Effective Workplace Writing

Effective Workplace Writing

Beth Camp
Linn-Benton Community College

IRWIN
MIRROR PRESS
Chicago • Bogotá • Boston • Buenos Aires • Caracas
London • Madrid • Mexico City • Sydney • Toronto

EFFECTIVE WORKPLACE WRITING

Copyright © 1997 by the McGraw-Hill Companies, Inc. All rights reserved. Printed in the United States of America. Except as permitted under the United States Copyright Act of 1976, no part of this publication may be reproduced or distributed in any form or by any means, or stored in a data base or retrieval system, without the prior written permission of the publisher.

This book was printed on recycled paper containing 10% postconsumer waste.

1 2 3 4 5 6 7 8 9 0 QB QB 9 0 9 8 7 6

ISBN 0-256-22016-6

Mirror Press: *David R. Helmstadter*
Elizabeth R. Deck

Marketing manager: *Carl Helwing*
Project supervisor: *Karen M. Smith*
Production supervisor: *Bob Lange*
Designer: *Larry J. Cope*
Cover illustration: *John Francis*
Prepress buyer: *Jon Christopher*
Compositor: *Electronic Publishing Services, Inc.*
Typeface: *11/13 Times Roman*
Printer: *Quebecor Printing Book Group/Dubuque*

Library of Congress Cataloging-in-Publication Data

Camp, Beth.
    Effective workplace writing / Beth Camp.
        p.   cm.
    Includes index.
    ISBN 0-256-22016-6
    1. Business writing.   I. Title.
    HF5718.3.C36   1997
    808'.06665—dc20   96-34762

http://www.mhcollege.com

To the Student

We are increasingly told that, without an education, we face careers with low pay and limited opportunity. We are told that people with some education after high school earn more money and have greater job security and job satisfaction than those with only a high-school diploma. We read daily about companies downsizing and investing in new kinds of workers—people who have high levels of technical skills, who can quickly adapt to new situations, who can work well with teams or alone without supervision, and who can be defined as "high performance" achievers.

Whether you are beginning your college education right out of high school or starting a new career, good communication skills will help you understand and respond to the changing demands of your career and the workplace.

This book invites you to strengthen your communication skills through:

- Questions that challenge your assumptions about working.

- Articles that help you think about the workplace.

- Writing exercises that immediately ask you to practice writing skills.

- Chapter discussions that highlight how employers communicate and what will be expected of you.

- An emphasis on teamwork—both how teams work and what skills are needed for you to function effectively as a team member.

- Writing and speaking assignments that will help you polish your writing to workplace standards.

Eastern philosophers say, "When the student is ready, the teacher appears." Many people will try to teach you how to do a certain task. You can learn from friends, co-workers, and family; but first, you need to "be ready" to commit yourself to learning.

The next steps to learning are not abstract ideas. These are the day-to-day expectations we have of ourselves. Do I read the assigned reading before class meets? Do I ask my instructor questions when I need information? Is my homework ready to turn in? Do I make up missing work or tests right away? Have I given this work my best effort?

Plan to be successful in this class and you will prepare yourself for continued success throughout your education and your career. Your best learning will occur when it is matched by your own initiative and willingness to work hard.

Beth Camp

To the Instructor

Teaching students how to write is challenging. National interest in educational reform has prompted much discussion over how and what we are teaching. We are challenged to prepare our students for the workplace of the 21st century, and we are beginning to see the impact computer technologies have on communications in classrooms and in workplaces of the future.

As a writing teacher, I often hear otherwise highly motivated technical, professional, and vocational students say, "I didn't have time to do the writing assignment because I had to finish homework for my major." Many students see writing courses as just another requirement on their way to gaining a degree.

But we hear a different story from employers, who assert that new workers are not prepared for the workplace because they lack basic communication skills.

Many new workers cannot write clearly or efficiently. They cannot give or receive critical feedback. They do not understand what criteria to follow to produce a "good" workplace document.

I wrote *Effective Workplace Writing* because I wanted to reach those students who were headed for the workplace with a two-year degree. At my community college, about half of our students will continue working toward a four-year degree, and many of these students must work while attending school. I felt these students needed workplace-focused writing instruction that would help them become more efficient writers and successful communicators.

I believe that reading, thinking, talking, and writing about workplace issues offers students important learning experiences. Students need both expressive and transactional writing skills that connect to

viii   To the Instructor

real audiences and that will develop their problem-solving abilities. I have structured the text in four parts to teach these skills.

Part I: Focusing on the Workplace introduces workplace writing processes, writing in teams, and writing with a sense of the criteria for "good" workplace writing. Students practice strategies for reading technical material critically, organizing research, and overcoming writer's block. Students work on writing topic sentences, developing paragraphs, preparing summaries, using citations properly when they use research from outside sources, and developing their own responses to information or ideas.

Part II: Improving Writing Productivity focuses on practical writing skills—describing and defining tools, equipment, and workplace processes, using forms, preparing informative and persuasive memos, and writing and responding to e-mail. Students polish their writing skills by working on clarity and document design to improve readability and writing efficiency. Students also begin to use graphics to improve the usefulness of their writing.

Part III: Building Communication Skills helps students to refine their writing skill using a variety of formats—from informative memos and letters, to instructions, to persuasive letters and proposals. The emphasis is on both the process (planning, writing, and revising) and the criteria (thoughtful content, logical organization, a clear and concise writing style, and readable format) that result in effective workplace writing. This section ends by highlighting strategies for successfully participating in and leading meetings.

Part IV: Writing about Your Career invites students to develop a career plan and to prepare a resume and cover letter that can be used to find a job or to move up within an organization.

Features

Chapter Previews introduce the concepts of each chapter. A list of questions follows the preview to explore and challenge students' knowledge about communication in the workplace.

Chapter Readings begin each chapter with current, topical articles relating to workplace issues. Students analyze each reading in discussion questions following each article. Questions can be answered in class discussion, in group settings, or as journal entries.

Section Exercises throughout each chapter let students practice evaluation or writing skills they studied in each section. Many of these exercises are designed for collaborative work, but they can also be used individually.

Writing Checklists in selected chapters give students a plan to edit and revise their work, either independently or with a peer editor. Throughout the text, an emphasis on peer editing and collaborative writing develops workplace skills of providing feedback and writing reports that meet workplace standards.

What's Coming Next summarizes the skills learned in each chapter and relates them to the material students will study in the next chapter.

Concept Review encourages students to define key terms listed from the chapter to aid concept retention.

Chapter Summaries include several types of exercises:

- *"What's Wrong with This _____"* exercises ask students to analyze writing samples and explain the errors found in each.

- *Chapter Assignments* are longer activities appropriate for homework.

- *Chapter Projects* are multistep assignments that emphasize group work and writing to address a real workplace issue.

- *Chapter Journal Entries* encourage students to write a personal response to situations relating to each chapter.

Quick Format Guides to memos, letters, reports, footnotes, and bibliographies appear in the Appendices. These guides provide a useful reference for business writing skills. They can also be used to tailor the text for an introductory Technical Writing class.

Instructor's Manual

The Instructor's Manual contains the following features:

- Overview of text.

- Sample class plans/syllabi for quarter and semester format.

- Grammar and punctuation skills diagnostic test; writing skills test.

- Chapter review for instructors, with highlights of key concepts, student challenges, and answers to selected exercises.

Our goals as writing teachers are both practical and idealistic as we work to prepare our students for their future. My hope is that this book will give you a framework to achieve your instructional goals and to give our students the discipline and creativity they need to be successful.

This book continues a dialogue that began in my classroom with students eager to learn how they could be more successful in the workplace. I welcome your suggestions and invite you to let me know what is most effective with your students.

B. C.

Acknowledgments

What began as conversations with my students, my husband, Allen Dorfman, friends, and between colleagues continues in a different form in this text. My corporate and classroom experiences helped me to think about what specific writing skills students need when they begin to write professionally. A book like this one is just one part of the dynamic learning that can occur when two or more people work together. It has been a learning experience for me, and I would like to thank those who gave so generously of their insights.

Thank you to my colleagues and friends Natalie Daley, Linda Spain, Linda Smith, and Jane White at Linn-Benton Community College for their valuable suggestions and support, and to my students over the last twelve years, whose questions and comments have been a faithful guide, especially Tony Falso, Tim Haag, Sung Lim, Joel Ewing, Betty Ann Viviano, Eric Schilling,

Teresa Trueba, Karen Fuller, and Roy Emery.

Helpful comments from reviewers Donna Alden, New Mexico State University; Heather Eaton, Daytona Beach Community College; Michael Hassett, Boise State University; Debra Journet, University of Louisville; Carolyn Plumb, University of Washington; Juanita Stock, Haywood Community College; Howard B. Tinberg, Bristol Community College; and Carol Wershoven, Palm Beach Community College, helped me probe more deeply into what professional writing involves.

I would especially like to thank those people who shared documents and their time in talking about "real" workplace writing: Hugh Vanderhaul and Chris Alexander, Hewlett-Packard; Joan Edwards, Overseas Private Investment Corporation; Jackie Paulson, Nursing, Jerry Phillips, Criminology, Linn-Benton Community

College; Kelly Duron, Pacific Corp; and Marcia Chapman of the Society of Technical Communications. My mentors—Simon Johnson and Lisa Ede, Oregon State University, and John Gage, University of Oregon—have continued to influence my teaching and thinking about writing.

Finally, a special thanks to my colleagues at Irwin and at Mirror Press for their dedication and support, especially Bruce Powell, whose interest in my students sparked the idea for this book; and David Helmstadter, whose early support led to my working with Carla Tishler and Bess Deck, *editors extraordinaire*. Their flexibility and tenacity have helped me throughout this process.

B. C.

Contents in Brief

Part I:
FOCUSING ON THE WORKPLACE

Chapter 1: Writing in the Workplace 2

Chapter 2: Developing a Writing Process for Work 22

Chapter 3: Reading and Researching for Workplace Writing 54

Part II:
IMPROVING WRITING PRODUCTIVITY

Chapter 4: Writing about Tools and Equipment 86

Chapter 5: Using Forms 128

Chapter 6: Communicating with E-Mail 172

Part III:
BUILDING COMMUNICATION SKILLS

Chapter 7: Writing for Co-Workers 192

Chapter 8: Writing for Customers 240

Chapter 9 Writing for Teams and Managers 276

Part IV:
WRITING ABOUT YOUR CAREER

Chapter 10: Writing to Get the Job You Want 314

Chapter 11: Writing for Promotion 360

Appendix A: Letters and Memos 394

Appendix B: Reports 402

Appendix C: Writing Skills Review 451

Index 503

Contents

Part I:
FOCUSING ON THE WORKPLACE
Chapter 1: WRITING IN THE WORKPLACE 2

Chapter Preview 2
Chapter Outline 3
Chapter Reading 3
Analyzing the Reading 5
Understanding Yourself as a
  Workplace Writer 6
  Observe How Other Workers Make a
  Commitment to Professionalism 8
Describing Your Writing Process 10
  Workplace Conditions That Can Affect
  Your Writing 11
  Understanding the Value of Team Writing
  in the Workplace 12
  Strategizing with Group Roles and Group
  Process 14
What's Coming Next? 19
Concept Review 19
Assignments 19
Project: Refining Your Writing Process 20
Journal Entry 21

**Chapter 2: DEVELOPING A WRITING
PROCESS FOR WORK 22**

Chapter Preview 22
Chapter Outline 23
Chapter Reading 23
Analyzing the Reading 28
Developing a Writing Process for
  the Workplace 28
Analyzing a Workplace Writing Process 29
  How Does a Writing Process Really
  Work? 30
  How Does My Individual Writing Process
  Fit in With Team Writing? 32
Writing Strategies: What Makes Good
  Workplace Writing? 33
  How Is the Workplace Writing Context
  Different from Writing for School? 35
  What Are the Criteria for "Good"
  Workplace Writing? 38
Tips for Overcoming Writer's Block 38
Writing Strategies: Writing with Topic
  Sentences 41
Writing Strategies: Developing Ideas by
  Using Facts and Examples 44

Contents **xv**

What's Coming Next? 48
Concept Review 49
Summary Exercise 49
Assignments 50
Project: Checking Content and
 Organization 51
Reading a Map 52
Journal Entry 53

Chapter 3: READING AND RESEARCHING FOR WORKPLACE WRITING 54

Chapter Preview 54
Chapter Outline 55
Chapter Reading 55
Analyzing the Reading 57
Reading for the Workplace 58
What Is a Trade Journal? 58
How to Read Technical Articles 60
How to Improve a Technical Vocabulary 61
How to Find Technical Articles 63
How to Save Time in the Library 64
How to Prepare a Bibliography 67
What Is a Summary? 71
How to Write a Summary 75
Editor's Checklist for a Summary 76
What's Coming Next? 77
Concept Review 77
Summary Exercise 78
Assignments 80
Project: Exploring Your Library Resources 81
Journal Entry 83

....
Part II:
IMPROVING WRITING PRODUCTIVITY
Chapter 4: WRITING ABOUT TOOLS AND EQUIPMENT 86

Chapter Preview 86
Chapter Outline 87
Chapter Reading 87
Analyzing the Reading 90
Describing Tools and Mechanisms 90
What Is a Tool or Mechanism? 93
How to Write a Technical Definition 95

Writing Parenthetical Definitions 95
Writing Formal Definitions 96
Writing Expanded Definitions 97
Preparing Stipulative Definitions 101
How to Describe a Tool or Mechanism 101
 Using an Organizing Pattern 102
How to Describe a Process 104
How to Do Graphics in Technical
 Descriptions 106
 Photographs 107
 Line Drawings 108
 Exploded Drawings 109
 Cut-Away Drawings 109
 Computer-Generated Graphics 110
Helpful Hints for Designing Graphics 110
Writing Strategy: Using Document Design to
 Improve Readability 112
Editor's Checklist for Technical
 Description 114
Writing Strategy: Preparing
 Specifications 115
 *Some Suggestions on Preparing
 Specifications 116*
Writing Strategy: Writing Persuasively 118
Editor's Checklist for Writing Memos 119
Editor's Checklist for Writing
 Descriptions 121
Concept Review 122
Summary Exercise 122
Assignments 123
Project: What Happened at Albany
 Motors? 125
Journal Entry 126

Chapter 5: USING FORMS: WRITING ON THE FLOOR AND IN THE FIELD 128

Chapter Preview 128
Chapter Outline 129
Chapter Reading 129
Analyzing the Reading 132
Writing with Forms 132
Writing Strategies: Working with Forms 133
Preparing Forms for an Audience 141

xvi   Contents

Writing Strategies: Reporting Raw Data   144
  Using Lists to Highlight and Clarify
    Information   144
  Using Headings to Direct the Reader's
    Attention   145
  Using Tables to Present Data   146
  Introducing and Interpreting Tables   146
  Interpreting and Summarizing
    Raw Data   148
  Reporting on Performance   151
Writing Strategy: Writing Coherently   153
Writing Strategy: Writing Concisely   155
Writing from the Field   157
Writing Strategy: Using Specialized
  Forms   159
  Forms for Tracking a Schedule   160
  Forms for Analyzing Tasks   160
  Forms for Managing Projects   165
What's Coming Next?   167
Concept Review   167
Summary Exercise   168
Assignments   169
Journal Entry   170

**Chapter 6:   COMMUNICATING WITH
              E-MAIL   172**

Chapter Preview   172
Chapter Outline   173
Chapter Reading   173
Analyzing the Reading   175
Writing with E-Mail   176
  What Is E-Mail and How Does It Work?   176
Working with E-Mail Day to Day   180
  Problems Related to Receivers   180
  Problems Related to Senders   182
  Problems Related to E-Mail
    Technology   183
What Writing Skills Are Needed to
  Use E-Mail?   184
Editor's Checklist for Writing E-Mail
  Messages   185
What's Coming Next?   186
Concept Review   187
Summary Exercise   187

Assignments   188
Journal Entry   188

**Part III:
BUILDING COMMUNICATION SKILLS
Chapter 7:   WRITING FOR
             CO-WORKERS   192**

Chapter Preview   192
Chapter Outline   193
Chapter Reading   193
Analyzing the Reading   197
Writing for Co-Workers   198
Writing Strategy: Writing Memos
  that Inform   200
Understanding the Parts to an Informative
  Memo   203
Editor's Checklist for Writing Informative
  Memos   204
Writing Strategy: Preparing Effective
  Instructions   206
Selecting a Format for Instructions   207
Preparing Instructions in a Narrative Style   207
Understanding the Parts of Instructions   209
Putting Together Document Design for
  Instructions   210
  Headings   213
  Lists   214
  Notes, Cautions, and Warnings   216
Writing Strategy: Planning Instructions   219
  Planning the Stages for Instructions   219
  Planning the Graphics for Instructions   221
  Deciding on a Basic Layout   221
Using a Process for Writing Instructions   226
  1. Draft the Instructions   226
  2. Have Your Instructions Tested by a
    Novice Reader   226
  3. Have Your Instructions Tested by an
    Expert Reader   226
  4. Revise Your Instructions   227
Anticipating the Context in Preparing
  Instructions   227
  Defining Your Readers' Needs   227
  Considering Where Instructions Will
    Be Used   229

Contents **xvii**

Protecting Readers from Hazards 229
Editor's Checklist for Writing Instructions 229
Designing an Effective Checklist 232
Editor's Checklist for Writing Checklists 234
What's Coming Next? 235
Concept Review 236
Summary Exercise 236
Assignments 237
Project 238
Journal Entry: Decoding Instructions 239

Chapter 8: WRITING FOR CUSTOMERS: INFORMING AND PERSUADING 240

Chapter Preview 240
Chapter Outline 241
Chapter Reading. 241
Analyzing the Reading 244
Writing for Customers 245
Writing Strategy: Writing Letters to Customers 246
Understanding the Parts of a Letter Format 248
   Deciding on a Conventional Letter Format 250
   Writing with a "You" Focus 252
Writing to Inform or Confirm 254
   Responding to Difficult Requests 255
Editor's Checklist for Writing an Informative Letter 258
Writing to Persuade 259
Writing Strategy: Responding to Customer Complaints 261
Writing Strategy: Writing to Collect Money 262
Editor's Checklist for Writing a Letter that Says No 265
Writing Strategy: Writing to Sell Products 267
   Understanding What Makes People Buy Products 267
Editor's Checklist for Writing a Persuasive Sales Letter 271
What's Coming Next? 272

Concept Review 273
Summary Exercise 273
Assignments 274
Project: We Can Do a Better Job! 275
Journal Entry 275

Chapter 9: WRITING FOR TEAMS AND MANAGERS: MEETING AND PRESENTING 276

Chapter Preview 276
Chapter Outline 277
Chapter Reading 277
Analyzing the Reading 281
Transformation in the Workplace 282
Teamwork and Continued Change 282
   Changes in Manufacturing 282
   Changes in Inventory 283
Working with a Team 284
   Understanding Learning Styles 285
Understanding Corporate Culture 286
Developing Meeting Skills 288
   Why Are Meetings Necessary? 288
   What Happens in a Meeting? 290
Managing the Meeting 291
   Preparing for a Meeting 292
Refining Your Communication Skills 304
Editor's Checklist: Planning and Following Up Meetings 305
What's Coming Next? 306
Concept Review 307
Summary Exercise 307
Assignments 308
Project: Who's on First? 310
Journal Entry 311

Part IV:
WRITING ABOUT YOUR CAREER
Chapter 10: WRITING TO GET THE JOB YOU WANT 314

Chapter Preview 314
Chapter Outline 315
Chapter Reading 316
Analyzing the Reading 318
Writing to Get the Job You Want 319

xviii Contents

Completing an Application 320
Preparing a Resume 326
  Picking a Resume Format 327
  Picking a Resume Type 327
  Selecting a Resume Layout 332
Checklist for Editing a Resume 338
Writing an Application Letter 340
  Drafting an Application Letter 340
Checklist for Editing an Application
  Letter 344
Preparing for the Employment Interview 346
Different Kinds of Interviews 348
  Types of Interviews 348
  *What Kinds of Questions Will Be
    Asked? 350*
  What Kinds of Questions Do I Ask? 353
Responding to a Job Offer 354
What's Next? 356
Summary Exercise 356
Concept Review 357
Assignments 357
Project: What's Out There for Me? 359
Journal Entry 359

**Chapter 11: WRITING FOR
              PROMOTION 360**

Chapter Preview 360
Chapter Outline 361

Chapter Readings 361
Analyzing the Reading 368
Writing for Promotion 369
Clarifying Personal Goals 370
Clarifying Professional Goals 371
Developing a Career Plan 372
After Accepting the Job, Anticipate
  Evaluation 375
Understanding Performance Evaluations 376
Understanding Appraisal Forms 377
Correct the Past or Develop for the
  Future? 381
Checklist for Writing Performance
  Evaluations 385
What's Next? 386
Concept Review 389
Summary Exercise 389
Assignments 390
Project: Writing a Performance Appraisal 391
Journal Entry 392

Appendix A: LETTERS AND MEMOS 394

Appendix B: REPORTS 402

Appendix C: WRITING SKILLS REVIEW 451

INDEX 503

FOCUSING ON THE WORKPLACE

Part I introduces workplace writing processes, writing in teams, and writing with a sense of the criteria for "good" workplace writing.

You will practice strategies for reading technical materials critically, organizing research, and overcoming writer's block. You will work on writing topic sentences, developing paragraphs, preparing summaries, and using citations for outside sources. You will also begin to develop your own responses to information and ideas by keeping a class journal and completing writing exercises.

1

Writing in the Workplace

Chapter PREVIEW

This chapter introduces you to workplace writing—as seen from your point of view as an employee working with co-workers and supervisors on projects and teams.

About the class journal. Writing in a class journal can help you clarify your thinking, improve your writing fluency and prepare you for class discussions and assignments. Your journal entries should emphasize what you think and your focus should be on expressing your ideas rather than polishing your writing to its final form.

You may select journal entries from suggestions at the beginning and end of each chapter, or your instructor may assign journal entries. Plan to discuss these questions with your instructor to clarify how you will use this important tool.

- What is the purpose of this class journal?
- What goes into a journal entry (date, chapter title, an introduction that lets the reader know what the entry is about)?
- How many entries a week should I write and how long should they be?
- Is a special notebook (for example, a three-ringed binder or a spiral-bound notebook) required for the journal?
- Should I bring my journal to class?
- How and when will the journal be graded? What criteria will be used?

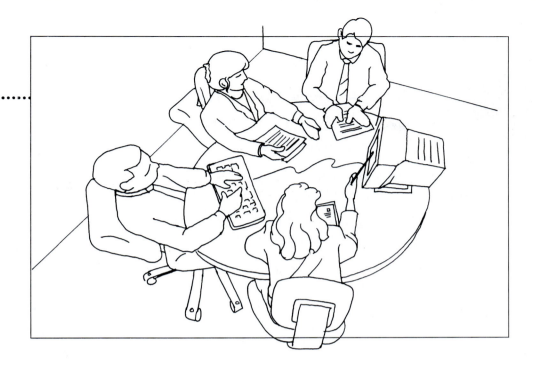

Journal ENTRIES

Begin thinking about writing in a workplace context by writing a short, one-paragraph answer to any of the following questions in your class journal or by preparing answers for class discussion, as directed by your instructor.

1. What qualities do businesses look for in their employees?
2. What reading, writing, thinking, and math skills do you think are most important for success on the job?
3. How is any workplace different from a classroom?
4. What kind of a process do you use for writing? How might this be different for the workplace?
5. What skills do you think are most important in working with groups, teams, or quality circles?

Chapter READING

What makes us successful in the workplace? Eric Lacitis suggests that well-polished communication skills can help you get the job you want. As you read his article, consider **what** the writer has to say and **how** he says it.

WHY JOB SEEKERS DIDN'T MAKE THE CUT

Eric Lacitis

Hello, kids! And their moms and dads who want to see them grow up to be successful!

You've heard about the "global economy," haven't you? It means fewer jobs, and lots more competition for those fewer jobs.

So you should make sure you're at your competitive best.

What do you think would be a valuable skill to have as we enter the 21st century and as we all talk to each other through computers? How do you think people communicate through computers?

And, kids, when you're adults, what kind of computer skills will bosses look for? Do you think they'll care that you were the best at playing Super Mario Brothers or Mortal Kombat or Street Fighter II? Or would they want someone who can write a simple sentence without fouling it up?

They Didn't Make the Cut.   While we think about that, let's try to figure out why the following job seekers—many of them college graduates—didn't make the "cut" when they recently applied for employment at a Seattle company.

The job was to be the company's accounting manager, and one qualification was that the person had to be a "great communicator." Some 300 people applied for the job.

The controller at that company let me look at the resumes. I don't think these examples were particularly unusual, and you could find them in job applications at any company.

Here's one: "I am a very hard worker and nonsmoker. I am presantly without any of car of my own so my job site needs to be acessable by buss."

All right, here's another resume, from a college grad:

"This letter constitutes as an application with a strong interest towards becoming an valuable asset for your firm."

Kids, would you like a good-paying job putting together video games worth millions of dollars and writing the rules to play them? Do you think the college grad I quoted could get such a job?

Let's try another resume. ". . . as you can see my qualifications fit nicely for the position mentiones . . . I am detailed oriented and like to be organezed in each facet of my position."

That was written by a graduate of one of our public high schools. She was looking forward to hearing real soon from the company. Can you figure out why she's going to be waiting a long, long time?

It won't be easy trying to make it in this global economy. In just about every one of those 300 resumes, people said they knew how to run various kinds of computer programs.

Chapter 1: Writing in the Workplace 5

That's fine, but once you're on that computer, you'll have to start doing something else. You'll have to . . . to . . . what's the word I'm looking for? You'll have to COMMUNICATE with other people at work who have computers. And the way people using computers communicate is actually kind of old-fashioned.

They have to write sentences, and paragraphs, and long memos.

Can you guess why a company might toss out the following resume? It had sentences such as:

"I Have flied Pacified Ocean five times."

How about someone who wrote about the "challanges" he wanted?

Or how his experience included meeting with customers to discuss how "there product" would be manufactured?

In the Minority.   Kids, parents, do you know why a personnel manager might wince at the spelling in the two sentences above? You do? Then you're in a minority of Americans.

It's kind of ironic, isn't it? Fewer and fewer people read much anymore. Book publishers, magazine publishers, newspaper publishers, they're all trying to figure out how to keep the readers they have.

But in this new electronic age, writing a coherent sentence is more important than ever.

Don't believe me, kids? Well, don't worry. Go back to the video.

In the global economy, there's also a place fer peeple that cen't write a semple sintence. At $4.25 an hour.

Source: Eric Lacitis, "Why Job Seekers Didn't Make the Cut," *The Seattle Times,* September 24, 1993. Copyright 1993, The Seattle Times Company. Used by permission.

Analyzing the READING

To build your reading skills, use the following questions to analyze the content of the article and to develop your reaction to it. Write a one-page journal entry in your class journal or prepare your answers for class discussion.

Reading for Content

1. What two main ideas does Eric Lacitis want you to remember?

2. What kind of facts, statistics, stories, or examples did Lacitis use to support his main ideas? Find one example of each type of support. Which are the two most important types of support and why?

3. What does the term *global economy* mean? The writer says there will be "lots more competition" for "fewer jobs." Do you believe the writer? What could the writer have done to prove this assertion? Does this assertion seem to be true for your area? For your field? How do you know?

6 Part I: Focusing on the Workplace

4. Who is Lacitis writing for? What does he want the reader to conclude after reading this article? Is his purpose clear from the introduction? Why or why not?

5. What do you think of this author's somewhat sarcastic tone? Does his informal style work for his intended audience? For you?

Reading for Reaction

1. What does this article mean to an employer? To a recent high school graduate who is looking for a job?

2. Do you agree with Lacitis about the quality of communication skills today's high school graduates have? Do you agree based on either your own experience or that of your friends?

3. What responsibility do schools have to prepare workers for the workplace? Is this the only responsibility schools should have? Do you feel the schools you have attended have met their responsibility to you?

UNDERSTANDING YOURSELF AS A WORKPLACE WRITER

In the workplace, you will be writing many different kinds of documents— from "just-the-facts" kinds of descriptive writing that reports on equipment, customers, or problems that need to be solved, to filling out routine forms, to writing more analytical and persuasive proposals, memos, and reports. Writing is only one of the important communication skills you will need to be successful on the job.

Communication skills (writing, speaking, and listening) are very important foundation skills that are linked together. If you are a good speaker, but are less confident about your writing, with practice, you can transfer your good speaking skills to paper. The same is true if you are a better writer than a speaker. Practice **can** improve your ability and your confidence, and that is exactly what this textbook is intended to do for you: improve your communication skills for job success.

In the workplace, **how** you carry out a task is as important as **what** you know. Whether you plan to work at a large company or a small one, understanding how all the pieces of the company fit together—and how the company fits into the larger economy and community—can help you be a more effective worker. You are hired for any job because of your

- **Technical skills** (skills and knowledge of your field).

- **Social skills** (communication skills and your ability to get along with others).

- **Operating skills** (ability to get tasks done and fit in with the work processes already in place at your company).

Assessing Your Skills   Exercise 1.1

Think of any job you have held. Make a list of the technical, social, and operating skills (using the previous definitions) that you needed to have or you gained **after** you had worked at a company for several months. Put at least three items under each heading.

Technical Skills	Social Skills	Operating Skills
_____	_____	_____
_____	_____	_____
_____	_____	_____
_____	_____	_____
_____	_____	_____
_____	_____	_____
_____	_____	_____
_____	_____	_____

Share your draft list with your discussion group and make any changes, additions, or deletions you think are needed.

After your list has been revised, write a sentence in response to each of the following questions.

1. Which skills do you feel most prepared in? Which skills may need more work? Which category of skills do you think was most important for your getting hired? For your personal success on the job after you were hired?

2. Did you notice any improvements in any of these skill areas because of your work experience? What and why?

3. Did your understanding of how the company did its business change after you began to work? In what ways?

4. Which category of skills do you think is most important for the overall success of the company? Why?

5. As you think over all your responses to this exercise, what writing or communicating skills do you feel would help you to be more successful in acquiring or improving your technical, social, and operating skills? Direct this response to your instructor.

Plan to share your findings with your discussion group and to turn in your list and your answers to your instructor.

The ability to improve your skills in each of these areas is essential. Let's get started by understanding your strengths and weaknesses in each of these areas.

You will improve your technical or operating skills through classes you are taking in your field. Speech and psychology classes will be important for building your social skills. This text will help you improve one aspect of your social

Part I: Focusing on the Workplace

skills—your **communication skills as a writer.** Some of the same ideas that help you organize and refine your written communication skills will help you to strengthen your speaking skills.

In part, you were hired for the job because of your **communication skills.** By using a combination of written and oral skills, you persuaded your employer to hire you instead of someone else.

Your communicating skills will be crucial after you are hired—to understand what work needs to be done (listening or reading), to let others know what you need or would like to have done (speaking or writing), and to let others know what you have done (speaking or writing).

Sometimes you will need to make decisions about communicating quickly and intuitively, without having the information you need, and often you will have many questions about writing in a workplace, for example,

- What kinds of writing will I be expected to do?

- Whom will I be writing to?

- What kinds of questions will my readers have?

- How do I know which is best: calling someone on the phone, dropping by his or her office, calling a meeting, or writing a memo or report?

- What processes can I use to write more efficiently?

Think about these questions as you read this chapter. We will begin to answer these questions in the coming chapters, and you will explore your ideas and build your communication skills through journal entries, class activities, and writing assignments. Before you continue, do Exercise 1.2.

Observe How Other Workers Make a Commitment to Professionalism

Once you are on the job, your employer will still be investing in your training. The best employees take advantage of this training by observing how others work, by asking questions to clarify what is expected, by doing their best, *and* by asking for feedback on what works best.

You will be constantly preparing yourself for the next project or the next promotion if you follow the suggestions listed here.

Notice how others complete jobs, what tools they use, how much time they take, and what criteria they use to decide the project is complete.

Keep accurate records for the company. These may include files on clients, work orders, shift reports, status reports, or work-in-progress forms.

Describing Your Workplace   Exercise 1.2

Write a paragraph of 5–7 sentences that describes the kind of workplace you would like to work in or plan to work in. Consider what field you will be working in and whether you will be working for a small or large company. Be as specific as possible. Additionally, answer the following questions in your paragraph:

1. How will your co-workers most likely communicate with each other? One-on-one? In meetings or work groups? Through memos?
2. What are two of your communication strengths? How does the workplace setting you've described match your communication strengths?
3. How will the size of your employer affect how information is shared? Do you think information will travel formally (through memos and letters) or informally (through conversations and other networks, sometimes called a grapevine)? In both ways?
4. How does this workplace description affect your decisions about classes you plan to take?

Share your draft paragraph with your discussion group, make any revisions that are needed, and turn it in to your instructor.

Add to company files by keeping precise notes when needed. One purpose of keeping a file is to create a legal record of what happened when and why. Accurate, focused, and concise records can help your co-workers do a better job in serving your customers and in analyzing the completed work.

Build personal networks with your co-workers (and with peers in your specialty who do not work for your company) that will help you understand how problems are solved and what resources are available.

Use a team approach for major projects. Pulling others in on key projects means more resources to get the work done and more support for new ideas.

Volunteer for more complex jobs to expand the skills you already have.

Ask for feedback tactfully on each job you complete. You may get honest feedback more easily from a peer than from your supervisor, but you can improve your work by always asking yourself: What worked well on this project, and where could I make improvements?

Stay current in your field by reading technical magazines and newspapers and by planning for your own professional development. Attend conferences, build your skills at workshops, or take formal classes to get the skills or training you need.

All of these suggestions will help you approach your work professionally. They can support good communications by adding to what you know and by improving how you work.

DESCRIBING YOUR WRITING PROCESS

You have been writing assignments and papers for school for many years. You may also have a few years or many years of experience in writing for the workplace. Already you have developed a process for how you **plan, draft,** and **revise** any writing assignment.

Frequently in the workplace, writing must be done quickly, without a great deal of time for planning or revision. Some companies operate in a crisis mode. This means that your skills need to be at their highest level. If a crisis occurs, you are then ready to do the work that is needed.

As you think about conditions that affect how you write in the workplace, it is important to analyze the exact steps you take (the writing process you now use), so that you can work on improving your writing efficiency.

Exercise 1.3 Describing Your Writing Process

Think about what you do when you need to complete a writing assignment. List the steps you generally follow when you complete a writing assignment for school or work.

Use the grid below to describe how you plan, research, draft, revise, and review a typical writing assignment for school or work. This could be an essay, a letter or memo, a research paper, or a formal report. Use the "comment" column to analyze which writing steps work best for you. Where do you want to see improvement? Which tasks do you feel very well prepared for? Which tasks (or skills) do you think will need more work? Please be as specific as possible.

Work Log Grid

Writing Steps	What I Do	Comment
Plan	Talk to friends Make an outline	I usually spend most of my time thinking about the writing problem and how I will solve it. The hardest part for me is to plan exactly what support is needed.
Research		
Draft		
Revise		
Review		

Share your "steps" with your discussion group. Make final revisions in your grid by adding more detail if needed. Conclude this writing exercise by listing two goals for you for this writing class that you think are important because of your analysis of how you currently plan, write, revise, and review your writing.

Workplace Conditions That Can Affect Your Writing

We might imagine a perfect workplace where there are few interruptions and the work goes smoothly. However, workplace conditions are almost never ideal. We can anticipate frequent interruptions, changes in communication technology, changes in our role of representing our employer (and an almost daily testing of personal integrity), and changes in who we work with to complete a specific task.

Interruptions Are Common.

You can expect frequent interruptions. One study reports the average manager has about 10 minutes of uninterrupted work time—on a good day. Not only are there many different projects going on at the same time, each with its own deadlines, but crises and breakdowns can occur at any time, affecting your ability to meet deadlines.

Knowing what are peak times for interruptions may help you to plan your workday. Understanding what your customers need, how your co-workers work, what kinds of problems occur most often, and what goals your supervisor has set will also help you plan your workday and to anticipate problems.

Communications Technology Will Continue to Change.

Your skill in using communication technology also may be very important. Will your future employer now or in the next five years use voice mail? E-mail? An electronic inventory control system? A computerized accounting system? Today, even very small companies are using computer software to lower the highest expense facing most businesses—salaries for workers.

Your ability to comfortably use and communicate through computer software that is appropriate for your field can give you a little more job security.

Your Role in Representing Your Employer Will Increase with Authority and Responsibility.

Another factor that can affect how you write in the workplace is the issue of who signs the order, the memo, or the letter. Frequently, new employees write for their supervisors. Only after going through a period of training, with your work closely reviewed, will you gradually be allowed to sign your own work and send it to the intended audience without your supervisor's approval.

As you gain experience at your job, you'll notice that the relationship between the writer and the reader affects **what** is written and **how** it is presented. You'll notice how some co-workers analyze their audiences to determine if their written work will be received enthusiastically or if it will have to face a hostile audience. People don't always get along with one another. Some people are protective of the ideas they develop or sponsor. Your social skills will be needed to define what "authorship" means to the people you work with.

Who owns what you write when you are working for an employer? The answer is the employer does. Because you have been hired to do a specific job, any writing you do that carries out that job is the absolute property of the employer, unless you have a specific agreement otherwise.

Your signature on documents going outside your employer is particularly important because in certain situations (for example, when you confirm an order or sign a contract), you are legally acting as an agent for the company, committing the company to whatever you have written.

Even on the telephone, when you represent the company, you are the company's agent. For example, an entry-level clerk at a major international bank quoted the daily investment rate to a major customer interested in investing hundreds of thousands of dollars. The clerk misquoted the daily rate. The customer asked, "Are you sure this is the correct rate?" The clerk quickly rechecked her calculations and restated the same rate. The bank lost several thousand dollars, and the employee was reprimanded. People make errors all the time. What is important is recognizing the potential costs of making errors—and working to prevent them.

Your Personal Integrity Will Be Tested. As you continue working for a particular employer, your supervisors, your co-workers, and your customers will come to trust you because you are able to deliver what you say you can deliver—on time. This idea can be described as your personal integrity, and it has great value, even if no one ever says a word about it.

Exercise 1.4   Describing a Workplace Promise

Think about a time when an employee of a company made a promise to you and did not keep it. Perhaps a clerk promised your order would be ready by a certain date, or perhaps you have eaten at a restaurant that promised quality service and did not deliver. What effect did this have on your immediate reaction to the situation—or to how you felt about the company? How do these kinds of experiences relate to your own actions as an employee?

Use a list to make some rough notes on your experiences (if needed) and write a rough draft of two or three paragraphs that describes what happened and what implications you think this has for the workplace.

Share your experiences with your discussion group or your class. Then, make any final revisions to clarify what happened or what implications you want to highlight. Be sure to include an introduction, a discussion, and a conclusion section. Turn your finished paper in with all notes and drafts to your instructor. Your instructor may request that you use a memo format; use the Quick Format for Letters and Memos section in Appendix A.

Understanding the Value of Team Writing in the Workplace

Frequently in larger companies, you will be asked to be part of a team of people working on a project. Or you may work with one or two others in your department to produce important memos or reports. The more important a project is

Chapter 1: Writing in the Workplace   **13**

to your employer, the more likely it is that more than one person will be working on that project.

In a classroom, your instructor may assign you to a permanent group for the term, or you may work with a new group on each project you complete. Part of the writing you will do in this textbook includes the hidden goal of helping you to build your skills in working with a group.

A project team in a larger company may be made up of either specialists from several different areas in the company or the people you work with every day. You may have been asked to work on the team because of your technical skills, your communication skills, or your writing skills.

Whenever you are working in a group, there are three key concepts to keep in mind: what roles people can play in groups, how a group functions over time or during its life cycle, and how to keep a group working as effectively as possible.

What Roles Do People Play in a Group?   If you were to play a piano, you would need to use more than a few keys from the keyboard to make good music. So, too, when you are working in a group, you need to know the different roles and "play" them when the time is right. A few of the more common group roles, divided into "lower" risk and "higher" risk roles, are listed in Exhibit 1.1.

Common Group Roles   Exhibit 1.1 ∎

"Lower" Risk Group Roles

Sleeper	Does not talk or participate in any work
Gopher	Serves the group with coffee or snacks or runs errands
Note Taker	Takes extensive notes of every meeting
Researcher	Gathers information that the group needs
Typist	Quickly volunteers to type all work or to coordinate the production work

"Higher" Risk Group Roles

Writer/Editor	Serves as group editor on all writing
Facilitator	Moderates the group's discussion and keeps the group focused on the task
Debater	Presents both sides of nearly all issues
Saboteur	Works to undermine the work of the group
Devil's Advocate	Questions all ideas, frequently with good reasons; may represent the "other side"
Presenter	Makes the formal presentation of the project

Each of these roles (and several other variations) can have a positive or negative impact on the work of the group. People may also play different roles at different times in the life of the group, depending on the dynamics between group members and the interest members have in the project.

Strategizing with Group Roles and Group Process

To facilitate the work and focus of your small group, you will need to set aside time to think about the roles people play in groups and the stages groups go through as they complete their work. Exhibit 1.2 summarizes the stages a small group goes through. As you can see, different roles are more important at different stages of a group's life cycle. For example, since bonding is an important result of the "forming" and "norming" stage, you would place high value on socializing to help new group members know one another. When the group is in the "storming" stage, you probably would want to support facilitating skills so that everyone understands the issues involved.

How Do I Anticipate Individual Attitudes in a Group?   Each person joins a work group with certain expectations, social skills, and technical skills that will support, or hinder, the group's goals. Careful listening, planning, and discussion can help you understand or clarify why people behave the way they

■ Exhibit 1.2   **Life Cycle of a Group**

Stage	Group Activities
Form	Select group members **Main purpose of this stage:** Identify group strengths and weaknesses by observing roles and communication skills Define group task and assign work Set meeting times 　　**Group focus:** Set up the team 　　**Group mood:** Exploring
Norm	Set norms for meetings; clarify goals, communications, and performance expectations **Main purpose of this stage:** Clarify group tasks, quality of work expected, and deadlines (or who does what, when, how, and how much) 　　**Group focus:** Clarify tasks 　　**Group mood:** Questioning
Work	Gather information and set up feedback Begin to evaluate performance **Main purpose of this stage:** Begin substantive work on goals 　　**Group focus:** Complete the task 　　**Group mood:** Producing
Storm	Resolve disagreements over performance, quality of work, or deadlines **Main purpose of this stage:** Analyze quality of work being done 　　**Group focus:** Check quality 　　**Group mood:** Evaluating
Perform	Present work to intended audience **Main purpose of this stage:** Complete tasks assigned 　　**Group focus:** Present the final project 　　**Group mood:** Achieving

do. Knowing that someone is quiet, for example, may lead you to get that person more involved by asking questions.

The leadership skills you use can help people who currently may not be productive to become more useful to the group. They can help to improve your group's overall productivity, without sacrificing morale or the quality of work.

You will also need to anticipate how different stages in the group's process will affect individual group members. Regardless of the attitudes of each group member, most people will feel discouraged during the storming stage, and most will also feel a sense of accomplishment during the performing stage when the group's task is completed (see Exhibit 1.3).

In any meeting these same group roles and stages of a group are in effect. Observing who does what and when it is done can give you valuable information about what your own role should be from start to finish in any group situation.

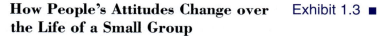

How People's Attitudes Change over the Life of a Small Group Exhibit 1.3 ∎

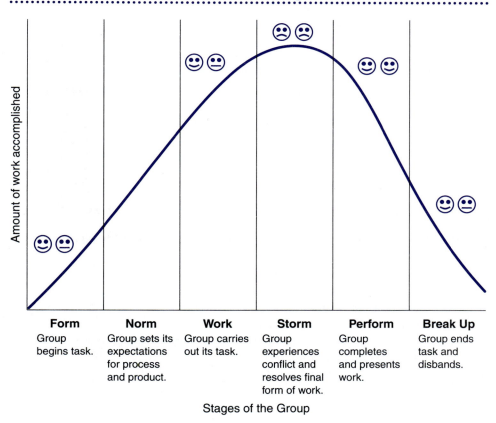

Exercise 1.5   **Analyzing Group Roles**

Think about the groups you have participated in—from sports, to classrooms, to social activities. Consider how people acted and review the list of common types that appear in groups. Answer the following questions in your class journal or share your ideas with your discussion group.

1. What factors make a particular role "low" or "high" risk to the individual? To the group? Why, for example, is a "saboteur" a high risk role?
2. Which roles do you think are most important for group success? Why?
3. Which roles are you most comfortable in? Which do you tend to take? What would you need to do to undertake some of the other roles?
4. Have you been a facilitator? A devil's advocate? Would you change your style after reading this section?
5. What group roles have you seen most frequently? What advice could you now give to someone who had trouble working with a group?

Understanding Group Process. Researchers disagree over exactly how many stages a small group will go through from beginning to end, but they do agree that all groups go through a process with similar activities happening at each stage. The life cycle of a group, adapted from the work of Robert Coles, has five stages (see Exhibit 1.2).

How Do I Encourage Good Teamwork? Ideally, every team member has a responsibility to work together to help the team be as productive as possible. Encouraging everyone to play an active role in your group is a first step toward successful group work. Some of your co-workers or classmates who start as sleepers or gophers may need to be encouraged to take on more responsibility.

Sleepers and saboteurs have much in common. Sometimes politics enters into a group with someone who does not want to participate or who will actively try to sabotage the group's efforts. If this occurs, do not postpone action! Allowing someone to have a "free ride" means more wasted time and perhaps outright hostility among group members.

Talk to the person involved to clarify the problem and possible solutions. If the person is not responsive within a specific time, ask for advice on how to follow up from your instructor or supervisor. It may be necessary to remove the person from your group. In the workplace, such steps could lead to demotions or the person losing his or her job. In the classroom, when a student is dropped from a group, the student must then complete the work alone. Usually, this action is enough of an incentive for someone to become a productive group member.

Devil's advocates do more than argue. Although they may be abrasive, devil's advocates often play very valuable roles because they clarify arguments that may be used against the group's project later. Groups who have a devil's advocate frequently are more analytical because group members are pressed to defend ideas more completely.

Assessing Your Group Skills   Exercise 1.6

Discuss the following questions in your small group or write a journal entry that explores these issues.

1. Which group roles do you think are most important at each stage of the life cycle of a group?
2. What steps could you take if a group member is very argumentative or refuses to do the assigned work?
3. If a group leader tried to do all the work by himself or herself, what would you do during the next group meeting? Why? How should work be divided fairly?
4. Consider a variety of groups you have been a part of. Does your experience match the theory being summarized here? Why or why not?
5. What changes would you make in how you work in a small group after reading this section?

Working in a group that has an outspoken devil's advocate means more careful preparation and analysis by group members. Overall, such interaction leads to a higher quality of work. Nurture your devil's advocate skills so you can help the group to focus on issues, rather than put people into a defensive, argumentative mood.

What Do I Do with a Team That Is Having Problems?   Many people are uncomfortable with conflict, especially if it occurs in a group where there are others to watch what happens. Conflict is inevitable when the project a group is working on could have an impact on the next promotion or the success of any person's career. The higher the stakes in a particular project (to an individual, a department, or the company), the more likely there will be conflict.

The best way to manage conflict is *not to squash it, but to channel it*. Refocusing the group's efforts away from personal attacks to the common interest of the group is one good way to minimize the negative effects of conflict. Everyone will have personal goals within your group, and politics can play a nasty part in the dynamics of any group. The best defenses are to be aware of the political implications of any task and to try to build consensus in the group about the group's goals.

Sometimes people become discouraged about working in groups, partly because of the possibility of conflict, and partly because working in groups can take more time than if people work alone. However, we can gain experience and knowledge from working in groups through:

- Using the skills and experiences of more people.
- Having more resources for critical analysis of the project and to complete the work.
- Building support for a project by having more people involved.

Group members can work very effectively together if each one, regardless of his or her role or understanding of the group process, wants the work of the group to be successful.

Part I: Focusing on the Workplace

What Ideas Will Help My Team To Write Effectively? People may think that writing is not collaborative. They may agree that planning and research are activities that can be done effectively in small groups, but they may think that writing is essentially an activity that needs to be done alone.

However, not every writing assignment is one that can be done alone. Some writing projects are so large that one person cannot complete them effectively. When you are working on a writing project that involves more than one person, group members should:

- Plan a work schedule that allows sufficient time for planning, researching, drafting, revising, and reviewing.

- Divide work fairly among group members, giving each one clearly defined tasks and deadlines.

- Allow enough time for reviewing and revising work before the final deadline.

- Allow reasonable amounts of time for production, especially if word processing or graphics for final documents will be produced outside the work team.

- Discuss a fair way to motivate or remove unproductive people from your team.

Experienced managers suggest doubling the time you think you will need to complete a writing project, until you have more experience in estimating how much time you will actually need.

WHAT'S COMING NEXT?

This chapter has asked you to analyze your communication skills as they relate to your success in the workplace. You've also described your writing process and have begun to think about what makes an effective team. In the next chapter, you'll further develop your writing process, define the concept of a workplace audience, and begin work on writing strategies, starting first with topic sentences and paragraphs.

■ CONCEPT REVIEW

Can you define the following key concepts from Chapter 1 without referring to the chapter? Write your own definitions in a journal entry or review them out loud.

communication skills
(speaking, reading, writing)

production or operations

authorship

acting as an agent for your employer

on-the-job training

workplace writing

writing process (plan, research, draft, revise, review)

collaborative writing

group roles

group process

WRITING SKILLS REVIEW ■

An important goal of this textbook is to help you express yourself in writing without distracting errors in sentence structure, punctuation, grammar usage, or spelling. To build your skills in editing and proofreading, begin a systematic review of *Appendix C: Writing Skills Review*. Your assignment:

__ Read "Understanding Sentence Structures, Parts of Speech" in *Appendix C.*

__ Complete exercises C.1 through C.11 in "Parts of Speech," checking your answers at the back of *Appendix C.*

__ Prepare a paragraph summarizing your progress for your instructor, or use the form your instructor provides to report your progress. Be sure to highlight any areas that you need to further review.

ASSIGNMENTS ■

Assignment 1: Interview a Professional in Your Field about Communications.

Prepare a list of questions you could use to interview someone working in your field. Write two or three questions about each of these categories: writing, speaking, reading, and math skills. Interview someone either face-to-face or over the phone. Write up your findings in memo form for your instructor. Conclude your memo by answering this question: What did you learn from this interview about communications requirements in your field?

Work with an editor from your discussion group to review your work, considering content, organization, style, and document design. Make final revisions and turn your assignment in to your instructor, including all rough drafts.

Assignment 2: Analyze Your Communications Strengths and Weaknesses.

After thinking about how important communication is to your success on the job, write a letter or memo of 4–5 paragraphs to your instructor describing the strengths and weaknesses of your own communication skills. Conclude your paper by recommending that your instructor include two or three key skills you want to acquire or improve in your writing class.

20   Part I: Focusing on the Workplace

Work with an editor from your discussion group to review your work, considering content, organization, style, and document design. Make final revisions and turn your assignment in to your instructor, including all rough drafts.

Assignment 3: Analyze a Small Group.   Observe a small group for two or three sessions. Keep a record of the roles people play and the process they go through. Were you able to identify all four stages of a group's life cycle? Would you say this group was a functional group or a dysfunctional group? Why or why not? Did you give the group any advice? If yes, was the advice helpful? In what ways? Talk over your findings with members of your discussion group.

Write up a summary of 4–5 paragraphs that reports on your observations and what you learned. Work with an editor from your discussion group to review your work, considering content, organization, style, and document design. Make final revisions and turn your assignment in to your instructor, including all rough drafts.

■ PROJECT: REFINING YOUR WRITING PROCESS

Work with one or two members of your discussion group to compare and discuss your grids from Exercise 1.3 that described your individual writing process. Your task is to collaboratively write a letter of advice to incoming college freshmen on how to improve their writing process so that they will be successful in their college work and will use their time wisely.

For this letter, you will need to

- Define a writing process.

- Identify common pitfalls and strengths for each stage.

- Conclude with personal and perhaps motivational advice.

Limit yourself to $1\frac{1}{2}$ to 2 pages. Divide the work evenly among your group members and allow time for peer review of the final document before it is turned in.

■ JOURNAL ENTRY

You have worked with your discussion group several times now. In this journal entry, analyze your interactions with your discussion group.

Describe briefly the stages your group has gone through so far.

Describe the group roles that each member of your discussion group has played so far. Note the communication strengths of each group member. For efficiency, you may want to use the following planning grid:

Chapter 1: Writing in the Workplace 21

Group Member	Group Roles	Communication Strengths

Write brief responses to the following questions as part of your journal entry:

Report What Has Happened with Your Group

1. How have you interacted with the group?

2. How would you describe your group's motivation?

3. What roles have you assumed so far?

4. How have you helped the group complete its tasks?

5. Have you identified any barriers to your group's success?

Analyze What Has Happened and What Could Happen with Your Group

1. How would you evaluate your contributions to the group?

2. What have you done to help the group be more productive?

3. Are there any changes you could make to help your discussion group work more effectively?

4. How does this group work relate to the skills you will need to work with your colleagues in your chosen field?

2

Developing a Writing Process for Work

Chapter PREVIEW

This chapter continues your work on defining a writing process, one that you can adapt to any writing project. This chapter also introduces you to the concept of audience. Writing to your audience and solving problems for your audience are two of the most important shifts from academic to workplace writing.

Journal ENTRIES

Begin thinking about your writing process and writing to an audience by writing a short, one-paragraph answer to any of the following questions in your class journal or by preparing answers for class discussion, as directed by your instructor.

1. Describe how you completed a simple writing project and a difficult writing project. Did you use different methods or resources for each?

2. When you write a letter, who is the audience? What about a writing assignment for school? A letter to the editor?

3. How does your personal writing style change when you write to a different audience? Give two or three examples.

4. What does *multicultural* mean to you? Do you communicate differently with people who are different from you in age, cultural background, position, gender, or nationality? Have you worked with a wide diversity of people?

5. Have you worked for the general public before? What lessons could you share with the class?

6. How do you know when a document is ready to be mailed?

Chapter READING

In what ways are we unique in the workplace? Sharon Nelton concludes that well-polished communication skills can help you work effectively with a wide range of people. As you read her article, consider **what** she has to say and **how** she says it.

NURTURING DIVERSITY

Sharon Nelton

Even if affirmative action is dismantled, diversity of the workforce is here to stay. Business owners and managers, experts say, will still need to maintain or step up efforts to recruit and advance minorities and women in the year 2000 and beyond. That's because having a diverse work force and managing it effectively will simply be good business.

"This is the great wave of the future," says George Henderson, chairman of the Department of Human Relations at the University of Oklahoma, in Norman. You can resist it, he says, "but you'll still be going with the wave."

One business leader who is happy to go with it is Michele Luna, president of Atlas Headwear Inc., a Phoenix company that manufactures military and sports hats. For Luna, the future is already here. She estimates that 94 percent of her employees are Asian and Hispanic, and many are immigrants. At least 80 percent are women.

When Atlas looks at employment candidates, says Luna, it never wonders whether a woman can do a particular job or if a person with an accent is going to fit in. "We've always considered people on their ability to get the job done," she says.

The company's promote-from-within policy has produced a senior management team that includes women and minorities, and Luna says she has proof that the firm's diversity is good for business. Atlas expects revenue of $8 million this year, up from $7.1 million last year. In the first three months of 1995, the company added 49 employees, bringing its total to 230 workers.

It has been eight years since *Workforce 2000,* a report prepared by the Hudson Institute for the U.S. Department of Labor, stunned the business world with its announcement that white males would make up only 15 percent of the net additions to the labor force between 1985 and 2000. White males were already in the minority, representing only 45 percent of America's 115.4 million workers in 1985.

The biggest gains, said the institute, a policy-research organization in Indianapolis, would be made by native white women, while other increases would come from minorities and immigrants.

New figures bear out the Hudson Institute's finding that white males are becoming more of a minority in the work force. A report issued in March by the Federal Glass Ceiling Commission, a 21-member bipartisan body created by the Civil Rights Act of 1991, says that 57 percent of those in the workforce today are women or minorities or both.

"Women and minority men will make up 62 percent of the workforce by the year 2005," according to the report. However, it says, "the world at the top of the corporate hierarchy does not yet look anything like America." The study shows that 97 percent of the senior managers of the nation's largest corporations are white and at least 95 percent are male.

Chapter 2: Developing a Writing Process for Work **25**

A newer survey suggests that smaller businesses have been more successful than larger ones in promoting minorities into upper management. The survey, released in April by the American Management Association, shows that in businesses with fewer than 500 employees, 19 percent of the senior managers are minorities, compared with 13 percent for businesses with 500 or more employees.

"Smaller firms are more likely to be minority-owned, and a single minority manager has a greater *statistical* impact when percentages are figured," the report explains.

Managing diversity goes "far beyond" meeting the legal requirements of equal employment opportunity and affirmative action, according to a joint survey of 785 human-resources managers conducted by the Society for Human Resource Managers and Commerce Clearing House. The 1993 SHRM/CCH report defines it as "the management of an organization's culture and systems to ensure that all people are given the opportunity to contribute to the business goals of the company."

This means including employees not only without regard to the obvious differences of race, sex, and age but also without regard to such "secondary dimensions" of diversity as marital or family status, sexual orientation, and disabilities, the report says. It also means including white males.

George Henderson, who has written a number of books on cultural diversity, puts it this way: "Diversity means optimizing the productivity of all the people that you have in your organization."

On the other hand, affirmative action, says author and teacher Susan D. Clayton, is any attempt to increase the representation of people from groups that have been traditionally underrepresented, which would be typically minority groups and women." Clayton is an assistant professor of psychology at the College of Wooster, in Wooster, Ohio, and the co-author of a book on women and affirmative action.

As small companies approach the year 2000, there are some compelling reasons for expanding their diversity, according to business leaders and experts. Among their assessments:

Employers Can Increase the Quality of Their Workforce. "It would be a mistake for small businesses not to embrace diversity, because they will be missing out on some tremendous talent," says Charles I. Story, president of INROADS, Inc., a nonprofit organization in St. Louis that places minority youths in internship programs in business and industry.

For example, Richard G. Cortez thought he was doing Hispanic college students a favor when he began hiring them for part-time work in his Boise, Idaho, precision sheet-metal business, Metalcraft, Inc. Cortez, who is Hispanic himself and is committed to diversity in his 40-employee company, quickly found the students were eager to learn and very adaptable to computer systems. He realized, he says, "I've got something here that's worth gold!"

Customer Bases Are Becoming Even More Diverse Than the Workforce.
The Glass Ceiling Commission notes that two out of every three people in the U.S. are females or minority-group members or both. Says the SHRM/CCH survey: "An organization with diverse employees can better meet the needs of diverse customers."

The companies she works with, says Ann M. Morrison, a Del Mar, Calif., diversity and leadership consultant, are looking at "the spending power of formerly niche-market groups like Latinos, like—in some cases—women, like blacks, and they're seeing that when you total up the so-called minority marketplace in the U.S., it's something like the GDP [gross domestic product] of Canada."

We're Becoming a Global Marketplace. Other countries, George Henderson forecasts, will provide impetus to U.S. contractors and subcontractors to increase the diversity of their employees. "It's inconceivable to me, as an illustration, that companies large or small that deal with South Africa can ignore the ethnic composition of the workforce," he says.

More and More Businesses Are Owned by Women and Minorities. An estimated 7.7 million U.S. businesses are owned by women. New figures on minority-owned businesses won't be available until later this year, but there were 1.2 million minority-owned businesses in 1987, up from 743,000 five years earlier.

The owners of these businesses may find it easier to sell to and more desirable to buy from businesses where women and minorities are included at management levels.

For all its benefits, diversity poses some problems and challenges. What's not often talked about, says Henderson, is the fact that different ethnic groups within a working environment can be very competitive with and antagonistic toward one another.

Communication takes extra effort. At Atlas Headwear, for example, bilingual employees help bridge the gap with employees who do not speak English.

Backlash—that is, resistance to diversity, particularly by white males—is another problem. The SHRM/CCH report says that white male resentment may be a result of narrow definitions of diversity that have excluded white males as well as a perception by some of them that diversity means preferential treatment for some groups. Charles Story attributes much of the white-male anger to the fear of losing jobs as a result of the downsizing of American corporations and of the recession.

Most small businesses can't match the resources that larger companies can draw upon to help them meet the challenges of—and enjoy the benefits of—a more diversified work force. Nevertheless, experts say there's much that owners and managers of small businesses can do for little or no cost to nurture diversity now and to prepare for the increased diversity that's on the way. Here are some suggestions:

• Educate yourself. Start with the Federal Glass Ceiling Commission's detailed report, *Good For Business: Making Full Use of the Nation's Human Capital,* available for $17 from the U.S. Superintendent of Documents, (202) 512-1800, or at regional Government Printing Office bookstores.

Other good sources of information and direction are *Cultural Diversity in the Workplace: Issues and Strategies,* by George Henderson, available for $19.95 plus $3.50 shipping from Greenwood Publishing Group, 1-800-225-5800; and *Justice, Gender and Affirmative Action,* by Susan D. Clayton and Faye J. Crosby, available for $13.95 plus $3.50 shipping from the University of Michigan Press, (313) 764-4392.

Charles Story notes that some local and state chambers of commerce offer diversity-education programs; if yours doesn't, you might want to urge it to do so.

• Start thinking of diversity as a strategic business issue, advises the SHRM/CCH report. Review the business reasons for expanding diversity that are listed above, and begin exploring what they mean to your company.

• See if changes might be needed in your corporate culture. Wooster College's Clayton suggests you examine your company's practices to see if you are unintentionally excluding members of certain groups. If so, she says, you can adopt changes to make hiring and advancement more inclusive. Like Atlas Headwear, for example, you can recruit through government agencies, community and church social-services groups, and refugee-resettlement programs.

The debate over affirmative action may rage on, but Charles Story says he expects "calmer heads will prevail," reaffirming its positive aspects and eliminating the negative ones. Business people need to strip this argument of its divisiveness, he says, and "steer it toward what is the best way to embrace people who want to work."

In any case, Susan Clayton says, it will be easy to tell when affirmative action is no longer necessary: "When you look around the workforce and see that members of all groups are being employed and that they are being employed at the high levels as well as at the lower levels, then we won't need it anymore."

Maybe we're getting there. The American Management Association study showed there have been increases in the share of management jobs held by minorities since AMA first studied the topic in 1992. And corporate recruitment of minority college seniors represented 24 percent of total campus hires last year, an all-time high, according to a study by the Hanigan Consulting Group, of New York City.

The increasing presence of women and minorities on boards can alter the way companies look, too. Frederick A. Miller is the only African American on the board of Ben & Jerry's Homemade Inc., in Waterbury, Vt. Asked if he influenced the ice cream company's recent selection of Robert Holland Jr., a black, as its president and CEO, Miller, an Albany-based management consultant, replies that Ben & Jerry's was committed to considering women and candidates of color as well as white men. Having a person of color on the board, he says, "let the search firm know that the organization was not joking about this."

Some critics—and some business owners, too—think diversity is just a fad. "Some people looked at computers that way, too," Ann Morrison says. "Remember that? 'Computerization? It'll go away one of these days.'"

Source: Sharon Nelton, "Nurturing Diversity," *Nation's Business,* June 1995, pp. 25–27. Excerpted by permission, *Nation's Business,* copyright 1995, U.S. Chamber of Commerce.

Analyzing the READING

To build your reading skills, use the following questions to analyze the content and to develop your reaction to the article. Write a one-page entry in your class journal or prepare your answers for class discussion.

Reading for Content

1. What is the main idea (or thesis) of the article?

2. What kind of facts, statistics, stories, quotations, or examples does Nelton use to support her main ideas? Find one example of each type of support. Which are the two most convincing types of support and why?

3. Can you divide the article into an introduction-discussion-conclusion pattern? What part of the article is informative, and what part is persuasive?

4. Nelton says there are "compelling reasons" for encouraging diversity in the workplace. What does *diversity* mean? What does the term *affirmative action* mean? Does the writer define these terms in the article? Why or why not?

5. Who is Nelton writing for? What does she want the reader to conclude after reading this article? Is her purpose clear from the introduction? Why or why not?

6. What do you think of Nelton's tone? Does her informal reporting style work for the intended audience? For you? How does it compare with Lacitus' tone in the reading appearing in Chapter 1?

Reading for Reaction

1. In what specific ways could employers "increase the quality of their workforce"?

2. What reasons are there for ending and for continuing affirmative action?

3. Does discrimination still exist in the workplace? If yes, in what ways? What should be done to counteract it?

4. How does diversity in the workplace affect communications? What additional skills or strategies do people need for them to work well with co-workers who come from different ethnic backgrounds, cultures, or religions?

5. Have either you or someone you know experienced discrimination? Think about sharing your experience with the class. How was the situation resolved? Do you believe this was fair? Why or why not?

DEVELOPING A WRITING PROCESS FOR THE WORKPLACE

The goal in developing a writing process for the workplace is to know what needs to be done so well that any routine kind of writing can be handled very quickly, and more complex kinds of writing can be tackled with confidence because you know **what** to do and **how** to do it.

Analyzing Your Writing Process   Exercise 2.1

In your class journal, or as a class exercise, review the description of your writing process you prepared for Exercise 1.3 in Chapter 1 and answer the following questions:

1. How does your writing process allow for interruptions?
2. How does technology affect your writing process?
3. In what ways does your writing reflect your personal integrity?
4. Do you always follow every step in your writing process? Where are you most likely to cut corners and why?
5. Do you sometimes leave key steps out because you need to meet a deadline? In what ways could this be prevented?
6. How does your writing process change when you work in a group? What parts of your process become more important when you are working in a group? Less important?
7. Which parts of your writing process do you think are most important to writing well?

Revise your description of your writing process (if needed), review this description with your discussion group, and turn it in to your instructor, including your analysis and all rough drafts when final changes have been made.

Your routine writing may include completing orders, filling out customer forms, correcting errors made by your company or by your customers; giving information to customers; processing orders, exchanges, returns of merchandise, and payments; or resolving routine customer complaints. These routine kinds of writing will be discussed in more detail in Chapter 8: Writing for Customers.

Depending on your technical specialty, your routine writing tasks may include completing patient charts, shift logs, or progress reports; describing equipment breakdowns; preparing specifications; or writing field reports; as well as some of the writing described above.

ANALYZING A WORKPLACE WRITING PROCESS

To anticipate the tasks involved in workplace writing, you can use the six-step writing process shown in Exhibit 2.1 to organize and manage the work you do for any writing task—small or large.

One reaction we might have to reading this list carefully is, "No wonder writing is so tough! There's so much involved!" Many of these decisions we make intuitively, based on how important the writing project is to our day-to-day job. However, as a new employee, you'll have the added responsibility of figuring out what your company's unwritten expectations are about each stage in the workplace writing process.

You will find that the time allocated to each stage of the writing process will change according to the size and importance of the specific job and its deadlines,

Part I: Focusing on the Workplace

■ Exhibit 2.1   Describing a Workplace Writing Process

Plan
Clarify writing purpose (to inform, to persuade)
Identify audience needs
Clarify type of document needed (memo, letter, report, presentation)
Identify resources to complete project (include people, time, and information needed)
Set or confirm deadlines
Draft content outline and research questions
Develop project work plan or proposal

Research
Review company files
Gather data (statistics, graphics, survey, interview)
Do library research if needed (include Internet)
Interview colleagues and specialists
Talk to everyone affected by problem
Take notes and record sources
Review and think about collected materials
Plan graphics needed to highlight and explain
Refine content outline
Decide length and appropriate format

Draft
Develop problem statement (thesis)
Further refine content outline
Draft ideas and support on paper
Weed out unnecessary information
Try to follow an introduction, discussion, conclusion order

Revise
First level of revision
Check purpose:
Does this writing solve the problem for this audience?
Check content:
Are main ideas clear?
Is there enough supporting evidence for this reader?
Are numerical calculations and graphics correct?
Are conclusions drawn from the data reasonable?
Check organization:
Are ideas presented logically?
Is the sequence of ideas most effective for this reader?

Second level of revision
Check style:
Are ideas expressed as clearly and concisely as possible?
Check mechanics:
Are grammar, punctuation, and spelling as correct as possible?
Check tone:
Is the style courteous?
Check document design:
Do page layout, headings, graphics, paragraph length, choice of memo, letter, or report enhance readability?

Review
Give to peer or supervisor for review with specific questions from first and second level of revision

Produce
Allow sufficient time for production (final word processing, proofreading, and photocopying for distribution)

the resources available, and the abilities of the people working on the project. You'll also find that these stages blend into each other. Most of the time, we cannot simply begin at one end and work through to the end. Real workplace writing projects can quickly become complicated and messy! As long as you have a clear picture of the overall process and your writing goals, the process will seem an ordinary one.

How Does a Writing Process Really Work?

It may be comforting to believe that everyone follows a writing process by starting at the beginning and working straight through all the stages until coming to the final document. Actually, you may find yourself digging for one more bit of last-minute research at the same moment that final production has begun.

Analyzing Your Workplace Writing Process   Exercise 2.2

Consider the workplace writing process highlighted in Exhibit 2.1. Write responses to the following questions in your class journal or prepare notes for your discussion group.

1. Do you agree with the order of this writing process? What would you change and why? Are there important tasks that should be added? What and why?
2. Which of these processes or steps do you think takes the most time and effort? Why? Allocate percentages to each step so that the total adds up to 100%. Did doing this change your original idea of how much time each step takes? Compare your percentage breakdown with other members of your discussion group. Were there significant differences? Where and why?
3. Which of these steps is most difficult for you? Why? Which is easiest?
4. Will your answers to these questions change if you are working in a group to produce a major report? Consider how and why for each step. Would working in a group change your percentage breakdown?

Share your answers with your discussion group, and turn in a summary of your group's discussion to your instructor with your notes attached.

Now that we have computer technology, making final copy changes is easier—although you will alienate production people if you continually push them with very tight deadlines.

Our minds are capable of working on several jobs at the same time, even if we don't realize it. So, while you are drafting, your brain may also be evaluating and revising at the same time—and interrupting your drafting. The flowchart in Exhibit 2.2 shows how this may work.

As you "read" the flowchart in Exhibit 2.2, think about your own writing process and how many times you can work step-by-step or how many times you loop back. Some writers can produce fine writing in a step-by-step fashion. Some of us can only work step-by-step on simpler writing projects.

■ **Exhibit 2.2** **Thinking and Writing Tasks Overlap and Repeat! Writing Can Be Messy!**

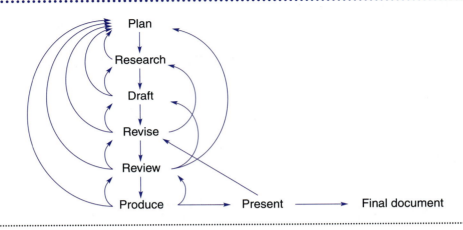

However, the more complex the writing project is, the more likely you will find insights about the meaning of the information you gather after you have written your rough draft or after you have talked to your peer reviewers.

Now, imagine how this flowchart changes when you write with one, two, or three other people.

How Does My Individual Writing Process Fit In with Team Writing?

When other people are depending on you to write a part of a document, your writing process must meet the team's expectations and schedule. In this case, what is required that is different from when you write alone? Flexibility, dependability, consistency, timeliness, good listening skills, and good writing skills all are necessary. Your ability to ask questions and clarify the team's expectations and needs as well as your knowledge about group roles and processes are added strengths.

WRITING STRATEGIES: WHAT MAKES GOOD WORKPLACE WRITING?

The last section concentrated on how you can improve your writing process whether you work alone or as part of a team. But how will you know when you have written something that is "good"? You can answer this question by considering four key factors: **writer, audience, purpose,** and **content.** Each time you work on any writing, consciously or not, you are using these four key factors to refine your writing.

Writer. Who you are as a person, your attitudes, your skills, and your experiences—all of these affect what and how you write. Additionally, your writing will be affected by your position in the company, your department, and even by your employer. All of these factors contribute to your writing.

Audience. Your audience also has a personal side. You will need to consider your audience's background, attitudes, skills, and experiences. Additionally, who the audience is—their titles, department, employer, industry, and purpose in needing your information—all of these factors will affect what you write and how you write, as shown in the diagram below.

Your audience may be the **primary reader** if the person has directly requested the information or is a decision maker. You will want to know what this reader needs and why.

You will also have many different kinds of **secondary readers;** these are people who may use the document, read parts of the document, or be affected by parts of the document. It's also important to anticipate these readers' needs and reactions.

Some researchers say the relationship between the writer and the audience is the most important relationship of all in workplace writing. As a technical specialist, many times you will have to translate what you are writing about for a less informed audience—even if your audience is a supervisor or senior manager. Here are some additional questions to consider about your audience:

- Is your audience inside or outside your company?
- Is your audience a novice or an expert? A technician? A specialist? A generalist?
- Is your audience a decision maker?
- Who besides your primary audience will read this? How many different kinds of readers will you have?
- Why does your audience need this information?
- How will your audience use this information?
- Where is your audience in the organization? A senior manager? A technician? A middle manager? A staff or line employee? Your co-worker? Your staff?
- Are there cultural differences between you and your audience?

How will all this background on your audience affect the information you need to provide? Over time, you will consider the needs of your audience almost intuitively, that is, every time you write.

Exercise 2.3 Identifying the Range of Your Audience

Think about a job you have held (if you have no work experience, think of any club activity or a class you have taken) and the people you have been in contact with at this job.

List as rapidly as you can four or five of these people, who they were, what their *functional role* in the company was (their job or position title), and how you communicated with them most of the time—in writing (by letter, memo, or forms) or speaking (face-to-face, in meetings, or by phone). Add a brief word or two about their personal background, perhaps their education or their cultural identity. You might want to use a grid to help you organize your information.

Person	Title	Functional Role and Background	How I Communicated
_____	_____	_____	_____
_____	_____	_____	_____
_____	_____	_____	_____
_____	_____	_____	_____
_____	_____	_____	_____

Look over your list. What do you notice about the range of your audience's needs? Was one communication mode (writing, speaking, listening) preferred over another? Did you adapt your communication style to your audience? List two or three reactions you have to this list and share it with your discussion group. Summarize your group's findings in a group memo to your instructor.

Exercise 2.4 Finding Primary and Secondary Audiences

In your class journal or as a class activity, identify the audience for each of the following. Who is the primary audience? Who is the secondary audience? Are there other readers? Jot down some notes for discussion with your group.

The Document	Primary Audience	Secondary Audience
1. A report card.	_____	_____
2. The yellow pages in a phone book.	_____	_____
3. Instructions for a new computer program.	_____	_____
4. An employment application.	_____	_____
5. A personnel evaluation.	_____	_____

Chapter 2: Developing a Writing Process for Work   **35**

6. A company's annual report.   _____   _____

7. A client's account.   _____   _____

8. All employee pay records.   _____   _____

9. A proposal.   _____   _____

For further discussion, answer this question: Does the tone of your writing change when you write to your supervisor? To senior management? To your peers? To your staff? Should it?

How Is the Workplace Writing Context Different from Writing for School?

When you write an assignment for a college class, your audience typically is made up of your instructor and perhaps your classmates. In the workplace, your audience is generally larger than a single reader—even for material marked "personal and confidential."

Think of the times so-called in-house reports or confidential memos have been leaked to the press to the embarrassment and sometimes increased legal liability of companies. This means every time you write something, even if what you are writing is just for the file, you will need to consider how other readers might respond—both inside and outside your company.

Written documents also travel inside companies: A report intended only for your supervisor might be attached to another report for senior management. A confidential assessment of a project with poor performance might float through the company. Writers need to be tactful as well as honest.

When you write an assignment for a class, you usually represent only yourself. When you write for the workplace, the writing itself will be evaluated to see if it meets the needs of your supervisors, your co-workers, or your customers. You represent the company as well as yourself in workplace writing.

Although you may need occasional library research to support classroom writing, in the workplace you will need to consistently support your workplace writing with specific information, dates, times, places, amounts, dollars, hours expanded, etc. You may find this information through research—sometimes in libraries, with computers, or in company files.

Your workplace drafts often must be reviewed more thoroughly than school assignments. Colleagues will work closely with you during revision and final production of any document. In the workplace, who does the *actual writing* is less important than the usefulness, the quality, and the timeliness of the writing.

Purpose.   Your purpose in writing also affects your audience and the content you select. You may be writing **to inform** or **to persuade.** In the workplace, you may also be writing to instruct or teach someone about a task or to direct someone to do something.

When your purpose is primarily *to inform*, you may want

- To give someone information.
- To tell someone how or why something works.
- To tell someone to do something.

When your purpose is primarily *to persuade*, you may want

- To sell something.
- To analyze or evaluate information for someone.
- To change someone's mind.
- To make a recommendation.

Rarely will you write to entertain, although this is a key strategy used by journalists to keep their audiences reading. Most workplace audiences are already motivated to read your document; they need your writing to get their work done.

Workplace writers need writing that is clearly and concisely expressed. Most business people, particularly those outside an organization, are too busy to wade through inarticulate or poorly expressed communication. Further, poorly written documents can result in miscommunication, which then leads to the need for *more* communication. Whenever a memo, letter, or report needs clarification in the workplace, money and time are wasted.

Notice how your purpose will control what content you include. Your purpose will also be influenced by how you envision the audience's reaction. Is your intended audience in favor of, neutral to, or opposed to what you write?

Every time you write, what your purpose is and who you are writing for will affect what you write and how you write it, as shown in the next diagram.

Exercise 2.5 Analyzing the Writer's Purpose

In your class journal or for class discussion, identify the purpose in writing each of the following documents. Are there instances where you will have more than one purpose?

1. Describe a new piece of equipment.
2. Prepare a training manual.
3. Write a progress report.

4. Complete a personnel evaluation.
5. Record an order for supplies or materials.
6. Propose a solution to a problem.

After you have identified a purpose for each of these writing situations, now consider how this document might need to be changed if it is directed to either a receptive or a hostile audience.

Discuss your findings with your group, and consider at what point in your writing process you analyze audience and purpose. What specific questions do you ask? Summarize your findings in a short group memo to your instructor. Include all rough notes.

Content. The content of your writing is the subject you are writing about. Here you will need to decide exactly what the subject is, how much supporting information to include, and in what order the information should appear.

You'll also make some decisions about logical organization, readable style, and document design (whether or not to use graphics or other visual aids, how to use headings, and how long your paragraphs will be). We can summarize these factors affecting content into the following four key categories:

- Thoughtful **content** = information your audience needs.
- Logical **organization** = information presented in an order that meets audience needs.
- Clear and readable **style** = information that is written clearly, correctly, courteously, and concisely so the audience can understand it easily.
- Clear and readable **document design** = information is presented on the page by manipulating page layout, graphics (especially numerical data), headings, and paragraph length so the audience can read it easily.

In some ways, thinking about your background as a writer, your audience's needs, and your writing purpose are activities that you carry out during the planning process. But working on *content, organization, style,* and *format* are activities that are crucial to your actual writing. Exhibit 2.3 shows how interconnected these writing factors are. These writing factors are so important, we will be working on them throughout this book.

Factors That Affect Your Planning   Exhibit 2.3 ■

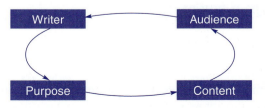

Exercise 2.6 Revising Your Writing Process

Consider the writing process we have been talking about: planning, researching, drafting, revising, and reviewing. Make notes to answer the following questions either for your class journal or for class discussion.

1. Would the writing process change for any of the purposes of writing (to inform, to direct, or to persuade)? If so, how or why?
2. Which of the writing factors (writer, audience, purpose, and content) would be most important in each step of the writing process? Why or why not?

Plan	Revise
Research	Review
Draft	Produce

3. In what situations would one step in the writing process be more important than other steps? Why or why not? How do you know?

Review your notes with your discussion group. Come to a consensus and turn in a group response to your instructor.

What Are the Criteria for "Good" Workplace Writing?

The best writing meets your reader's needs. If you follow the guidelines discussed here, you will develop your own process that will include some work in each of the steps outlined here: planning, researching, drafting, revising, reviewing, and producing.

Your writing will be most effective when it:

- Is relevant for its *audience* and fulfills its *purpose.*
- Helps the reader solve the intended problem (*content*).
- Is clearly and correctly expressed (*style*).
- Is logically presented (*organization*).
- Is easy to read (*document design + proofreading*).

TIPS FOR OVERCOMING WRITER'S BLOCK

Workplace audiences are very practical readers. They need the information you have, and they usually have many questions for which they will want the answers right away!

Sometimes when a writing project is important to you or you are working with tight deadlines, you may experience the dreaded *writer's block*. Writer's block occurs when you want to write, but somehow it is impossible to get your thoughts down on paper.

Here are some suggestions that may help you get started if you experience writer's block. Use any of these that seem helpful. Do not worry about using exactly the right words, the order of your ideas, whether your ideas are complete, or even spelling! These techniques are used just to get you started.

1. Write a Discovery Draft.

Set a timer and write nonstop for 15 minutes, putting anything you think of down on paper. Do not stop to evaluate or reread what you have written. Just keep writing. After your timer goes off, go over what you have written. Underline any key ideas that you think will be useful. Set your timer again and write about the underlined ideas.

2. Clarify Why You Are Writing.

Start by writing down the problem you are trying to solve: Jot down your audience's needs, and then start writing definitions for key terms. Next, use one of the other planning or prewriting techniques listed here to get more ideas down on paper.

3. Make a List of Key Questions You Need to Answer.

Write one sentence that describes the problem you are trying to solve, review your audience needs, and then write a list of questions. Start each question with one of the following key words:

What... How... Why... When... Where... How much... Who...

Try to write at least three questions for each key word. **What, how, when, where, who,** and **how much** tend to prompt descriptive writing. Asking **why** leads to analytical writing. After you can't think of any more questions, begin to organize your work by putting similar questions together, combining questions, or deleting questions that seem repetitive. Use your problem statement to decide what is useful.

4. Talk to Key People, a Colleague, or a Friend.

Discuss your ideas or the project. Sometimes as we explain our ideas to others, what we are thinking about becomes clearer to us. New insights or ideas on how to approach the planning, researching, or drafting steps occur to us as we speak!

5. Make a Map.

Some people are visual thinkers. They can "see" better if they draw a picture or a map of their ideas. Try this out to see if this technique works for you. A content map can be very close to an outline, only with a freer format.

■ Exhibit 2.4 A Student's Map

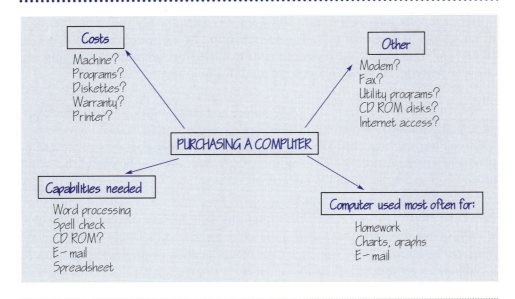

An example of a map is shown in Exhibit 2.4. This student started in the middle with her general topic and then branched out to the main subjects supporting her main topic. She did not put ideas in any particular order, but at this point she can "see" the main sections of her project.

Once you have finished your map, you may notice that some areas include more ideas and supporting information than do others. A map like this doesn't really show which idea should come first or second. Ask which ideas are most important, and then organize and write your first draft.

6. Use a Formula. You can also use formulas (like a five-paragraph theme) to get started. For example, using an introduction-discussion-conclusion pattern helps you know what you need to write in each section of a memo. Remember to revise your writing so the "bones" of the formula don't show!

7. Consider the Context for Your Writing. You may need to clarify who really has final authority over the document. Sometimes the revision process you use within the workplace (for example, sending your draft to several colleagues or supervisors) can drastically change your original document. However, if the project changes after the document has been reviewed, the writer's job is to make the needed revisions.

8. Think About Your Expectations for Your Own Writing. You may not be happy with your first draft. Very few writers produce final copy on their first draft.

Sometimes we care so much about the quality of what we are writing that we have difficulty in getting our ideas down on paper. Linda Flower, a researcher on writing techniques, suggests we "satisfice" when we are drafting. This means turning off that internal critic and accepting what you have written for now, knowing you can revise it later. Inspiration takes only a moment, but good writing requires revision!

WRITING STRATEGIES: WRITING WITH TOPIC SENTENCES

Most workplace readers skim what they read; they do not have the time to read everything that is given to them. Sometimes, however, workplace readers need to read very technical information slowly and carefully to memorize the information and to use it.

You can improve your writing and make it easier for your readers if you organize your writing so that your reader can grasp the main ideas quickly.

Topic Sentences. These sentences introduce the main idea to the reader. A topic sentence generally has two parts: *a main idea* and *the opinion of the writer*, as shown below

Each word in the topic sentence "controls" what must be developed in the paragraph (or in the longer piece of writing). Notice which words in the sample topic sentence predict what the writing will need to cover.

Topic sentence: Homeowners in Phoenix, Arizona, prefer solar power because of its low cost and its lack of pollution.
Main idea: Solar power in Phoenix, Arizona
Opinion: Homeowners prefer it
This paragraph should answer the following questions:

What is the low cost of solar power?

How does solar power avoid pollution?

How does solar power compare to other types of power in Phoenix?

What does all this information mean to the reader?

The writer working with this paragraph would have to prove why solar power is preferred. The writer could choose to support this point of view by using facts,

Part I: Focusing on the Workplace

examples, or statistics. Which of these he or she chooses and how much support is needed will depend on how receptive the audience will be to this idea.

Just as the topic sentence controls what goes into a paragraph, so too the main idea (sometimes called a thesis) should control what goes into any longer piece of writing such as a memo, letter, or report. Notice how the main idea (or thesis) is introduced in the first paragraph from our chapter reading:

Main idea Even if affirmative action is dismantled, **diversity of the workforce is here to stay.** Business owners and managers, experts say, will still need to maintain or step up efforts to recruit and advance minorities and women in the year 2000 and beyond. That's because having a diverse workforce **Writer's** and managing it effectively **will simply be good business. opinion**

Notice that this opening paragraph lays out some serious writing challenges for this writer, Sharon Nelton. Not only will she have to show why a diverse

TIPS FOR WRITING PARAGRAPHS

Here are some suggestions that can help you write focused paragraphs and understand how paragraphs fit together in longer documents.

- Try to put the **main idea** into **one sentence** (called a topic sentence).

- Decide whether the **topic sentence** should be placed **first** in each paragraph.

- Make sure that **each paragraph** discusses only **one** idea.
  The **topic sentence** controls what goes into each paragraph just as a **thesis statement** controls what goes into a longer document.
  The rest of the paragraph develops and supports the main idea of the paragraph, following the guideline of one main idea to each paragraph.

- Decide if your paragraph should end with a **summary statement** that draws a conclusion from the information presented or a **transition statement** that leads the reader to the next paragraph or section in your writing.

- Try to **limit the length** of your paragraphs to 6–7 lines of text for readability.

- Check that all paragraphs in any piece of writing relate to the **purpose** of the writing and the needs of your **audience.**

As you'll notice in the paragraphs we study in this chapter, writers do not always follow these guidelines. For example, journalists frequently write very short paragraphs for emphasis or so that readers will keep on reading. Writers make exceptions to these guidelines to make their writing easier to read or understand —or to emphasize a point. Paragraph writing is an art, not a science!

workforce is here to stay, but she will have to explain how this will benefit employers. If Nelton wants to help business owners achieve these goals, she will also need to show them how. Does the resulting article achieve these goals?

Analyzing Topic Sentences Exercise 2.7

As your reading and writing will show you, thesis statements are rarely that first sentence in a longer document; however, topic sentences for paragraphs are often placed first in informative writing to introduce the content of that particular paragraph.

Practice analyzing the parts of a topic sentence for your class journal. First, answer the following questions for each of the topic sentences below:

- Underline the words that introduce the **topic**.
- Circle those words that suggest the **writer's opinion** or point of view.
- List the **questions** that each paragraph would have to answer.

Finally, decide which of these topic sentences would not work well as a topic sentence. Why?

1. Operations at the Smithville plant can be improved by controlling costs and by limiting overtime.
   a. Topic.
   b. Opinion.
   c. Questions this paragraph needs to answer.
2. Multinational companies face many challenges in hiring and developing technicians who can work effectively overseas.
   a. Topic.
   b. Opinion.
   c. Questions this paragraph needs to answer.
3. Profits are good.
   a. Topic.
   b. Opinion.
   c. Questions this paragraph needs to answer.
4. Emerson Electronics laid off 25,000 workers by July 30, 1994.
   a. Topic.
   b. Opinion.
   c. Questions this paragraph needs to answer.
5. Entry-level workers must complete three kinds of training before they can begin work in their new unit.
   a. Topic.
   b. Opinion.
   c. Questions this paragraph needs to answer.

After completing this exercise, write a definition of a topic sentence to share with your discussion group.

As this exercise has shown you, some topic sentences can be either too general or too narrow. Sometimes if the topic sentence is left out of a paragraph, the reader must work hard to find the meaning of the paragraph. Your goal is to write topic sentences that precisely introduce and limit the topic to be discussed.

44   Part I: Focusing on the Workplace

Exercise 2.8   Writing Topic Sentences

Practice writing topic sentences for each of the following general topics. Narrow your topic so that the discussion would be one paragraph in length. Include controlling words that limit your main idea.

- Job hunting
- A promotion
- Education

Share your topic sentences with your discussion group. Check to make sure that each topic sentence includes a sufficiently narrow subject that will fit both a paragraph **and** the writer's opinion to shape the development of the paragraph. Make any revisions that seem necessary, and turn it in to your instructor or put it into your class journal.

WRITING STRATEGIES: DEVELOPING IDEAS BY USING FACTS AND EXAMPLES

Workplace audiences prefer facts to feelings. Few will approve a new project because the writer thinks it's a good idea without reviewing the benefits, the costs, and the plans that such a project might involve. Workplace writers work hard to pick out the most important facts to support their proposals. Learning how to use paragraphs to support your main idea is an important writing skill.

Here's an example of two **supporting paragraphs,** taken from the middle of the chapter reading, that emphasize facts. Notice how each fact helps the reader gain a clearer picture of the main idea for that paragraph.

Topic sentence   New figures bear out the Hudson Institute's finding that white males are becoming more of a minority in the workforce. A report issued in March by the Federal Glass Ceiling Commission, a 21-member bipartisan body created by the Civil Rights Act of 1991, says that 57 percent of those in the workforce today are women or minorities or both.

Supporting facts

Topic sentence   "Women and minority men will make up 62 percent of the workforce by the year 2005," according to the report. However, it says, "the world at the top of the corporate hierarchy does not yet look anything like America." The study shows that 97 percent of the senior managers of the nation's largest corporations are white and at least 95 percent are male.

Supporting facts

Here's an example of another section that uses a supporting quotation and an explanation to clarify the main idea.

Topic sentence	Managing diversity goes "far beyond" meeting the legal requirements of equal employment opportunity and affirmative action, **according to** a joint survey of 785 human-resources managers conducted by the Society for Human Resource Managers and Commerce Clearing House. The 1993 SHRM/CCH report defines it as "the management of an organization's culture and systems
Supporting quotation	to ensure that all people are given the opportunity to contribute to the business goals of the company."
Explanation	**This means including** employees not only without regard to the obvious differences of race, sex, and age **but also** without regard to such "secondary dimensions" of diversity as marital or family status, sexual orientation, and disabilities, the report says. **It also means including** white males.

Using Transitional Words and Phrases to Clarify Meaning. Notice how **transitional words** (shown in boldface above)—actually words, phrases, sentences, or even paragraphs—connect one part of the writing to another. Review the list of transitional words in Exhibit 2.5 to help you see what tools writers use to connect their ideas at the word and phrase level.

Transitional Words and Phrases   Exhibit 2.5 ■

Repeating key words and using transitional words helps the reader understand how ideas are connected.

Accordingly	However	Otherwise
After all	In addition	On the other hand
Again	In fact	Perhaps
Also	Meanwhile	Still
As a result	Nevertheless	Then
Finally	Next	Therefore
For example	Now	Too
Further	Of course	Yet

Finding Transitional Words   Exercise 2.9

Read over the following paragraph and underline any transitional words or phrases, using the list in Exhibit 2.5 to help you.

    Advances in technology allow a clearer look into the earth. Supercomputers, for example, now provide geophysical images in three dimensions, crunching numbers that sum up how scientists believe the earth's insides work. A team led by UCLA geophysicist Paul Tackley used such computer power to simulate two billion years of mantle movement.

Such computer models need information about buried rock. To search for clues, geophysicists study the rocks themselves. With a diamond anvil, they replicate deep-earth pressures and heat and observe how the mineral content of the rocks changes.

Earthquake vibrations offer another view of what lies underfoot. As they travel through the earth, these seismic waves pass more quickly through cooler areas and more slowly through warmer ones. Monitors around the world record their arrival. Computers then turn arrival data into images resembling sonograms.

This technique, seismic tomography, shows one moment from cycles lasting millions of years, and that presents a challenge. "It's almost like trying to figure out the plot of Humphrey Bogart's *The Big Sleep* by looking at a single frame of the movie," says Michael Wysession, a Washington University seismologist.

Source: Keav Davidson and A. R. Williams, "Theories of Motion." Illustration by Chuck Carter and Allen Carroll. *National Geographic*, vol. 189, no. 1 (January 1996), p. 108.

Using Comparisons and Metaphors to Clarify Meaning. Another writing strategy that Keav Davidson and A. R. Williams used in the excerpt in Exercise 2.9 is comparing something the reader already knows to something that is new or unfamiliar. Using **comparisons** (also called **similes** and **metaphors**) can help the reader understand new ideas more quickly and clearly. For example, comparing the cycle of earthquake waves to one frame from a popular movie helps the reader to understand the "seismic waves" as part of a larger system of constant waves moving through the earth's crust.

A **simile** explicitly compares two different things and is usually introduced by *like* or *as*.
Example: He eats like a bird.

A **metaphor** implies a comparison between two different things, without using *like* or *as*.
Example: He is a bird when it comes to eating.

These kinds of comparisons can be helpful to the reader, especially when they are concrete; however, with the example above, even though most birds "pick at" their food, birds also eat several times their weight each day. Which comparison do you want your reader to think of?

Exercise 2.10 Analyzing the Writer's Support

Read this informative paragraph on hot spring cyanobacteria, then answer the questions that follow in your class journal or prepare for class discussion.

Hot spring cyanobacteria are wonders of life at high temperatures. Some live in waters as hot as 167 degrees F (75 degrees C). At this temperature, they are usually yellow, but become darker—orange, rust, or brown—as the water cools. Between 115 to 130 degrees F (46 to 54

Chapter 2: Developing a Writing Process for Work   **47**

degrees C), other species may appear which will modify the colors even more. Certain varieties are scientific curiosities because they are extremely specific for their environment. They may be found around the world living only in hot spring waters.

Source: The Yellowstone Association (1995).

1. What is the topic sentence?

2. What is the writer's main idea? Can you find the writer's point of view?

3. Can you predict what should be in this paragraph just from reading the topic sentence? Is anything missing? What and why?

4. What is a fact? Can you describe the difference between a fact and a conclusion? Find two facts and two conclusions the writer used. Which, in your opinion, is more effective?

5. Did the writer use any comparisons or metaphors to clarify new information?

6. Do you feel the writer provided enough support for the main idea? Too much support?

7. Does this paragraph have a conclusion? If yes, where is it? If not, why not?

EDITOR'S CHECKLIST FOR REVISION

Use this checklist to help you in your revision before you turn in a writing assignment. Ask someone in your discussion group to act as your peer editor and to answer the following questions for your rough draft.

Encourage your peer editor either to set aside about 20 minutes to work through the list with you, discussing each question, or to write out sentences in response to each question.

Check Audience and Purpose:

_____ 1. Can I quickly tell *who the audience is* for this writing?

_____ 2. Is the writing *purpose descriptive or persuasive*?

_____ 3. Is the *topic too large* for this document?

Check Content and Organization:

_____ 1. Do *paragraphs begin with topic sentences*? If this is a longer document, can I easily *find a thesis statement* in the introduction?

_____ 2. Does *each topic sentence have a main idea and an opinion*? Are any topic sentences too narrow or too general?

_____ 3. Does the writer *use facts or examples to explain the main idea* sufficiently? Is there enough support for the main idea?

_____ 4. Do any *words need to be defined*? Would *comparisons or metaphors* help clarify new information?

_____ 5. Does *each paragraph end with a conclusion* that either summarizes the main idea or interprets it? Would you as a reader like more information or more interpretation?

Part I: Focusing on the Workplace

_____ 6. Is *any information missing*? What and where?

_____ 7. Are *ideas presented logically*? Are they in the same order as the topic sentence (or thesis) that introduces them?

Check Style and Document Design:

_____ 1. Do you see any places where *ideas could be expressed more clearly*?

_____ 2. Are *ideas presented as concisely as possible*? Are there any extra sentences or words that could be deleted?

_____ 3. Are *any paragraphs too long* for easy readability?

_____ 4. Does the writer *use transitional words* so that sentences flow smoothly from one to the next? Are transitions used between paragraphs? Can you suggest any changes?

_____ 5. Can you spot *any spelling, grammar, or punctuation problems*?

WHAT'S COMING NEXT?

This chapter has emphasized understanding your own writing process as well as reviewing some of the factors that affect the planning of your writing—including audience and purpose. You have read about (and practiced) writing strategies for writing topic sentences, writing paragraphs, and developing ideas by using facts and examples. You can use the previous editor's checklist to review any document.

In Chapter 3, Reading for the Workplace, you'll see how research prepares you to be a good workplace writer. We generally do research in school to support a particular assignment; in the workplace, you'll be researching formally and informally nearly all the time to solve a variety of problems. You won't have time to go to the library for background materials. You'll need to absorb this information from your professional reading that you do at home each week, from observations you make as you watch others work, or from talking to your colleagues. Research is nothing more than the information you need to understand what's happening around you and how you can do your job more effectively.

■ CONCEPT REVIEW

Can you define the following key concepts from Chapter 2 without referring to the chapter? Write your own definitions in a journal entry or review them out loud.

multicultural communications	topic sentence
thesis statement	writing process
prewriting	rewriting
page layout	writing style

Chapter 2: Developing a Writing Process for Work   **49**

multiple audiences

secondary reader

persuasive writing

organization

similes

primary reader

informative writing

document design

transitions

metaphors

WRITING SKILLS REVIEW ■

An important goal of this textbook is to help you express yourself in writing without distracting errors in sentence structure, punctuation, grammar usage, or spelling. To build your skills in editing and proofreading, begin a systematic review of *Appendix C: Writing Skills Review.* Your assignment:

__ Read "Understanding Sentence Structures, Combining Sentences" in *Appendix C.*

__ Complete exercise C.12 through C.15, including the Summary Exercise, in "Combining Sentences," checking your answers at the back of *Appendix C.*

__ Prepare a paragraph summarizing your progress for your instructor or use the form your instructor provides to report your progress. Be sure to highlight any areas you need to further review.

SUMMARY EXERCISE ■

What's Wrong With These Paragraphs?

Prepare a memo to the writer of these two paragraphs listing strengths and weaknesses for each paragraph along with suggestions for improvement. Include a list of the questions you think the paragraph needs to answer. Turn in your finished memo (with drafts) to your instructor or complete this exercise in your class journal.

Paragraph 1
Extremely hot water has properties important in the development of a hot spring's plumbing system. First of all, its lower density allows it to rise easier through small channels. Second, it is a much better solvent than cooler waters; it dissolves astounding amounts of silica, a common component of volcanic rocks. In this way, some small channels are enlarged while others are soon clogged with new deposits.

Paragraph 2
Silica, sometimes called sinter, is the grayish white deposit that lines the bottom of this spring, forms terraces along the runoff channels, and gives the spring its name. Silex Spring's water supply is so great that it usually overflows throughout the year; the overflow provides a restricted hot environment for mats of bacteria and algae. These living mats are food for several kinds of flies that live in and on the hot water. These flies, in turn, are food for mites, spiders, other various insects, and bird life.

Source: Yellowstone Association, "*Fountain Paint Pot and Firehold Lake Drive*" (June 1994).

Part I: Focusing on the Workplace

■ ASSIGNMENTS

Assignment 1: Analyze Writing. Use the editor's checklist to analyze any longer piece of writing or term paper you have completed for another class. If you don't have such a writing sample, use a newspaper or magazine article on a topic you are interested in. Prepare a memo for your instructor that also answers the following questions.

- What was the writer's purpose?
- Who was the writer's intended audience?
- Did the type of writing have an impact on the content, organization, style, or document design?
- Was this particular writing successful? Why or why not?
- What specific editing suggestions would you make to this writer?

Ask someone from your discussion group to read and respond to your memo. The following are some questions your reviewer may ask: Does the introduction clearly introduce the purpose of this memo? Is there enough supporting detail so that the memo can "stand on its own"? Is there a conclusion? Turn in all rough drafts with the final memo and a copy of the writing you analyzed.

Assignment 2: Analyze Your Revision Skills. After thinking about how important efficient writing and revision skills are to your success on the job, write a memo to your instructor describing the strengths and weaknesses of your writing process as it relates to your future career.

Read through the Writing Skills Review section in Appendix C. Complete two or three of the exercises in areas where you feel you may need more work. Report on the results of your study and conclude your memo by recommending that your instructor include two or three key revision skills you want to acquire or improve in your writing class.

Assignment 3: Clarify a Professional Writing Process. Reread Exhibit 2.1 (page 30), which describes the writing process and review it with a working professional or instructor in your field. Does the writing process summarized in this chapter reflect what really happens in the workplace for your field?

Summarize the results of your conversation for your discussion group and the class in a short memo that highlights the most important steps in this process. Identify any missing parts. Make recommendations to the class on how they could prepare themselves for the workplace, based on your reading of this chapter and your discussion.

Use the editing checklist to have a peer editor from your discussion group review your work. Make final revisions and turn the memo in to your instructor.

Chapter 2: Developing a Writing Process for Work   51

PROJECT: CHECKING CONTENT AND ORGANIZATION ■

Work with a partner to analyze the following reading. Your memo to the writers will highlight their writing strengths and your unanswered questions. Answer the following questions and use the editor's checklist to complete your review.

- How could you describe the writer's purpose?
- What do you know about the intended audience?
- Can you divide the short article into introduction, discussion, and conclusion sections?
- Can you find a thesis in the introduction? What is the main point of this article?
- Because this writing is informative, does each paragraph begin with a topic sentence?
- Are transitions used effectively to connect ideas throughout?
- Are any unusual terms defined?
- Do the writers use any similes or metaphors? Where and why?
- Are you satisfied with the writer's use of facts and conclusions?
- Does the article's conclusion answer the question implied in the thesis?
- What is your overall opinion of the effectiveness of this article?

THEORIES OF MOTION

Keav Davidson and A. R. Williams

Push South America and Africa together, and the fit leaves no room for argument. They were locked together as part of a large landmass, Pangaea, until some 200 million years ago, when it began to break into the mobile continents of modern times.

Ever since, the Atlantic Ocean has widened along a hot, crust-producing seam that runs through Iceland toward Antarctica. This process of sea floor spreading adds about an inch a year between the Eastern and Western Hemispheres.

The currents driving these vast tectonic changes sweep through the underlying mantle rock about as fast as fingernails grow. The idea that solid rock can flow is hard to grasp because rock seems immutable in the human time frame. Yet over millions of years, rock can move like glacier ice.

Heat makes the rock in the mantle flow. Most comes from decaying radioactive isotopes in the layer itself. Some comes from the outer core, as from the bottom of a double boiler.

Experts agree on all these points. Yet as Caltech geophysicist Don Anderson points out, "The paradigm behind the science of geology is, 'Life is a mess.'" And scientists have yet to sort out completely the mess in the mantle.

52   Part I: Focusing on the Workplace

Its pattern of movement remains the most contentious issue. Some researchers believe that the upper and lower mantles convect separately, exchanging little or no material. Others think currents sweep through the whole mantle. Increasingly, though, scientists see in the newest studies evidence for both patterns. Richard Carlson, a geochemist at the Carnegie Institution, compares the earth to a layered drink. "It's like alcohol and orange juice. If you stir them gently, they will form layers, but with a lot of stirring, they'll mix."

Source: Keav Davidson and A. R. Williams, "Theories of Motion." *National Geographic*, vol. 189, no. 1 (January 1996), pp. 106–107. Reprinted by permission.

■ READING A MAP

Which do you "read" first—an excerpted article or a map? Most readers would be drawn first to the map—even when a map is shown in black and white. When visuals are used in a longer report, it's quite likely the reader will thumb through the pages, looking for illustrations.

That's because we understand visuals much faster than we understand text. Somehow we get the information all at once by seeing the picture as a whole (and its relationships), rather than having to read the words one at a time and then put the pieces together.

Look at the black-and-white map of ocean currents in Exhibit 2.6 to answer the following questions—either for your class journal or for class discussion.

- Can you tell what the illustration is about by looking *only* at its title?

- What is the purpose of this visual?

- Who is the intended audience?

- Can you understand the visual *without* reading some discussion?

- How is this visual organized? How do you think most people would read this map—top to bottom? Inside to outside? Left to right?

- Are the titles and subtitles descriptive enough so the visual can stand on its own?

- What do you think is the most important part of this visual? Why?

- What visually reinforces your understanding about how ocean currents move?

- Is there anything you would change about this visual?

Ocean Currents Exhibit 2.6 ■

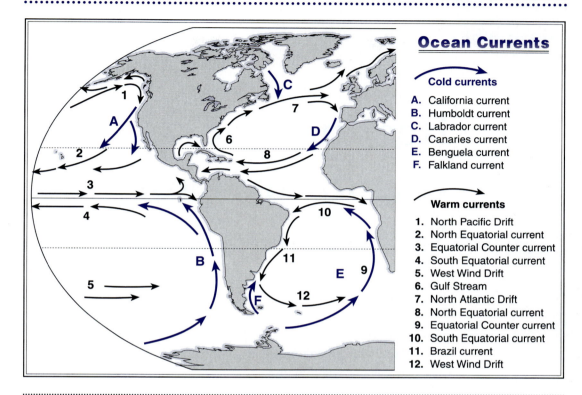

JOURNAL ENTRY ■

In this journal entry, experiment with metaphors. Describe an object or a process that you know very well in one or two paragraphs. Your audience is someone who is totally new to this object or task. Now, go back over your description and add at least three comparisons (similes that use "like" or metaphors).

If your intent were to entertain, these additions might look like: "Sprinkle salt like there is no tomorrow" or "The freshly popped popcorn looked wilted, like yesterday's rose." However, your intent is to clarify exactly what the reader needs to know—so write concretely with precision, as scientists measure liquids for an experiment!

3

Reading and Researching for Workplace Writing

Chapter PREVIEW

This chapter introduces you to ways to help you improve your reading skills for technical materials. You'll also learn why research is important and how to find information quickly in the library or by using computer technology such as the Internet.

Two writing strategies will be introduced: *writing summaries* (condensing the main ideas from your reading) and *writing from research* (weaving other people's work into yours and crediting them appropriately).

Journal ENTRIES

Begin thinking about your experiences in writing research papers and how writing from research might be different in the workplace. Write a short, one-paragraph answer to any of the following questions in your class journal or prepare answers for class discussion, as directed by your instructor.

1. How do I read technical information?
2. How do I know what's happening in my field?
3. How fast is information changing in my field?
4. How have I used the library in the past? Which library resources am I most familiar (and comfortable) with?
5. How do I summarize and interpret research?
6. Why is a bibliography important? Why should I use some form of footnotes?
7. What is plagiarism, and how do I avoid it?

Chapter READING

What kinds of skills are needed for success in the workplace? As you read Mubarak Dahir's article, think about how your experiences in and out of school prepare you for the workplace.

EDUCATING ENGINEERS FOR THE REAL WORLD

Mubarak Dahir

How well are engineering schools preparing students for the working world? To find out, the National Society of Professional Engineers (NSPE) recently conducted a two-year survey of some 1,000 employers in industry and government.

More than a third of the respondents said the current engineering baccalaureate programs do not produce engineers who can meet their company's basic needs. And nearly half said they have to provide additional training for new engineers beyond "on-the-job" experience.

Industry leaders say technical know-how alone is no longer enough. "When we interview college graduates, we do little to analyze their technical skills," says Stephen Tucker, program manager of university recruiting for General Electric. "We assume all the graduates from the schools where we recruit have those skills." Instead, GE looks primarily at five other qualifications: communication and interpersonal skills, analytical ability, self-confidence, personal initiative, and the willingness to adapt to change. Unfortunately, says Tucker, "engineering schools are not covering any of those areas." As a result, two-thirds of GE's new engineers undergo special in-house educational programs before they start their jobs.

Indeed, 8 out of 10 employers surveyed placed a high value on teamwork, but only 1 in 4 felt graduating engineers were well prepared in this area. Similarly, 7 out of 10 highly valued leadership, but only 1 in 10 thought graduates had developed this quality. When asked which areas of study would merit more attention in a revised curriculum, more than 60 percent said students need to improve communication skills, while about 45 percent said they should get more practical experience.

Engineering educators vary in their response to such criticisms. At one end of the spectrum are reformers like Alfred Ingersol, chair of the NSPE study and professor emeritus at UCLA's Department of Civil Engineering, who wants students to complete either a five-year undergraduate program or a master's program before awarding them their first professional engineering degree. He believes the extra time is needed for teaching additional courses in humanities and ethics, which he feels will not only address employers' concerns but also help engineers become true professionals rather than "glorified technicians."

However, industry manifests little support for increasing the number of years of engineering school. Only 1 in 4 of the industry leaders said they would favor a five-year undergraduate degree program, just 1 in 10 agreed that a master's should be the first professional engineering degree, and fewer than 1 in 4 said their firms would increase salaries for entry-level engineers from longer undergraduate programs.

Some industry representatives say that 5-year programs would actually hurt the quality of engineering graduates since many students would opt for other 4-year disciplines. They also point out that while the current engineering curriculum is theoretically only 4 years, it is so demanding that the average engineering student takes 4.7 years to complete the curriculum. Adding another year of course work could lengthen the average program to nearly 6 years.

Chapter 3: Reading and Researching for Workplace Writing   **57**

A second group of educators favors significant but less draconian changes in the curriculum, the most popular of which entails adding a co-op program. This approach, the only five-year schedule industry supports, would allow students to use an internship to develop leadership and teamwork skills. Dow Chemical, for one, not only recruits co-op students more heavily but also offers them significantly higher salaries.

"The trend toward co-op programs is a recognition of the fact that engineering at its core must have a feel for practical applications," says Woodrow Leake, deputy executive director for the American Society of Engineering Education. Schools must do more than simply throw the student into the workplace.

Blending Theory and Practice.   Before 1950, Leake explains, engineering was seen primarily as a pragmatic vocation. But educators have since grabbed onto the idea of "engineering sciences," emphasizing the more theoretical and intellectual aspects of engineering at the expense of practical applications.

The task now, Leake says, is to incorporate the down-to-earth aspects of engineering in the classroom without sacrificing the theoretical. For example, besides merely understanding the complex concepts of electromagnetic waves and quantum theory, electrical engineering students need to know how to design working circuits and components as well as understand the cost, safety, and environmental trade-offs of the various design options. Integrating these aspects of engineering will require relating conventional textbook exercises to design projects and ethical discussions.

Some educators defend the current system against criticism either from industry or university reformers and say they are battling an image problem as much as an academic one. "Engineering schools do a much better job than the survey implies," says Donald Rathbone, dean of engineering at Kansas State University. He points to the 92 percent of respondents who say that they would continue training new graduates even if the engineering curriculum were revised.

Even some reform advocates, such as Constantine Papadakis, dean of engineering at the University of Cincinnati and a strong supporter of five-year co-op plans, feel that not all the criticisms are valid. "It's impossible in either 4 or 5 years to prepare someone for everything they're going to need in a 40-year career," he says. "Giving people the tools and the desire for lifelong learning—that's the key."

Source Mubarak Dahir, "Educating Engineers for the Real World." *Technology Review* (August/September 1993), pp. 14–16. Reprinted with permission.

Analyzing THE READING

To build your reading skills, use the following questions to analyze the article and to develop your reaction to it. Write a one-page entry in your class journal or prepare your answers for class discussion.

58   Part I: Focusing on the Workplace

Reading for Content

1. What are the two or three main ideas of this article? What are the five areas that students need more training in? List them.

2. What is the source of the information in this article? Do you think it is trustworthy?

3. What is the reaction of employers? How is this different from the reaction of educators?

4. What is in the best interests of students?

Reading for Reaction

1. What is your personal reaction to this article? Was there one fact or statement that surprised you? Which statement and why?

2. What should educators do about this survey? Do you think that companies should continue their training programs? Why or why not?

3. Reconsider the five critical skills that the employers have identified. How have your classes or work experiences prepared you in these areas? Based on your experience, which of these are most important and why?

4. Regardless of your field, how could you use the information in this article?

READING FOR THE WORKPLACE

We don't always think that reading skills are essential to our success on the job, but they are. You will need to be able to read quickly and with accuracy, summarizing the essential messages that you'll find in a variety of materials, reports, and forms—from company sources to more technical resources you typically find in a library.

How do you keep current with what's happening in your field? Survey after survey shows that most working professionals read between 4 and 10 hours each week to keep up with changes in their field. They read newspapers, books, technical reports, and trade journals (magazines that report on trends in a particular industry or field).

Knowing what to read and how to use the information you read will help you prepare for your profession. This chapter will reinforce these skills.

WHAT IS A TRADE JOURNAL?

As you become more involved in your field, you will need to be familiar with trade journals—magazines that report on key events or changes in a particular field. Through these magazines, not only can you stay up-to-date on new products or processes, but you can also find out about trends that will affect your industry, professional associations, job openings, and new products.

One of your goals as a working professional is to constantly improve your understanding of your field. After leaving school, most of your learning takes place on the job. You are always working toward more responsibilities and toward the next promotion. Finding out which trade journals are most important to your field and starting to read one or two regularly can help you to find out about new products, changes in government regulation, or trends in hiring or selling patterns.

Professional trade associations do more than publish trade journals. They offer working professionals a way to network with others in the same field. Perhaps most important, trade associations sponsor local, regional, and national meetings, conferences, and workshops. Trade associations also support research, publish studies, often maintain a speakers' bureau, and gather statistics useful to a particular field.

Because there may be several trade journals that are important in your field, the next exercises will help you find out what they are and how to evaluate them.

Surveying Professional Reading Habits Exercise 3.1

Find someone who is working as a technician or professional in your field by asking friends, family, or by calling up local employers. Plan to ask three or four questions about what kinds of magazines or journals the person reads to keep current in your field, how much time is spent reading, and how the magazines are obtained. Ask if the person belongs to any professional trade associations and note the names, if any. Ask the person which trade journals are the most useful and why.

Make a list of the magazines or journals that you discover from this survey and look at them in your local or school library. If you have difficulty in finding copies of the journal, ask for help from your instructor in your field. Note which journals look most useful to you and why. Summarize what you learned in a journal entry or in a memo to the class.

Finding Professional Associations Exercise 3.2

Go to your library and use the *Encyclopedia of Associations* to find out what professional organizations (also called trade associations) are listed under your field (and related fields). Take some time to get to know this resource. Read through the sections relating to your field and pick out three associations that you think are important to your field.

Read each entry and take notes on three facts about each association. Be sure to notice if the association publishes a trade journal. Check to see if the library carries this publication. If not, consider writing a short letter to the association saying that you are majoring in a particular field and requesting a review copy. (Use the section "Letters and Memos" in Appendix A to set up your letter format.)

Be prepared to tell your discussion group which association looks most interesting, why it is interesting, and what you learned from this exercise.

HOW TO READ TECHNICAL ARTICLES

Not everyone has time to read every article that may have useful information. What is helpful is to be able to read the article quickly, to find out if it has any ideas or information that may be useful to you, and to be able to restate in your own words the most important information from the article.

Sometimes what the writer thinks is most important and what you need are two different things. Sometimes your supervisor will ask you to summarize a report for your co-workers, but the process of reading and summarizing information stays the same.

When you need to read a technical article or a formal report in order to prepare a summary, plan to read actively by making notes and asking questions about the material you are reading. Make a photocopy of the article so you can write on the margins.

If you need to report to someone else on what information is contained in the article, you will need to write a summary that restates the main ideas—without adding your own opinion.

If you are reading the article to gain understanding—whether for yourself or others, you will need to summarize the main ideas and then interpret them (describe your own personal reaction to the information). After you understand what is most important, you can apply this new information. Begin reading by asking the following questions:

- **What** is here?   **Summarize**
- **Why** is it important?   **Interpret**
- **How** can I use this information?   **Apply**

Exercise 3.3   Thinking About How You Read

In your class journal, write a paragraph that describes how you usually read. Consider what kinds of reading you do, where you are when you read, what kind of noise is around you, if you take notes or underline, and how much time each week you spend reading. Add any other information about your reading habits you can think of.

Now, pick out some reading that you have to do for a difficult class, perhaps accounting or anatomy. As you read the article or chapter from your textbook, follow the guidelines shown in the next box, "How Do You Read a Technical Article?" exactly.

After you have finished your reading, go back to your journal and write a second paragraph that tells about your experience. Were there any differences? What did you change about how you read and why? Plan to discuss what happened and your reaction to this exercise in class.

Chapter 3: Reading and Researching for Workplace Writing **61**

HOW DO YOU READ A TECHNICAL ARTICLE?

How you read often depends on your purpose for reading. If you are going to read an article that has technical information that you must understand and remember, and perhaps use, try the following steps:

- **Preview** the article by reading the title, the abstract (a short summary of the article that usually appears at the beginning), any pictures or charts, and the conclusion.

- **Skim** read the article quickly for an overview of what's in the article, noticing which sections will be most useful. Underline any technical words you notice that need to be defined. Look these words up after you finish skim reading. Use a technical dictionary, handbook, or textbook. Jot down key words or notes for each technical word.

- **Reread** the article, paying special attention to the introduction (where the writer usually introduces the main ideas of the article) and the conclusion (where the writer restates what is most important).

- **Make notes and comments** in the margin as you read so that you can quickly find the purpose, main ideas, or conclusion. Ask questions or write your comments as you read. This is particularly important if you are going to write a summary or if you have to remember the information in the article.

- **Reread** those portions you have highlighted or where you have made notes. Come up with your own answer to these two questions:

  What is this writer saying?   **Summary**
  What does this mean to me?   **Reaction**

- **Practice summarizing and reacting** to what you read any time you do any reading, whether it's a newspaper article, reports or memos from your workplace, or your homework. You can improve your problem-solving skills by making these questions routine.

HOW TO IMPROVE A TECHNICAL VOCABULARY

Part of being able to summarize information clearly from a technical article depends on your understanding of the information being presented. Most of the time, you will need a very precise definition of key terms. Notice how your technical textbooks define key concepts, tools, and processes.

You can use a technical dictionary or a handbook to define key concepts and terms for your field. For example, the *Machinery's Handbook* (published by Industrial Press) provides hundreds of formulas, drawings, tolerance tables, and definitions for such topics as threads, gears, and plastics.

Once you are out in the workplace, you will continue to learn new words that describe how people work, what tools they use, and how they carry out operating processes. Your on-the-job training and day-to-day experiences will keep you learning new vocabulary through what you see and do and through conversations with co-workers. You won't necessarily have to make long lists of new words. You will pick these up naturally on the job.

The longer you work in your field, the more comfortable you will become with its technical vocabulary, often called **jargon** because outsiders don't know what these technical terms mean. For example, while most people are familiar with *postoperative* as the recovery stage following surgery, the term *perioperative management,* which refers to what happens before, during, and after surgery, should be used only when nurses, technicians, and doctors are talking to each other.

Sometimes technical terms are translated into **slang**—words that save time; thus, *postoperative* becomes *postop*, a word that has not yet made its way into a dictionary. Many workers like to use slang because it is informal. For example, our budget request went "down the tubes."

Your strategy in using either jargon or slang would be to consider the audience: What level of vocabulary does my audience already understand? Will using a technical term clarify or confuse my audience? Will my audience understand slang—since it's not often in the dictionary?

Often at work, you must translate very technical information to people who are less knowledgeable than you. Imagine explaining how a clutch works to a car owner who is mostly worried about how much it will cost to replace that

Exercise 3.4 Defining Key Terms

Find an article from a technical journal or a chapter from a textbook that reports on a new process or new materials used in your field and that is of interest to you.

Read through the article or chapter. Are there any difficult concepts or words you don't understand? Find four or five key terms and look them up in a technical dictionary or handbook (find these dictionaries or handbooks in the reference section of your library). Some textbooks have a glossary of terms (shortened dictionary) in the back of the book.

Copy the definition for each word. Study these definitions by underlining key words and then rereading the definition several times. Take a moment to rewrite the definition *in your own words*, and then return to the article.

Did your understanding of the article change once you reviewed the definitions of terms? Write a paragraph for your journal about what you did and what you learned. Discuss your answers with your group and the class.

Defining Technical Terms in Your Field   Exercise 3.5

Pick your hardest technical class. Start a list of technical terms from this class and keep it for two weeks, adding new terms after every class.

For each technical term you list, write a definition in your own words and a sentence that includes this new word. Review this list of technical terms every time you study. At the end of the two weeks, write a paragraph in your journal about what you learned from this exercise.

---

clutch. Or you might need to explain to a group of seasoned truck drivers how an on-board computer collects detailed information on their driving habits for the main office.

When you must translate the information for a less informed audience, always begin by asking: **What** does my audience already know? **Why** do they need this information? **How** are they going to use or react to this information? These answers will help you sort out how to best "translate" and present the technical parts of your discussion.

HOW TO FIND TECHNICAL ARTICLES

Learning about new materials, processes, or needed skills in your field is absolutely crucial to your success as a technician. Some companies subscribe to technical journals so that their employees can easily find this information. If you work for a small company, however, you may have to take extra steps to stay well informed about changes in your field.

Technical journals are not always easily available. They aren't on the newsstand right next to *Newsweek* or *Time*. Your local or school libraries may subscribe to only a few technical journals—unless you can go to a college or university library. While you are in school, another good source for technical articles in your field is your instructor. Once you are working, you may need to subscribe to the key trade journals in your field, if your employer does not provide them.

More and more people and companies are using the Internet to stay current with new technical information. You can also use CD ROM databases in your library's computerized research center to find technical articles that are useful to you. CD ROM databases allow you to sit down at a computer, type in a few key words, and locate the articles you need quickly. Some libraries provide printers so you may print out the entire article right there.

If you cannot get the whole article from a CD ROM database, consider how badly you need the information. You may be able to order the article you need through your library's interlibrary loan service.

This next section will help you save time in the library by summarizing what library resources are available and suggesting how and when to use them.

64   Part I: Focusing on the Workplace

HOW TO SAVE TIME IN THE LIBRARY

If you know what resources are available in the library and make a plan before you go to the library, you can save time—whether you work alone or with the librarian. This next section highlights how you can use different resources in the library—including the librarian! After completing some of the exercises in this section, you should be able to use:

- The *Library of Congress Subject Index* to find key words.
- The "call number" and card catalog to find related books, videos, or cassette tapes.
- The reference section of your library.
- The CD ROM computer search database.
- The journals related to your field located in your school library.
- Internet resources.

Before you go to the library, ask:

1. **What** am I looking for? A book? A magazine article? Some background information?
2. Do I have a **List** of key words or questions to guide my search?
3. **Where** am I most likely to find this information?

Using the *Library of Congress Subject Index.*   Sometimes you'll need information from the library, but you're not sure where the information has been filed. It's important to use the same key words that librarians use to index (or file) books, magazines, and articles. For example, information on older people may not be under senior citizens, the elderly, or retirees, but you will find much information under "the aged."

The Library of Congress puts out a giant, three-volume subject index (or list) of all the key headings used by librarians everywhere. You can use this valuable book to look up your key words in alphabetical order and find out what subject headings the library has used to file the information. The *Library of Congress Subject Index* is usually found near the reference section of your library.

Using Call Numbers and the Card Catalog.   Just as the librarian assigns any book, magazine, or other materials to a certain subject, each library item also has its own number that appears on the spine of the book or front of the magazine. This call number tells you instantly where this book, videocassette, recording, or other print materials can be found.

You can look up this call number by using the card catalog or a computerized card catalog that lists resources under author, subject, or title. Copy the call number down exactly! **Note:** Some libraries use the Dewey Decimal system to assign numbers to their books, while other libraries use the Library of Congress system to catalog their books with letters and numbers.

Once you have looked up the call number for your particular subject, you can go to that part of the library and "shelf browse" to see what else the library has in this area.

Using the Reference Section. Sometimes you need background information so you can understand your subject more fully and select the best key words to use. Go to the reference section of the library and use any encyclopedia (such as the *Encyclopedia Americana* or the *Encyclopeadia Britannica*) to look up your topic.

Although you may not be able to check out books kept in the reference section, if you use your call number here, you may discover hard-to-find information, general handbooks, and dictionaries related to your field. Plan to use a technical dictionary to define key terms.

Using the CD ROM Computer Search Service. Most libraries now have several computerized database services—like *InfoTrac* or *Academic Abstracts*—that allow you to use a computer to find articles in journals and magazines.

Sit down in front of the computer, type in your key research words, and press enter. The computer program "searches" and "matches" your key words with articles it already has filed on a massive bibliography stored on one compact disk, much like songs are stored on a CD (compact disk).

The computer screen will show a listing of those articles that match your key words. You can then read through the list to select and, in some instances print out, those articles you are interested in. With some services, you can press a button and see an abstract (also called a summary) of the article. And, in some instances, you can press a button and print out the whole article.

If you are not satisfied with what you find, use the keys to "step back" a page to start a new search with new key words. Remember to ask the librarian for help in refining your search.

Print out the summary of information about the article or write it down. This information will give you what you need to find the article. Unfortunately, most printouts do not show the call number for your magazine. This is because many libraries follow different systems to catalog and index their magazine collections.

Finding the Journal or Magazine. Sometimes you'll find journals and magazines right in the "stacks" (where the books are kept), arranged by call number; other times, the journals are arranged alphabetically in their own section.

Part I: Focusing on the Workplace

Once you have located your journal or magazine, skim read the articles to find out which ones best meet your needs. You can photocopy the best articles, or, if the library allows it, check out the magazines you are most interested in.

Using Interlibrary Loan.   If the journal you need is not available, talk to your librarian about using an interlibrary loan service that librarians use to request copies of your journal article or book from other libraries in your region or across the country. Interlibrary loan takes about 8–10 days to deliver the article, and there may be a small photocopying fee.

Using the Internet.   A few libraries are beginning to offer direct access to the Internet along with workshops or one-on-one help in using the Internet. Your school, local library, or employer may provide you with access to the Internet through a modem or a "host computer" that connects to the Internet (commonly called the "information highway" because so much information is available through it).

When computers are hooked together in an office, for example, so that information can easily be exchanged between different people using different computers, we say that the computers have been "networked" together. The Internet is actually a name for a "network of networks"[1] with millions of computers connected to each other—all around the world.

The Internet is a recent innovation. It was set up in 1969 by the Department of Defense to link some universities together who were doing military-funded research. The project was so useful that every major university in the country wanted to participate. By 1980, Internet use shifted from large computers to smaller desktop workstations run by individual users.[2] Today, people consider access to the Internet almost as a constitutional right—like the right to privacy or the right of free speech.

If your school or local university does not have student access to the Internet, you can subscribe to services like CompuServe or America Online, for an access account. The range of services—and information—available via the Internet is staggering! With double clicks of your mouse, you can

- Get job applications from major employers and background information on the company.

- Order or preview a wide range of products and services—from airline tickets to new compact disk recordings.

- Sign up to belong to a special interest group or "listserv" that allows you to send and receive e-mail to the whole group.

- "Talk" to other technical professionals.

---

[1] John R. Levine and Carol Baroudi, *The Internet for Dummies* (San Mateo, CA: IDG Books Worldwide, Inc., 1993), p. 8.
[2] Levine and Baroudi, *The Internet for Dummies,* pp. 11–12.

- Download, save, and print documents from research libraries, government agencies, or special interest groups (other people working in your field with similar questions or problems).
- Send and receive mail from anyone who has an e-mail account.

Using the Internet well takes time and skill, and it is fast becoming an important job-related skill, regardless of your field.

Regardless of which kinds of resources you use to find information, as a researcher and a writer, one of your primary responsibilities is to use the information you find ethically. This means providing a bibliography citation for each resource you use, whether you're using a newspaper article or a page from a newsgroup through the Internet.

Finding Out about the Internet Exercise 3.7

Run a key-word search through the Card Catalog at your library on the key word *Internet*. Sit down at one of the CD ROM databases and run a similar search on the key word, *Internet*. How much information did you find? Look up two or three articles on the Internet from the resources your search turned up. Which articles were most useful? Why?

Write a journal entry that describes how you found your information, which resource was the most useful, and what you thought of the articles you found. Did your library have the articles that interested you? Did the title and summary of the article in the CD ROM database, for example, actually match what you found? How easy was it to use the Card Catalog? The CD ROM service? Which was easier to use? What did you learn from this activity?

Review your journal entry and prepare a list of two or three ideas to share with your discussion group or the class.

Discovering Where You Can Get Internet Access Exercise 3.8

If your employer or school does not have Internet accounts available for students, try to find out if there are public access accounts through a local college, library, or the city you live in. List the types of services and their start-up and monthly costs. This will involve making phone calls or visiting the various offices to find out about these programs. Ask if there are any workshops or seminars on getting started with the Internet.

Your goal is to find out how easy or difficult it would be for you to have access to the Internet. Report your findings in a journal entry and share the results with your discussion group or the class.

HOW TO PREPARE A BIBLIOGRAPHY

A *bibliography* is an alphabetical or numbered list of articles, books, newspapers, interviews, or other materials that usually appears at the end of most technical articles and in books. Bibliographies are formal ways of presenting information. Each citation (one item on the list) must be complete. These next sections will show you how to prepare:

68 Part I: Focusing on the Workplace

1. A **bibliography** that shows all of the publication information (like the name of the author, the exact title of the book, the publishing company, and date of publication) about any one source.

2. **In-text citations** that refer to the bibliography and tell exactly what page this particular information came from.

All academic writers (scientists, technicians, psychologists, teachers, and researchers) follow a very formal system of in-text citations and bibliographies. Bibliographies can be useful to readers. They tell the reader where the writer got the information and where to go for more information. Sometimes writers prepare *annotated bibliographies,* which include a very brief summary of each book, article, or resource, plus their own opinion about the usefulness of that resource.

Many workplace writers use in-text citations sparingly, but they will use a bibliography to help their readers find more information. Workplace writers are also very careful about separating what they write from what they summarize from other sources. For more information on using in-text citations and footnoting, see the section on Footnotes and Bibliographies in Appendix B.

The key to putting together a bibliography is to consider **how** your reader will use this information and to **select** a format for bibliographies that is used in your field. Right now, you can choose from about 14 different formats for bibliographies, depending on your discipline, whether it is biology, engineering, journalism, law, medicine, music, or psychology.

What's important to remember is that your printout from the CD ROM resource or computerized card catalog will **not** be in proper bibliographic form. So you need to translate this information about your journal article or book into the proper form.

An example of a **printout of a journal article citation** from *InfoTrac* is shown in Exhibit 3.1.

■ Exhibit 3.1 **Printout of Journal Article Citation**
··

```
Heading:  SHORT-TERM MEMORY
          * Dictionary definition
     7.  How to aid short-term memory. by Steven Peterson
         v15 Total Health April '93 p50(2)
     TEXT / HEADINGS
TEXT (96 lines)
```

Chapter 3: Reading and Researching for Workplace Writing **69**

Printout of Newspaper Article Citation   Exhibit 3.2 ■

Copyright © 1995 by UMI Company. All rights reserved.

Access No: 9300020108 ProQuest – The New York Times (R) Ondisc

Title: MEDICAL BREAKTHROUGHS, ON-LINE

Authors: Miriam Schuchman and Michael S. Wilkes; Miriam Shuchman
             and Michael S. Wilkes, both medical doctors in
             California, are medical commentators on National Public
             Radio.

Source: The New York Times, Late Edition—Final

Date: Friday, Jul 7, 1995 Sec: A Editorial Desk p:25
             Length: Medium (593 words) Type: OP-ED

Subjects: MEDICINE & HEALTH; DISCLOSURE OF INFORMATION; RESEARCH;
             ELECTRONIC INFORMATION SYSTEMS; INTERNET (COMPUTER
             NETWORK)

Companies: NEW ENGLAND JOURNAL OF MEDICINE; JOURNAL OF THE AMERICAN
             MEDICAL ASSN

Abstract: Miriam Shuchman and Michael S. Wilkes say that top
             medical journals like *The New England Journal of Medicine*
             and the *Journal of the American Medical Association* are
             too slow in their efforts to inform people fast enough
             about developments that affect their health.

Copyright 1995 The New York Times Company. Data supplied by NEXIS (R) Service.

Exhibit 3.2 is an example of a **printout of a newspaper article citation** from *The New York Times,* and Exhibit 3.3 shows an example of a **printout of a book citation** from a computerized card catalog.

This next section shows you these three printouts rewritten into one of the three most commonly used formats for bibliographies: Alphabetical (used by academics and business), Numbered (used by scientists and technicians), and Author/Date (used by psychologists and researchers).

Choose the format that comes closest to fitting the needs of your field and **copy** the format **exactly** when you prepare your citation for a book or an article. In the following samples, note the various styles that are preferred. Choose one style and use it consistently.

70 Part I: Focusing on the Workplace

■ Exhibit 3.3 **Printout of Book Citation**

```
Title:          Every student's guide to the Internet/Keiko
                Pitter . . . [et al.].
Publication:    New York : McGraw-Hill, c1995.
Description:    viii, 183 p. : 24 cm.

Branch has:     TK5105.875.I57 E94 1995     Stacks: available

NOTE            Includes bibliographical references (p. 173-174)
                and index.
NOTE            Internet or bust—E-mail/listserv—Usenet newsgroups :
                bulletin board services of the Internet—Gopher :
                what's on the menu?—Telnet : warning! Leaving the
                primrose path—FTP : getting it from there to here
                (and back again)—WAIS : indexes and databases—World
                wide web (Mosaic) : bringing it all together.

SUBJECT:        Internet (computer network)
ADD'L AUTHOR    Pitter, Keiko M.
```

Formats for Journal Articles

Alphabetical   Peterson, Steven. "How to Aid Short-Term Memory,"
Total Health April 1993: 50–51.

Numbered   2. Peterson, S. How to Aid Short-Term Memory,
Total Health 1993; 15(4): 50–51.

Author/Date   Peterson, S. (1993). How to aid short-term memory.
Total Health 15,(4), 50–51.

Formats for Newspaper Articles

Alphabetical   Shuchman, Miriam, and Michael S. Wilkes. "Medical Breakthroughs,
On-Line." *The New York Times* 7 July 1995: A25.

Numbered   2. Shuchman, M. and M. S. Wilkes. Medical Breakthroughs, On-Line.
The New York Times 1995; July 7: A25.

Author/Date   Shuchman, M., and M. S. Wilkes. (1995, July 7). Medical breakthroughs,
on-line. *The New York Times,* pp. A25.

Chapter 3: Reading and Researching for Workplace Writing   **71**

Formats for Books

Alphabetical   Pittner, Keiko M. *Every Student's Guide to the Internet.* New York:
McGraw-Hill, 1995.

Numbered   3. Pittner, K. M. Every Student's Guide to the Internet.
New York: McGraw-Hill; 1995, 195 p.

Author/Date   Pittner, K. M. (1995). *Every student's guide to the Internet.*
New York: McGraw-Hill.

The easiest way to prepare a bibliography is simply to organize your citations in alphabetical or numeric order.

Most academic writers in the humanities set up their bibliography in alphabetical order, referred to as the MLA system after guidelines issued by the Modern Language Association. Most technicians and scientists in physics, chemistry, and biology use the numbered system guided by the Council of Biology Editors or CBE. Psychologists and social scientists follow the author/date system, called the APA system after guidelines issued by the American Psychological Association.

See the section on Footnotes and Bibliographies in Appendix B for more guidelines on preparing in-text citations and bibliographies.

....

WHAT IS A SUMMARY?

A summary simply restates main ideas in your own words. If you can summarize important information, whether from a book or from events happening around you, whether in writing or by speaking, you have gained a valuable skill to support your problem solving.

We're always selecting information from all that happens to us. We pick out the highlights and we think about what this information means.

Suppose your supervisor gives you a new task to do. This may be the chance for a promotion—or this may be a routine assignment. Your ability to summarize what kind of assignments your supervisor has given you, what opportunities there may be for promotion, and how this new task fits into the requirements for a promotion all come together very quickly. Your performance may be a critical factor in letting your supervisor know you are ready for that promotion. Or suppose a key piece of equipment breaks down. Your ability to summarize what's wrong and to say exactly what action needs to be taken could be critical in solving the breakdown quickly.

The same process applies whether you are writing or speaking. You need to gather the information, think about it, and then compress the information into its key elements by preparing a summary. You should think about it again, and then prepare your reaction and analysis.

Part I: Focusing on the Workplace

In writing a summary, you will need to know how to

- Summarize   Compress the chapter, book, entire article, or entire event down to a sentence or two—in your own words.

  Example of a *summary:*

  Top medical journals like *The New England Journal of Medicine* and the *Journal of the American Medical Association* are too slow in their efforts to inform people quickly enough about developments that affect their health.

- Paraphrase   Take someone else's one idea and translate it into your own words.

  Example of a *paraphrase:*

  Because major medical journals need three to six months for review of all articles before publication, the general public does not have speedy access to information that could be used for more effective treatments. The major medical journals want to protect the public, but people who are ill have a hard time waiting.

- Quote   Take someone else's words and repeat them *exactly* as they were expressed—inside quotation marks.

  Example of a *quotation* with a *lead-in:*

  According to Drs. Miriam Shuchman and Michael S. Wilkes, major medical journals like *The New England Journal of Medicine* " . . . must take all the time that is necessary before publishing findings to protect people from scientific misinformation and half-baked ideas—a worthy motive."

Notice that although the information is roughly the same, how it is presented depends on the writer's strategy. Summaries are most useful because they can help the reader understand the main idea from a longer piece without having to actually read it. Paraphrases are commonly used to translate technical information into wording that is easier to understand, and quotations are used to highlight exactly what the speaker or writer said—because it was said so well.

Regardless of which of these techniques you use, remember to give credit to the person or resource—every time.

Developing Your Reaction to Your Reading.   Sometimes people write summaries by using the notes they've written in the margin of an article or report. This may work for you. However, if the reading is very technical, it may help to take a more formal approach by listing the main points in your own words and by writing a response to each main idea to discover your reaction.

Exercise 3.9 Practice Paraphrasing and Quoting as Part of a Summary

Select an article from any journal on a technical or workplace topic of interest to you. Prepare a summary that includes a citation, two-three paraphrased ideas, two or three quotations (with a lead-in), and a summary of two sentences that restates the essence of the article.

Bring your draft summary and the original article to your discussion group. Review each draft, checking for completeness, suggesting changes, and reading carefully to avoid plagiarism. One form of plagiarism occurs if you use the same wording in your summary of the article without using a lead-in and some sort of quotations or in-text citation. Ask if the author is clearly given credit for any unique ideas. Are there any paraphrases that should be quotations? Identify the strengths and weaknesses for each draft summary that your group reviews. Revise if asked to do so by your instructor, and turn in all drafts to your instructor.

Two ways to write journal entries that will help you prepare summaries and develop your own responses to your reading are **reaction entries** and **double entries.**

Reaction entries in a reading journal, much like your class journal, can be a place to explore your ideas. They can be used as a basis for discussion, to explore ideas you are interested in, or to support writing papers or reports. A reading journal should help you think about what you have read by giving you a place to ask questions, answer questions, and record your reactions to what you read.

One reader's reaction to Mubarak Dahir's comment in paragraph 7 of this chapter's introductory article—that industry doesn't want students to spend more time in school, yet industry wants the curriculum expanded or changed—is shown in Exhibit 3.4.

Exhibit 3.4 A Reader's Reaction

Dahir says that industry wants better prepared students, but that industry wants students to spend less time in school. School is so expensive, and student scholarships and work-study programs are being cut back. Most of my friends are working and going to school part-time. Some of us are going to need longer time in school or will have to borrow really large amounts of money to stay in school full-time. Ted filed bankruptcy after he graduated. I'd rather get out of school quicker, but I want to stay out of debt and I'm already working 20 hours a week. I'm not sure I can keep up with my technical classes. Also, some students are not ready to begin their programs without taking basic classes in math, reading or writing. So how do we get caught up? Something seems wrong, but how do we fix it?

Dahir highlights five areas that students need more skills in—such as communication skills, the ability to work on teams, and greater flexibility. We can't get these skills by listening to lectures. Teachers need retraining too. How will we get the training we need if companies and schools can't work together?

Reaction to Mubarak Dahir's article, "Educating Engineers for the Real World" *Technology Review* (August/September 1993), pp. 14–16. Reprinted by permission.

Part I: Focusing on the Workplace

A **double-entry journal** can be very effective in helping you find the key ideas you need to write the summary and develop your own reaction as well.

To set up a double-entry journal page, just draw a line down the middle of the next page of your journal. On the left side of the paper, summarize the main ideas from the article. Immediately opposite on the right side, write your questions and reactions. Exhibit 3.5 is one reader's double-entry journal for part of Mubarak Dahir's article.

After you have prepared one or both of these kinds of entries, take some time to think about what you have read—and what you have written. This important "processing" time helps you sort through information and decide what is most important.

Using a variety of formats like this double-entry journal can help you analyze different aspects of a problem in a very powerful way. Other formats that can help you "see" information differently are maps, flowcharts, grids, diagrams, drawing, and doodling.

■ Exhibit 3.5 **Example of a Double-Entry Journal Page**

Author: Mubarak Dahir
Title of article: "Educating Engineers for the Real World"
Name of magazine: *Technology Review*
Date of magazine: August/September 1993   Pages: 14–16

Main Ideas and Highlights	**Questions and Reactions**
Graduating engineers not meeting basic needs of industry	Why not? What are the basic needs of industry?
Technical knowledge not enough; students need:	What other skills are being taught in my classes?
Teamwork	Where have I gotten these skills?
Communication/interpersonal skills	How important are these really?
Analytical ability	All of these skills will improve with practice.
Self-confidence and initiative	Are my teachers giving me enough practice?
Willingness to adapt to change	Which of these are most important?
	Which of these do I need to work on?
	I can see how flexibility is important to job success because we may need to change jobs four or five times over a career—and in more than one industry. There's a lot of job shrinkage going on as job requirements increase. I will need the best technical and communication skills possible.

Chapter 3: Reading and Researching for Workplace Writing   **75**

HOW TO WRITE A SUMMARY

Before you prepare an actual summary, think about **who** is going to read the summary, **why** they are going to read it, and **how** the summary will be used. Some of your readers will want a summary of the entire article; other readers will only want a summary of the part of the article that is most useful to them.

Before You Write the Summary

1. *Skim read* the article quickly to get an overview.
    a. Look for *advance organizers* like preview statements, introductions, headings, illustrations and their titles, and conclusions.

2. *Reread* the article slowly.
    a. Underline key points and key terms.
       (Use a dictionary to look up any terms you don't know.)
    b. Highlight, circle, or asterisk the *main idea* or writer's *purpose.*
    c. Note *key points* in the *margin.*
    d. Jot down *questions* as they occur to you. You can use these questions to start writing your reaction, if one is requested.
    e. *Label* the parts of the article (introduction, main points, conclusion).

3. *Reread* only the underlined or highlighted material. Notice which ideas are most important.

When You Write the Summary

1. List the *main ideas* from the article—without looking back too often at the article. Make sure to translate key ideas into your own words. This is called *paraphrasing.* You will use this information to write your summary.

2. *Reread* the article and check your list of main ideas for completeness.

3. *Reread* your list of main ideas and draft a *summary* of one, two, or three sentences that restates the main ideas of the entire article in your own words.

4. *Reread* the article and check it against the draft summary for completeness. Have you caught the essence of the article? Should any information be added? Remember, a summary should not include your opinion!

When You Write the Reaction

1. Prepare your *reaction* by writing a paragraph that summarizes your interpretation of this article. Do you have any new insights to add? Sometimes when we work with material a little longer, we find new things to say.

2. Expand your analysis by asking how this information is relevant to your situation or needs. Reconsider the *purpose* of the *audience.* Ask how this information will be used. Why is it important?

76   Part I: Focusing on the Workplace

3. When you have drafted your reaction paragraph, *compare* it to the summary. Have you responded to all important main ideas? Have you left out any important implications?

Before you complete your summary and reaction, reread the article one last time to check the following:

- *Content and organization*: Is your summary an accurate paraphrase of the contents of the article? Are they logically organized?

- *Style*: Is your tone objective in the summary section? Is your tone reasonable in the reaction? Is your writing concise?

- *Document design:* Have you followed conventional format for a summary: Proper citation form at the beginning? Short paragraphs? Headings if needed? Separate paragraphs for the summary and the reaction?

- *Proofreading*: Have you checked punctuation, grammar, and spelling for correctness?

■ EDITOR'S CHECKLIST FOR A SUMMARY

Use this checklist to help focus your revision before you turn in your summary. Ask someone in your discussion group to act as your peer editor, answering the following questions for your rough draft.

Encourage your peer editor to either set aside about 20 minutes to work through the list with you, discussing each question, or to write out sentence responses to each question.

Check Audience and Purpose

_____ 1. Can I quickly tell *who the audience is* for this writing?

_____ 2. Is the writing purpose *informative* in the summary section?

_____ 3. Is the writing purpose *persuasive* in the reaction section?

Check Content and Organization

_____ 1. Does the *summary* accurately *reflect the article?*

_____ 2. Is any *important information missing?*

_____ 3. Are ideas *presented in the same order* as in the original article?

_____ 4. Does the *reaction* section *emphasize analysis* and implications?

_____ 5. Does each section have a *clear introduction, discussion, and conclusion?*

Chapter 3: Reading and Researching for Workplace Writing **77**

Check Style and Document Design

_____ 1. Do you see any places where *ideas could be expressed more clearly?*

_____ 2. Are *ideas presented* as *concisely* as possible? Are there any extra sentences or words that could be deleted?

_____ 3. Is the *tone objective* in the summary section? Reasonable in the reaction section?

_____ 4. Are *paragraphs too long* for easy readability? Did the writer use separate paragraphs for the summary and the reaction?

_____ 5. Are *transitions smooth* between sentences and paragraphs?

_____ 6. Is the bibliographic *citation format correct?*

_____ 7. Can you spot *any punctuation, grammar, or spelling problems?*

WHAT'S COMING NEXT? ■

This chapter has highlighted reading and researching techniques that will help you continue to develop as a professional. You have practiced writing strategies to help you summarize your technical reading and edit for conciseness. You can also use summarizing skills to report what you observe on the job.

Preparing a formal bibliography and using some sort of in-text citations will probably be more important for academic writing than for workplace writing, but you may need these skills as you continue to use the library to find information related to your field. Knowing how to read and use such tools as bibliographies and article summaries in technical journals can save you research time. Also, technical articles are often written at a higher technical level than the reader expects. Using "advance preview" techniques to help you translate highly technical information and using journal entries to help you map out the most important ideas and your reaction to them can be very useful.

In Chapter 4, Writing About Tools and Equipment, you'll work on describing tools and mechanisms—what they are and how they work. You'll also design documents that include graphics to help readers understand more quickly what something is and how it works.

CONCEPT REVIEW ■

Can you define the following key concepts from Chapter 3 without referring back to the chapter? Write your own definitions in a journal entry or review them out loud.

Part I: Focusing on the Workplace

jargon	slang
technical or trade journal	quote
professional or trade association	reference section in library
bibliography	CD ROM database
summarize	computerized card catalog
paraphrase	Internet

■ WRITING SKILLS REVIEW

An important goal of this textbook is to help you express yourself in writing without distracting errors in sentence structure, punctuation, grammar usage, or spelling. To build your skills in editing and proofreading, begin a systematic review of *Appendix C: Writing Skills Review.* Your assignment:

___ Read "Building Punctuation Skills" in *Appendix C.*

___ Complete exercises C.16 through C.17 in "Using Periods, Semicolons, and Question Marks," and Exercises C.18 through C.23 in "Using Commas," checking your answers at the back of *Appendix C.*

___ Prepare a paragraph summarizing your progress for your instructor or use the form your instructor provides to report your progress. Be sure to highlight any areas you need to further review.

■ SUMMARY EXERCISE

What's Wrong With This Summary?

Read the article by Miriam Shuchman and Michael S. Wilkes on "Medical Breakthroughs, On-Line," which is reprinted in Exhibit 3.6. A student's draft summary and reaction is given below.

Use the editor's checklist to evaluate the following summary. List your editing suggestions in a memo to the student writer of the review. Bring your draft memo to your discussion group and decide what further suggestions this writer needs to follow. Either turn in the completed exercise to your instructor or add this to your class journal.

Summary: This article is about Lou Gehrig's disease and how people exchange medical information more easily by using the Internet. People are not informed fast enough about developments that affect their health. Editors agree that people need faster information about health, but the scientific peer review process takes too much time, often as long as two to three months.

Chapter 3: Reading and Researching for Workplace Writing 79

An On-Line Article   Exhibit 3.6 ∎

```
Copyright © 1995 by UMI Company. All rights reserved.

Access No:   9300020108 ProQuest - The New York Times (R) Ondisc
Title:       MEDICAL BREAKTHROUGHS, ON-LINE
Authors:     Miriam Schuchman and Michael S. Wilkes; Miriam Shuchman and Michael
             S. Wilkes, both medical doctors in California, are medical
             commentators on National Public Radio.
Source:      The New York Times, Late Edition—Final
Date:        Friday, Jul 7, 1995  Sec: A Editorial Desk  p:25
             Length: Medium (593 words)  Type: OP-ED
Subjects:    MEDICINE & HEALTH; DISCLOSURE OF INFORMATION; RESEARCH; ELECTRONIC
             INFORMATION SYSTEMS; INTERNET (COMPUTER NETWORK)
Companies:   NEW ENGLAND JOURNAL OF MEDICINE; JOURNAL OF THE AMERICAN MEDICAL
             ASSN

Abstract:    Miriam Shuchman and Michael S. Wilkes say that top medical journals
             like The New England Journal of Medicine and the Journal of the
             American Medical Association are too slow in their efforts to
             inform people fast enough about developments that affect their
             health.

Copyright 1995 The New York Times Company. Data supplied by NEXIS (R) Service.

Article Text:
  Many people who are paralyzed by Lou Gehrig's disease are unable to speak
intelligibly. They communicate through computer on-line services. Last fall,
they networked about Neurontin, a drug that helps prevent epileptic seizures.
Excitement ran high because anecdotal evidence indicated that the drug relieved
their suffering. Doctors nationwide were so inundated by inquiries that a
scientific study was initiated.
  If medical science establishes that Neurontin provides a safe relief from
the symptoms of the fatal muscle disease, it still may take months—too many
months—for the results of the study to appear in prestigious journals like
The New England Journal of Medicine and the Journal of the American Medical
Association. They are the most influential journals in their fields. They affect
what less powerful journals do.
  In the Internet age, the decades-old editorial policies of both big journals
are turning them into dinosaurs. They say they must take all the time that is
necessary before publishing findings to protect people from scientific
misinformation and half-baked ideas—a worthy motive. But the problem is they
are far too slow. People are not informed fast enough about developments that
affect their health.
  The editors of both journals argue that rapid dissemination of information
is less important than presenting valid information. They are right. But they
behave as if they are the sole legitimate sources of such information.
  Once science's peer-review process establishes the validity of a scientific
finding, the contents of the paper describing those findings should soon be made
publicly available. But because of the power of the journals they may not be.
The journals do not want scientists to take their findings directly to the
public and to the news media before they appear in their own pages. Career
```

80   Part I: Focusing on the Workplace

■ Exhibit 3.6 *(concluded)*

considerations then come into play. Researchers who divulge their unpublished work risk being labeled publicity seekers and having their articles rejected, even if the articles pass the journals' own peer-review process.

The New England Journal of Medicine has a fast-track procedure. This means speeding the information into print within two to three months. The Journal does consent to prepublication release of scientific conclusions—but, typically, only when a federal agency like the National Institutes of Health or the Centers for Disease Control declares the need for urgent dissemination of information.

The On-Line Journal of Current Clinical Trials, published by the American Association for the Advancement of Science, disseminates medical research articles which have passed peer review that it sponsors. It promises publication within 48 hours of the time the article passes its own muster.

Nobody thinks medical findings should be prematurely popped into the Internet: This would spawn false hope. Nonetheless, unpublished but fully reviewed research may be valuable to patients and their families.

In March, the press carried reports that calcium channel blockers, drugs that reduce blood pressure, could increase risk of a heart attack. The story was based on research presented at an American Heart Association meeting to which journalists had been invited. Patients across the nation called their doctors, but few details were known at the time. Here it is July, and the study has not yet appeared in a journal.

Medical researchers, most of whom receive public financing, feel a responsibility to communicate with the public about their work. They should not be thwarted by journals.

Source: Miriam Shuchman and Michael Wilkes, "Medical Breakthroughs, On-Line," *The New York Times,* July 7, 1995, p. A-25. Copyright © 1995 by the New York Times Co. Reprinted by permission.

Reaction: The government has a responsibility to make sure the public gets medical information faster, especially if the research is funded by the public. Services like InfoTrac's Health Reference Center are really helpful when someone you love has a disease like cancer. You can look up all the technical terms and prepare questions to ask your doctor. Even learning about survival rates can be useful, although it might be depressing.

Citation: Miriam Shuchman, and Michael S. Wilkes. "Medical Breakthroughs, On-Line." *New York Times* Section A (July 7, 1995), page 25.

The full text of the article is reprinted in Exhibit 3.6.

Chapter 3: Reading and Researching for Workplace Writing

ASSIGNMENTS ■

Assignment 1: Prepare an Article Summary.   Reread either Miriam Shuchman and Michael S. Wilkes' article, "Medical Breakthroughs, On-Line," or Dahir Mubarak's article, "Educating Engineers for the Real World." Prepare a formal summary (and reaction) to practice your skills in summarizing, paraphrasing, and interpreting. If you prefer, photocopy one article from one of the technical journals you've found that interests you and summarize it.

Follow the guidelines in this chapter on how to write a summary. Start by writing a sentence that defines your audience and purpose; then, read through the article several times, identifying key words and key concepts. Make notes in the margins (if you are using a photocopy) and complete a double-entry journal entry, using the format shown in this book. Set up the format so you can record the information you would need for a bibliography.

After you have completed your double-entry journal entry, prepare a one- or two-paragraph summary and a one-paragraph reaction. Make sure you include some paraphrasing, summarizing, and one quotation.

Review your draft summary with someone from your discussion group. Make any final editing changes and then turn in the final summary and reaction to your instructor. If you used a different article than the one in this book, turn in a photocopy of your article with your summary.

Assignment 2: Build an Annotated Bibliography.   Think of a topic from your field that is of interest to you. Go to the CD ROM center at your school library and use the key words to review and select 8 to 10 articles related to your topic.

Use the "browse" feature to explore related subjects if you are having difficulty finding good articles. You may have to narrow your topic or make it broader. For example, instead of *inventory*, you may need to look at *inventory management*. Print out the citations and, if they are available, print out the abstracts as well.

After you have your working list of citations, read through the material and decide which would be most interesting. Select five of the 8 to 10 articles and write a sentence or two on your own reaction to the topic, article title, or abstract. You have now created an *annotated bibliography*. Share this with your discussion group. Highlight for your discussion group what was most difficult about this project, summarize the group's discussion, and turn this project in to your instructor with all printouts.

Assignment 3: Prepare a Field Report on Technical Journals.   Go to your school library and compile a list of the technical journals and resources in the Reference Section that relate to your field (you'll need to find the call numbers for your field for this part).

Use the *Directory of Associations* to find a list of the trade associations in your field. Check to see if any of these trade associations publish a trade journal and find out the subscription information. Share your list of journals and resources with an instructor who teaches in your major and ask if any additions or deletions are needed.

Part I: Focusing on the Workplace

Write a memo to your instructor evaluating the library resources that relate to your field and recommend any changes (if needed). Review your draft memo with someone from your discussion group. Make any final editing changes and then turn in the final memo to your instructor. If you collected any printouts as part of your review of library resources, include them with your memo.

■ PROJECT: EXPLORING YOUR LIBRARY RESOURCES

Select a topic from a field you are interested in, such as employment opportunities or professional development in your field, or choose a topic like how nurses use computers or the Internet for beginners. Your goal is are to use the library to find two newspaper articles, two journal articles, and two books related to your topic. You may choose to work with a partner for this project.

Before You Go to the Library

1. Select a partner.
2. Select your topic.
3. Write down four or five key words related to your topic.
4. Write three or four questions you would like to answer about your topic.

While You Are at the Library

1. Use the *Library of Congress Subject Index headings* to find out what terms librarians have used for information.
2. Use the *CD ROM resources* to help you find two journals in your field. Print out a "working bibliography" of five or six citations you are interested in. Find two articles and make a photocopy if you cannot check the journals out. If you are working on a team, find four articles.
3. Use the *card catalog* to help you find two books (or other materials) on your topic. Shelf browse, using the call number you find from the card catalog.
4. Check the *reference section,* using your call number.
5. Ask the *librarian* about additional resources *after* you have collected as much information as you can.

After You Have Completed the Library Research

1. Write a *journal entry* that describes how you worked in the library. Which resources were most useful? What were the strengths and weaknesses of the CD ROM resource? The reference section? The card catalog? Did your library have journals that were useful to you? Books or other print materials? What strengths and weaknesses did you notice about the journal and book collections? Was it easy to use library resources overall? Why or why not? Conclude by noting what you learned from this activity.

Chapter 3: Reading and Researching for Workplace Writing

2. Prepare a *brief summary* and *reaction* to one magazine or newspaper article and one book. Include a citation at the top of each summary and reaction. Use a heading to start each section of your review and follow chapter guidelines on writing summaries and reactions.

For the book, first look through the book, skim reading a few pages here and there to get an overall sense of the author's style and the topic. Use the table of contents to describe what is in the book when you prepare the summary. You cannot read the book for this assignment. Your goal is simply to decide if the book is worth reading. So, in your reaction, respond to what is said in the introduction, the table of contents, and on the book jacket (if one is available).

3. Work with a peer editor to get some feedback on your first drafts, then revise and proofread before turning in the final summary to your instructor with a photocopy of the magazine or newspaper article.

JOURNAL ENTRY ■

In this journal entry, analyze your research process. Take a moment to describe your current research process. What do you tend to do first when you must complete a research project? Second? Use a list or draft a flowchart to describe each step you go through.

Try to identify which stages you feel most comfortable with and which ones you need more work in. For example, are you comfortable using CD ROM resources for research? What about the Internet? Which areas of the library are you most comfortable working in? Which presented challenges when you completed exercises in this chapter? What specific steps would help minimize problems? Finally, what kinds of research do you think will be most useful in future classes you will take? In your future career?

IMPROVING WRITING PRODUCTIVITY

Part II emphasizes practical writing skills—describing and defining tools, equipment, and workplace processes, using forms, preparing informative and persuasive memos, and writing and responding to e-mail.

You can write effectively by connecting what you write to what your audience needs. By choosing the appropriate format and by revising for clarity, you will help your audience understand the information more easily. In this unit, you will also explore how to use graphics to improve the usefulness of your writing.

4

Writing about Tools and Equipment

Chapter PREVIEW

This chapter introduces you to strategies workplace writers use to describe technical objects and processes. You'll practice how to analyze and define tools or processes so that others can understand and use them. You'll work on designing your documents so they are easy to read, and you'll analyze and prepare instructions.

Two key writing strategies will be introduced: (1) choosing some sort of logical pattern to organize the information being presented, and (2) carefully crafting the document design itself.

Journal ENTRIES

Begin thinking about how you show others to use tools or equipment. Write a short one-paragraph answer to any of the following questions in your class journal or prepare answers for class discussion, as directed by your instructor.

1. What is the difference between a tool and a mechanism?

2. What motivates people to use a new tool or process?

3. How important are pictures (for example, photographs or line drawings) to understanding what something is?

4. What do you remember about the driving manual you studied when you first got your driver's license? Its contents? Its style? Its overall document design?

5. What kind of tools or equipment does a technician use in your field? What kinds of tools or equipment might you be expected to use in the first three years of your job?

Chapter READING

Do we understand the tools we work with? Sometimes we don't really know what makes a tool work—even if we use it every day. This "black-box" approach puts the *tool* in charge of us, rather than *us* in charge of the tool.

We need to know how our tools perform and what the risks are in using our tools. As you read this next article, excerpted from David Macauley's *The Way Things Work,* consider how much we assume about the tools and technologies we use every day.

ELECTRICITY AND AUTOMATION

David Macauley

Introduction. The power behind electricity comes from the smallest things known to science. These are electrons, tiny particles within atoms, that each bear a minute electric charge. If a million million of them were lined up, they would scarcely reach across the head of a pin. When an electric current flows through a wire, these tiny particles surge through the metal in unimaginable numbers. In a current of 1 ampere, sufficient to light a flashlight bulb for example, 6 million million million electrons pass any point in just one second. Each electron moves relatively slowly, but the movement itself is passed from electron to electron at the speed of light.

If the nineteenth century was the heyday of mechanical machines, then the twentieth century belongs to machines powered by electricity. This does not mean that the age of mechanical machines is behind us. Machines that move will always be needed for doing work but machines that think, operated by electricity, are increasingly what controls them.

Exploiting Electrons. The machines in this part of *The Way Things Work* use electricity in a number of different ways. In many electrical machines, moving electrons and magnetic fields are intimately linked. As soon as electrons start to move along a wire, they create a magnetic field around them. Quite why they do this cannot easily be explained, but the fact that they do is borne out in all machines that use electric motors, because all these make use of magnetism. So too do electric generators, which produce our supply of electricity in the first place.

Electrons also produce electric fields, which have the same ability to attract and repel as magnetic ones. Some machines in the following pages, such as the photocopier and the air ionizer, work by shifting electrons about so that electrical attraction and repulsion comes into play.

Yet more machines, like the calculator and the computer, use electrons as a means to carry information. But despite these differences, the principles that govern the flow of electricity in all machines are just the same. The electrons always need energy to make them move. They always travel in a set direction (perversely from negative to positive) and at a set speed. Furthermore, they will always produce particular effects while they are on the move. One of these effects is heat; magnetism is another.

Electricity and Movement. As a source of power, electricity has no rival. It is clean, silent, can be turned on and off instantly, and can be fed easily to where it is needed.

Electric machines that produce movement are extra-ordinarily diverse. At first sight, there is little similarity between, for example, a quartz wristwatch and an electric locomotive. However, both use the motive force produced by the magnetic effects of an electric current—although the current used by a train is hundreds of thousands of times greater than that which flows in a watch.

Like all electrical machines, those that use electricity to produce movement

Chapter 4: Writing about Tools and Equipment

take only as much power as they need. An electric motor will only take a set amount of current. This means that one source can power many machines with each one taking only the current it needs and no more.

Electricity for Signals. In the same way that energy in waves can be made to carry information, so too can energy in the form of electricity. Like light and radio waves, electricity travels almost instantaneously, so the message arrives at its destination with little or no delay.

Machines like calculators and computers that make use of electricity to carry information are known as electronic machines. This means that they work by controlling the movements of electrons. Because electrons are so small, it is possible to make the components that control them very small indeed. The components of a computer can therefore be highly miniaturized and assembled in complex circuits to give the computer its extraordinary range of abilities. A single microchip can store the street map of any capital city—something big enough for you to get lost in. Yet it is so small that you could very easily lose it!

Machines that Control Themselves. The paramount importance of electric signals is to control machines—not just to switch them on and off but to provide information and instructions that govern the way in which they work.

Sensors and detectors are often the source of these signals. They can detect the presence of physical objects, like metal or smoke, and then can also measure quantities, such as speed. Linked with powerful computers, electrical machines can be used to control mechanical machines and process information faster, more reliably and more accurately than humans ever can.

An air journey, for example, shows how many tasks they have taken away from human hands. Even before you board a plane, a computer will have booked your place and probably printed your ticket, while scanners have checked your baggage. Although the aircraft itself is not basically an electrical machine—a paper dart can fly, for example—an airliner could not even leave the ground without electricity to start its engines and power its on-board systems.

Once in the air, you travel in safety because electrically operated radar enables air traffic controllers to guide the pilot, and you get to your destination mainly because the autopilot has kept the plane on course. The human crew will have trained on a computer-controlled flight simulator, learning how to cope with everyday matters and emergencies thrown at them by machine. Even mundane creature comforts—fresh air, light, music and movies—all require electricity.

More and more aspects of our everyday lives now depend in some way on the flow of billions of electrons flowing through countless switches in machines that can almost think. Just as in the last century when the development of mechanical machines seemed to be unbounded, so today the growth of electronic machines seems unlimited. With the computer and the robot, mankind is advancing towards the ultimate machine—one that will be able to perform any task required of it.

Source: David Macauley, "The Transducer," *The Way Things Work*. Boston: Houghton Mifflin, 1988, pp. 274–77. Compilation copyright 1988 by Dorling Kindersley, Ltd. Text copyright 1988 by David Macauley and Neil Ardley. Reprinted by permission of Houghton Mifflin Co. All rights reserved.

Part II: Improving Writing Productivity

Analyzing THE READING

To build your reading skills, use the following questions to analyze the content and to develop your reaction to this excerpt. Write a one-page journal entry in your class journal or prepare your answers for class discussion.

Reading for Content

1. What is the major difference between mechanical machines and electrical machines?

2. What are some of the advantages of electrically powered machines? Does Macauley identify any disadvantages?

3. Macauley says that electric machines can produce movement, carry information, and control other machines. Can you think of a machine for each of these three categories?

Reading for Reaction

1. Macauley says that when we link electrical machines with computers, the resulting machine "can be used to control mechanical machines and process information faster, more reliably and more accurately than even humans can."

   Because employers benefit from increased productivity and cost savings, we will increasingly work with computers. However, what happens when these machines "lock up"? What happens to the types and number of jobs? What happens if the programmer makes a mistake in the program the machine follows? Jot down your responses to these questions and consider your previous and future workplace. In what ways are machines changing your workplace?

2. Macauley says today we depend increasingly on machines that "almost think" and that we are moving toward "the ultimate machine—one that will be able to perform any task required of it." What do you think of machines that can "think"? What separates humans from machines? What are some effects of our dependence on machines?

DESCRIBING TOOLS AND MECHANISMS

One of the more exciting days in your training comes the first time you get to work with the actual tools and machines that are an essential part of your technical field. You may be learning how to use an oscilloscope, an electron microscope, a lathe, or a heart monitor.

However, no matter how skilled you are in using the tools and machines you are familiar with today, each advance in technology brings with it new tools and new equipment—and refinements in the tools and machines you are presently using.

How will you know how to use these tools and machines most effectively? You may have studied them in school. Your supervisor or co-worker may show you how to use the tool or machine. Someone may hand you a manual or a checklist and tell you to study it and then run the machine. If you are working on a shift with few people and many different machines and someone reports in sick, you may be asked to run complicated machines with little or no training.

Many employers emphasize ongoing training to minimize injuries. A community college student, working at night full-time to pay for school costs, was asked to clean up an assembly line saw, a machine he had not yet been trained to work on. A piece of wood got jammed in the saw; he reached back under the saw to pull out the jammed wood, without realizing two high-power saws were still running. A quick zing, and the tops of two fingers were gone. Sometimes we don't realize the risks we take as we get used to working in a particular place with familiar tools.

Most of the time, your supervisor will make sure that you know the safety precautions to take when you are working with the equipment on the job, but it is also your responsibility to make sure you know what the hazards are. Also, when you are describing tools and equipment for others, it is important that you highlight any risks so they are easy for the reader to see and understand. You'll practice writing *notes, cautions,* and *warnings* in Chapter 7, Writing for Co-Workers: Informing and Instructing.

Here are some questions you may have when you are learning to use new tools:

- How do I read a manual and get the most out of it?

- How do I know what a tool or machine can do?

- How do I understand its basic operating principle? Its parts and subparts? Its limitations? Its risks?

Here are some questions you may have when you are preparing forms, instructions, proposals, or reports on tools or machines.

- How do I describe a tool or machine, if I am asked to do so?

- How do I describe how it works?

- How do I prepare specifications?

These questions are the focus of this chapter. Keep these questions in mind as you read the chapter and work through the exercises. What is most important is that you have a sense of who your audience will be and how, where, and when the information you write will be used.

Exercise 4.1 Inventory Your Experience With Tools

Think over the tools and equipment you have used over the last several years, either on the job or with a hobby. Draft a list in your journal of as many of these tools as possible, considering whether they are simple or complex tools.

Can you think of a category of tools you have used? Power tools? Hand-held tools? Shop tools? Try to organize your list of tools into categories by considering how they were used (their functions), what they were made of (their physical attributes), and what they could be used for (their performance capabilities).

Write a one-paragraph summary that concludes your journal entry to highlight how you have used tools, noting which categories of tools have been most important to you or most challenging to work with. Share the results of your work with your discussion group.

HOW DO YOU READ A MANUAL TO LEARN ABOUT NEW EQUIPMENT?

Frequently when you work with new equipment or processes, you will be oriented by your supervisor or co-workers. You then may be handed a manual. You can always ask your co-workers questions, but you can also improve your understanding of oral instructions by working on your reading skills. Improve your use of technical manuals and instructions by following the reading process outlined here.

- *Do I understand* **what** *is here?* What is the key information? Define any terms or concepts you don't know.
- **Why** *is it important?* Why is this information included? How should I use it? How does this information relate to my problem or to using this machine or tool?
- **How** *can I use this information?* How do I use this information in this particular situation? Interpret key information as it applies to your situation as you read, checking frequently for problems.

Manuals and instructions are typically organized in a very specific order, with sections designed to meet the needs of novice or more advanced readers.

- **Preview** the manual or set of instructions to see how the writer has organized the material and what special sections may be useful (a glossary; a list of special materials, conditions, or hazards; or how steps are presented).
- **Read** headings and graphics only. Slowly read through the headings in the section you need. Read the graphics to see **what** is being highlighted. Look through the document to get familiar with it.
- **Reread** the entire manual or section, using a marker to highlight key information, especially key steps, special operating conditions, the troubleshooting guide, or any hazards in materials or operations. Make any notes that will be helpful to you.

- **Stop** to define key words and ask questions as you read. Double-check any terms, concepts, relationships, amounts, or key action steps in the process. Are there any gaps in the information being presented or in your understanding? List your questions and talk to your supervisor or a co-worker to answer your questions. As a novice, never assume!

- **Observe** someone using the equipment to verify the process you are to follow. Ask questions right away if any part of the procedure is unclear to you. If this is not possible, ask someone to monitor your use of the tool or machinery.

- **Take** some "think" time. Reading a manual can be confusing, especially if the terminology or the process is new to you. After you have read the manual, take some time to think about what you have read. What parts are most important to you? What parts do you need to study further? Do you need more information?

- **Refer** to the manual often as you work with your new equipment. Continue to use your manual as a first resource to answer any questions.

Somehow, you must balance between using your own common sense to solve problems and asking your supervisor or your co-workers how to use the equipment more effectively. Keep in mind that your performance will improve as you understand **what** you are doing, **why** you are doing it, and **how** you can best do it!

WHAT IS A TOOL OR MECHANISM?

A tool is most commonly defined today as any hand-held device that helps a worker get a job done. Tools can be very simple, with no moving parts (like a hammer) or very complex with many parts and subparts (like an electrical protractor, which is a surgical instrument used in heart surgery to hold the chest cavity open). Even a rock can be considered a type of tool if it is used to carry out a task, such as breaking open a lock.

Tools frequently have *mechanisms* that enable them to work. A mechanism can be defined as any thing or group of things that acts as a functional unit. For example, the starter to a car would be such a mechanism. However, mechanisms do not have to have moving parts. Therefore,

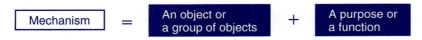

Mechanisms can be part of a tool or machine—or they can be complete in themselves.

Part II: Improving Writing Productivity

Frequently you may be asked to read or write technical descriptions of tools, machines, or equipment. If you must describe a tool or machine so that others can understand what it is, you are preparing a description of the object itself, that is, a description of the mechanism.

A *mechanism description* tells a specific reader:

- **What** the mechanism is made of by describing its **physical** characteristics, its parts, and its subparts.

- **Why** this mechanism is useful by describing its purpose or its **functional** characteristics.

- **How** this mechanism works by describing its **operating process.**

Mechanism descriptions frequently appear in operating manuals, instructions, and textbooks. Exhibit 4.1 is an example of a mechanism description that is used to teach artists how to draw the human neck.

■ Exhibit 4.1 **The Human Neck**

The Neck

Muscles of the Neck

1. Sterno-cleido-mastoid.
2. Levator of the scapula.
3. Trapezius.

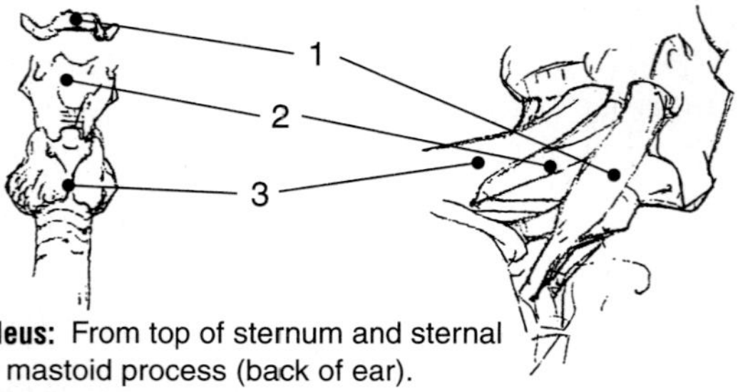

Sterno-cleido-mastoideus: From top of sternum and sternal end of clavicle to mastoid process (back of ear).
Action: Together, pull head forward; separately, rotates to opposite side, depresses head.

Levator of the Scapula: From upper cervical vertebrae to upper angle of shoulder blade.
Action: Raises angle of should blade.

Trapezius: From occipital bone, nape ligament, and spine as far as twelfth dorsal, to clavicle, acromion, and ridge of shoulder blade.
Action: Extends head, elevates shoulder, and rotates shoulder blade.

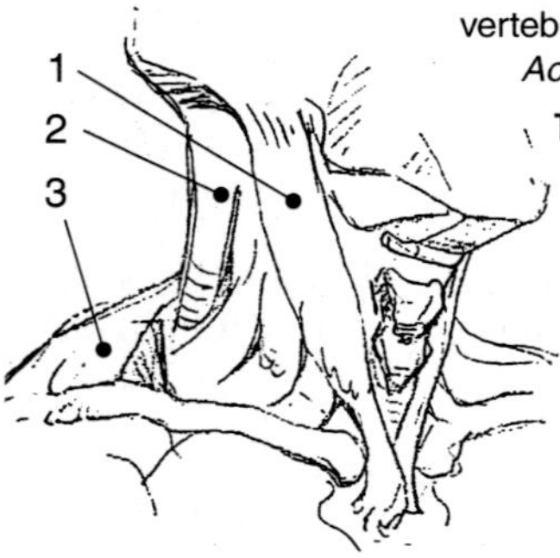

Source: George B. Bridgman, "The Neck," *Constructive Anatomy.* New York: Dover Publications (1960), pp. 74–76.

Analyzing a Mechanism Exercise 4.2

List two or three physical and functional characteristics for each of the following. Describe in one or two sentences its operating process.

   A pencil.

   A hammer.

   A photocopy machine.

Compare your list with the lists of members of your discussion groups. Were your lists of characteristics the same? Were your descriptions of the operating process similar?

Working with a partner or in your small group, summarize the difference between physical and functional characteristics in a short memo for your instructor. In what way are operating characteristics different from physical and functional characteristics? Consider which characteristics are more important for two or three different types of workplace audiences. Then, turn in a group memo to your instructor with your worksheets attached.

As you can see, graphics are an essential part of any mechanism description—regardless of the field of work involved—because they clarify and illustrate what an object is and how it works.

HOW TO WRITE A TECHNICAL DEFINITION

A technical description gives the reader precise and accurate information about **what** something is. Most technical descriptions begin with a formal definition.

Depending on the type of document you are writing, you can use any one of four common types of technical definition:

   Parenthetical definition

   Formal definition

   Expanded definition

   Stipulative definition

Writing Parenthetical Definitions

Parenthetical definitions define the term right in the text with a few words. Your reader can instantly understand a difficult concept. A parenthetical definition is frequently used in any description and in longer reports which contain special terminology that may not be understood by most readers.

The following sentence includes a parenthetical definition. Notice how the definition appears right after the word it defines—in parentheses.

Part II: Improving Writing Productivity

Because of its good thermal conductivity (the property of transmitting heat very readily), copper is used in radiators of motor vehicles. (Source: Marshall Cavendish, p. 198.)

Writing Formal Definitions

Formal definitions define the term with a few precise sentences. Formal definitions commonly appear in dictionaries, glossaries, and usually at the beginning of technical descriptions. Formal definitions also follow this formula:

TERM = CLASS + DISTINGUISHING FEATURES

The following formal definition describes an *object:*

A *hypodermic syringe* is a medical instrument used to inject fluids into the body or draw them from it. The syringe is fitted with a hypodermic needle and a cylinder that fits into a larger cylinder and moves under fluid pressure. The syringe acts as a piston, and displaces or compresses fluids, as in a pump. (Source: *The American Heritage College Dictionary*)

A formal definition can also describe a *process.*

Scrubbing is the method of removing impurities from a gas by allowing it to contact a liquid or solid with which the impurities will react or in which they will dissolve, often carried out using a column. (Source: *Dictionary of Teledyne Wah Chang's Language*, 1992).

To decide whether a formal definition is complete, use this widely accepted formula to analyze the next sentences.

Term	=	Class	+	Distinguishing characteristics
1. An orange		is a fruit.		
2. An orange		is a citrus fruit.		
3. An orange		is a citrus fruit		often used to make juice.
4. An orange		is a citrus fruit		that is reddish-yellow in color and is often used to make juice.

Notice how the definition becomes clearer as the class is more narrowly defined and the distinguishing characteristics are described more specifically.

Writing a definition by following a formula can present some challenges. However, by using all three parts of the formula, you can work toward a very precise description. Knowing the object well, of course, helps you to write with more precision.

Compare your analysis with these notes.

| Term | = | Class | + | Distinguishing characteristics |

1. An orange   is a fruit.
   Note: The class is too broad and there are no distinguishing features.

2. An orange   is a citrus fruit.
   Note: The class is more precisely defined, but because there are no distinguishing features, any citrus fruit could fit this definition.

3. An orange   is a citrus fruit   often used to make juice.
   Note: All parts of the formula have been used, but the distinguishing features may not be specific enough, because some people drink grapefruit juice or other citrus juices.

4. An orange   is a citrus fruit   that is reddish-yellow in color and is often used to make juice.
   Note: The definition is improved, but what about tangerine juice?

The goal is to write a definition that will describe only one term. As you can see, writing a clear, concise, and complete definition requires knowing the term to be defined very well as well as using a great deal of precision.

Writing Expanded Definitions

An *expanded definition* is used when the reader needs more information than a sentence or two to describe **what** something is and **how** it works. Very complex concepts can require a few paragraphs, a section, chapters, or even entire books to define the term or concept fully.

Analyzing Technical Definitions   Exercise 4.3

What's wrong with the following formal technical definitions? List what part of the formula is missing (TERM = CLASS + DISTINGUISHING FEATURES). Note whether the writer used functional features or operating features. Rewrite one of the following definitions. (List the name of your dictionary and the page number if you use one).

A *hammer* is used to hit nails.

A *motorbike* is a two-wheeled conveyance used for travel.

The *entectic point* occurs when two metals are mixed, raise each other's melting points, and change from liquid to solid state at the same time.

The *pineal gland,* a pea-sized organ buried in the brain, stimulates drowsiness by secreting melatonin. It also influences moods, hunger and metabolism.[1]

Bring your definitions to your discussion group for review. In your group, compare the revisions, commenting on strengths of each definition. Select one definition and write a brief statement why this definition was selected as the "best of the group" before turning this group project in to your instructor.

---

[1]"Bright Light Fights Winter Blues," *USA Today: Newsletter Edition,* 122, no. 2585 (February 1994), p. 15.

Find the *term*, the *class*, and the *distinguishing characteristics* in this formal definition of **sonar.** Does the supporting discussion clarify the distinguishing characteristics of sonar? Are the *physical features* of the term described as well as its *functional features* (its process)?

> Sonar, which stands for Sound Navigation And Ranging, is a sensing system that detects objects with sound waves. It is mainly used underwater, where other kinds of waves and rays do not travel so well. Ships employ sonar to measure the depth of water, to find shoals of fish and to detect wrecks. A transducer emits a pulse of sound, which travels down through the water and is reflected back. The transducer picks up this echo and the sonar converts the time it takes the sound to return into a value for the object's distance.[2]

What about this definition of a *transducer?* Can you find the *term,* the *class,* and the *distinguishing characteristics?* Do you have a clear understanding of the *physical* as well as the *functional characteristics?* Can you imagine understanding the transducer without the drawing shown in Exhibit 4.2?

> A *transducer* is a device that converts one kind of signal into another. In sonar, a transducer on the hull of a ship converts an electric pulse into a pulse of sound; it then converts the returning sound waves back into an electric signal. It is similar to a combined loudspeaker and microphone.[3]

■ Exhibit 4.2 **Transducer**

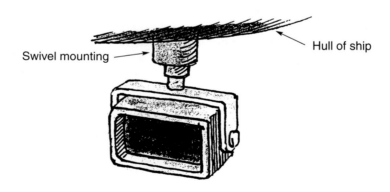

Source: David Macoulay, *The Way Things Work*, p. 318.

---

[2]David Macauley, "Electricity and Automation," *The Way Things Work*. Boston: Houghton Mifflin, 1988, p. 318. Compilation copyright 1988 by Dorling Kindersley, Ltd. Text copyright 1988 by David Macauley and Neil Ardley. Reprinted by permission of Houghton Mifflin Co. All rights reserved.

[3]David Macauley, "The Transducer," *The Way Things Work*. Boston: Houghton Mifflin, 1988, p. 318. Compilation copyright 1988 by Dorling Kindersley, Ltd. Text copyright 1988 by David Macauley and Neil Ardley. Reprinted by permission of Houghton Mifflin Co. All rights reserved.

WRITING STRATEGIES: USING EXPANDING TECHNIQUES

Writers can use a variety of expanding techniques to define complex terms. Each of the following techniques may be used by itself or in combination to add more information or analysis about the term being defined.

You can decide which of these to use by thinking about your audience and what their needs are: How will your audience use the information, how much does your audience already know, and why does your audience need this information?

Expanding Techniques That Primarily Describe the Term

Define the key term (formally or parenthetically).

Explain how this object works (its operating principle).

Describe the process (function) of the object.

Describe special conditions (if any).

Provide background on the concept or the object.

Give a history of the word itself (etymology).

Use drawings, photographs, or some graphic to illustrate a key concept or part of the object.

Expanding Techniques That Analyze the Term

Compare some aspect of this object with a familiar object (metaphor).

Describe what the object can **not** do (negation).

Divide the object into parts (partition).

Sort the object into categories (classification).

Show the causes or effects of the object.

Describe the implications of the object (its form or function).

Notice how these techniques are used in the reading below. This particular discussion of a process did not include any illustrations—photographs, line drawings, or computer-generated drawings. Would these have helped your understanding of the material?

Identifying Expanding Techniques   Exercise 4.4

Look through a textbook in your major to find an example of an expanded definition. Make a photocopy of the definition and list as many expanding techniques as possible that this writer used.

Bring your list to class for discussion with your group. See if additional expanding techniques can be identified in your group's examples. Write one or two summarizing sentences about what your group discovered in comparing these expanded definitions and turn in your group project to your instructor.

COMPUTERS TO AID BLOOD TYPING

Imagine yourself, five years from now, in an emergency room. You're losing a lot of blood, and the hospital staff is about to give you more. The type of desktop computer on hand at the blood bank may make the difference between a quick recovery and prolonged sickness—even death.

The job of a technologist is to match a patient's blood with that of a donor by finding which antibodies—defense agents of the immune system—are in the patient's blood. According to Phil Smith, professor of industrial systems engineering, Ohio State University, there are more than 400 other antibodies that also may be present, and identifying them requires expert detective work. "Though relatively rare—certainly less than one-tenth of the population have an unusual antibody—they must be taken into account when matching blood." If an antibody present in a patient's blood is undetected or misidentified, he or she may be given blood that triggers a rejection response.

Correctly identifying antibodies can be as difficult as diagnosing a disease. "By adding a small amount of blood serum to test cells and monitoring the reaction, technologists can narrow down the field of possibilities, but the process involves lots of high-level reasoning, logic, and intuition." With the help of the right type of computer, they may be able to improve their ability to identify antibodies. "But, as we [have] learned from other fields, it's important to find the best way for the computer to share its expertise with the operator so that overall performance can be improved."

Smith and graduate student Stephanie Guerlain found that blood bank technologists are more likely to make mistakes in typing people's blood when they use partially automated computer aids than those that just monitor and critique their performance. "A partially automated system uses a rigid set of rules and often encourages overreliance," Smith indicates. The alternative—a monitoring or critiquing procedure—is more similar to the systems used by commercial aircraft to avoid traffic. This type of system is designed to analyze a situation and periodically interrupt to give a warning and advice if it detects a problem. If the system disagrees with any of the conclusions of its user, it displays a brief error message.

The researchers found that there was a trade-off in performance when the technologists used the partially automated system. In simple cases, they performed twice as well as did those using the critiquing system—with an average of a 5.6% misidentification rate vs. 11.9%. However, in cases where more complex reasoning was required—when unusual antibodies were in the blood—the partially automated system led the technologists to make one-third more errors than those working with critiquing systems.

Source: "Computers to Aid Blood Typing," *USA Today: Newsletter Edition,* vol. 122, no. 2585 (February 1994), p. 12.

Chapter 4: Writing about Tools and Equipment **101**

PREPARING STIPULATIVE DEFINITIONS

A *stipulative definition* is used when the writer wants to limit the definition of a term for a specific situation. If you sign a rental agreement, for example, you will be signing as the lessee. A stipulative definition in such a rental agreement could read like this:

> Said aforementioned premises shall be occupied by no more than two adults and three children, hereinafter known as the lessee. (Source: Adapted from Stevens-Ness Law Publishing Co. Form 818 Rental Agreement)

We use stipulative definitions in the workplace to make sure the reader or listener understands the limitations affecting his or her use of a particular tool or piece of equipment. This wording was included as part of installation instructions for a dishwasher:

> Caution: When house wiring is aluminum, be sure to use U.L. approved anti-oxidant compound and aluminum-to-copper connectors. (Source: General Electric Company, Pub. 31-3494)

HOW TO DESCRIBE A TOOL OR MECHANISM

Although you can use a simple pattern of introduction, discussion, and conclusion to start your draft of a technical description, it may be helpful to follow a more detailed outline, adapting it to your mechanism, your writing purpose, and your audience's needs.

This Outline Could Include

A **title** that specifies the object you are describing.

An **introduction** that introduces the mechanism, summarizes the purpose of your description, and previews what's in this description.

A **formal definition** that explains what this mechanism is in dictionary terms, with any special terms defined parenthetically throughout your description.

A **list of parts and subparts.**

Illustrations that show **what** the object is and, if needed, **how** it works (its process), or **how** it is put together (its assembly), or **what** tasks it can do (its function).

A **description** of the object that includes its physical and functional characteristics. Operating principles may be included, if needed.

A **conclusion** that summarizes the purpose of this description or directs the reader to further information.

Part II: Improving Writing Productivity

You can see these features of technical description in the description of a hand-held pencil sharpener written by Paul for a class exercise (see Exhibit 4.3). Paul reported that the most difficult parts of writing this description were:

1. Understanding the object well enough to define it accurately. This came only with drawing and using the object.

2. Having the right words to describe and define the object. A dictionary was useful!

3. Using parallel form so that the description read smoothly.

4. Separating **what** from **how** so this object description only described what the pencil sharpener was (definition + parts + subparts).

5. Making and labeling the drawing.

As you read Paul's description, do you see any improvements that would help him? Did Paul leave any parts out of this description?

The more complex your mechanism is, the more you will need to use two key writing strategies: (1) *choosing some sort of logical pattern to organize the information* being presented, and (2) careful *crafting of the document design* itself—using headings, graphics, labels, and short lists and paragraphs. These writing strategies are essential in making your description easy to read and use.

Using an Organizing Pattern

You can organize your description of a tool or mechanism by using one of four commonly used organizational patterns. Each of these is best for one particular audience or a particular writing purpose.

1. Spatial Order.   Your audience needs to understand **what** this tool or mechanism is. Here, your emphasis is on the object itself, so you will describe **parts** and **subparts** and the relationship of these parts and subparts to the whole object. You will most likely describe these parts in a certain order, for example, from top to bottom, or left to right, or inside to outside, so that your reader can understand the parts in relation to each other.

You can describe simple objects quite well using spatial order. For example, an egg has these subparts (moving from inside to outside): the yolk, the white, the protective membrane, and the outer shell. Spatial order works best with objects that do not have moving parts.

2. Operating Order.   With this pattern, you want your audience to understand **how** the tool or mechanism works, so your discussion is organized around how it operates. Frequently, this type of description is called a *process description*

Description of a Hand-Held Pencil Sharpener   Exhibit 4.3 ■

A hand-held pencil sharpener is a small tool used to make a fine point at the writing end of the graphite-and-wood end of a pencil so that a writer may use the pencil to write characters, usually on paper. This plastic sharpener has three basic parts: a base, a razor, and a cup.

1. The *base* of 1" x 1-1/2" has four subparts:
    a. A flat *platform* on the bottom of the pencil sharpener so it can stand upright.
    b. A *mount* for the razor.
    c. A *round hole* matching the size of the pencil to be sharpened.
    d. A ridged *exterior* so the base can be grasped firmly when the sharpener is used to sharpen pencils or to open the sharpener to discard shavings from the pencil.
2. A flat, single-edged *razor,* mounted on the top inside of the base, with the cutting edge of the razor positioned to sharpen the pencil.
3. A clear, removable *plastic cup* of 1-1/2 " x 1-1/2" that fits tightly over the base and catches the shavings made as the pencil is sharpened. The cup has two thumb-sized indentations so that the pencil sharpener can be grasped with one hand while the pencil is being sharpened.

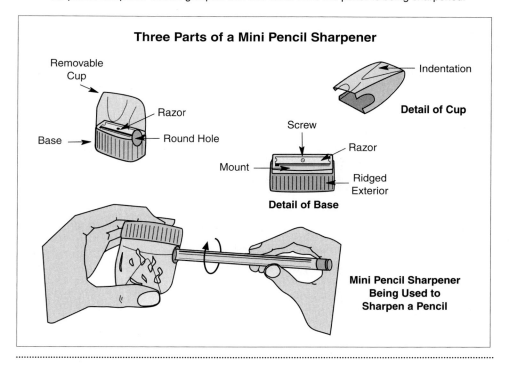

because it describes the actual sequence the mechanism follows when it is working. Sometimes a process description is divided into steps or stages (for example, step one, step two, etc.). An operating order works best with simpler mechanisms, like a can opener or a drill.

3. Order of Importance. On very complex mechanisms, your audience needs to understand not only **which system is most important** but also **what** it is and **how** it works. For example, if you were describing a bicycle, which would you describe first? The braking system? The steering system? The lighting system? Analyzing your audience's needs and then organizing your mechanism into major and minor parts or systems will help you decide what information to put first.

4. Order of Assembly. When your audience must put the mechanism together or wants to know how something is assembled, you would describe the mechanism in terms of **what parts are assembled first,** and then continue by describing how to assemble the object—in chronological order. Your key transition words would be *first, second, third,* and *finally.* Nearly everyone has assembled a child's toy; this order of assembly is very often used.

HOW TO DESCRIBE A PROCESS

When you are describing a process, you are explaining what happens from beginning to end or explaining what happens during one stage or one part of a larger process. For example, you might be describing a process as simple as how a meat thermometer works or a process as complex as how an anode rod works in an electric water heater.

Your *process description* can clarify:

- How something **works** (how an ear or a camera works, for example).

- How something is **made** (how dollar bills, glue, or airplanes are manufactured, for instance).

- How something **occurs** (such as what happens when a volcano erupts or how eyestrain occurs).

The following process description illustrates a concept:

To understand the difference between metals and non-metals, consider the concept of the metallic state. In a true metal, the identical atoms of the metal arrange themselves in an orderly way as closely as possible, rather like a lot of billiard balls in rows and layers. The outer free electrons then act as if they are interchangeable between atoms and their attractive force binds the atoms together. These outer free electrons act as a sort of cloud of glue. When an electrical pressure is applied, these free electrons are available to drift along. They make up the electrical current, and their movement also allows easy conducting of heat. (Source: Cavendish, *Dictionary of Technical Terms,* p. 199)

Notice where the writer moves from describing the mechanism to describing the process. What special technique does the writer use to make the concepts as clear as possible to the reader? Does this strategy help you visualize how the metallic state works?

Using comparisons or **creating a metaphor** helps the reader compare what is unknown to something that is known. In this example, Cavendish uses billiard balls, a cloud, and glue as concrete ways of showing the reader how this abstract concept works.

Often writers use metaphors to introduce new words. For example, in working with the Internet, you will find new words everywhere that come from innovation—new ideas, new processes, and new objects. Java is a new software program that people can use to write small programs or *aplets* on Web pages. Joe Kilsheimer points out that these aplets have "nothing to do with apples; an aplet is a small application, just like a wavelet is a small wave."[4]

In describing a process, your goal is to write with clarity, precision, and conciseness. Before beginning to describe the process, you will need to consider **who** is using this process description, **why** they are using it, and **where** they will be using it. For example, describing how to assemble a new type of bicycle would be very different for the following three audiences: consumers, bicycle repair shop owners, and manufacturers.

Your planning for the content, style, organization, and document design for the process description will change, depending on your audience, their background, and their purpose. If you have several different audiences to consider, you may have to organize different sections of your description as some manual writers do, creating a section for a novice and another section for an expert reader.

Describing a Process   Exercise 4.5

Follow the steps outlined above to write a two- or three-paragraph description of how you complete a particular task. Think of a simple task that you are already very familiar with, such as how you might wash your dog or how you might go about applying for a job. Emphasize how you do this task (process description), not how you want someone else to do this task (instructions). Include a definition as well as your description of the process.

Bring your draft process description to your discussion group for review. Together look at the expanding techniques discussed previously in this chapter to identify two additional writing strategies that could be used to clarify this process for your reader. Turn in your drafts to your instructor with notes that highlight the strengths and weaknesses of these drafts.

---

[4]Joe Kilsheimer, "Java: It's Going to Revolutionize the Internet," *Corvallis Gazette-Times* (January 8, 1996), p. C4.

WHAT'S THE DIFFERENCE BETWEEN A DEFINITION AND A PROCESS DESCRIPTION?

Most of the time, you will need to use both a **definition** to show **what** makes things happen and a **process description** to show **how** things happen.

This chart summarizes the key differences between these two types of informative writing. Both are important for workplace writing.

Definition	Process Description
What is this thing? Is it an object? A tool? A mechanism?	**How** does this thing work?
What are the key parts and subparts?	**What** are the key operating steps or stages?
Why is this definition needed? To help readers understand exactly what this object is.	**Why** is this process description needed? To help readers understand exactly how this process works.
Example: The **scrubber** is a column that removes pollution from a gas stream.	*Example*: **Scrubbing** removes impurities from a gas by allowing it to come into contact with a liquid or solid that dissolves or changes the impurity into a harmless emission.

HOW TO USE GRAPHICS IN TECHNICAL DESCRIPTIONS

The old saying that one picture is worth a thousand words is very true, especially when you must write about technical subjects. In fact, some training manuals have more pictures than words!

Learning from images (pictures or graphics) is a powerful way to reinforce concepts your readers have just read about. Actually, we retain more of what we learn when we not only read about, but can "see" it. We can remember complicated information more easily if we are given opportunities to discuss the information and to actually carry out the task.

Using graphics (and document design) can help readers understand objects, concepts, processes, and relationships. In many ways, whether we write with words or communicate with graphics, we are answering the following questions our readers have:

- **What** is it?

- **How** does it work?

- **Why** is it important?

This section will help you understand how photographs, line drawings, and computer-generated drawings, maps, or models are used in technical description. These kinds of graphics show **what** the object is (object) or **how** a process works (process). Subsequent chapters will cover other types of graphics.

When you are writing about a technical object or process, frequently your readers do not have a clear picture in their minds about what the object looks like or how it works. You can use photographs, line drawings, and computer-generated graphics to clarify the reader's mental image.

Photographs

A *photograph* shows exactly what an object looks like. However, photographs do not highlight key parts or subparts. Novice readers may also be confused by the amount of detail shown. This photograph of arrowheads gives the reader a clear sense of the range of styles and materials possible, much faster than through words or a drawing (see Exhibit 4.4).

Various Types of Arrowheads   Exhibit 4.4 ■

Source: Tony Stone Worldwide, Ltd./© Steve Elmore.

Line Drawings

A *line drawing* is often used in place of a photograph because the line drawing simplifies the object or the process being illustrated. Key parts can be highlighted and labeled more easily. Notice how the drawing in Exhibit 4.5 clarifies the process of manufacturing silicon chips.

■ Exhibit 4.5   From Silicon to Chip

The soaring worldwide demand for microchips – used in everything from hair dryers to computers – has fueled the Portland area's high-tech boom. Semiconductor companies will need thousands of workers over the next five years to help meet the demand. Here's how workers turn silicon into chips:

1 Silicon used by the semiconductor industry comes from the Earth's crust. Workers grow a pure silicon rod – called an ingot – by heating silicon crystals in a furnace.

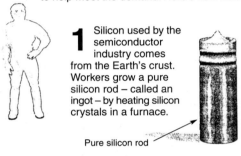

Pure silicon rod

2 Using high-tech machines, the silicon rod is shaped into a perfectly rounded cylinder weighing from 10 to 200 pounds. Next, it is sliced into thin wafers about the size and thickness of a compact disc. They are then cleaned and polished.

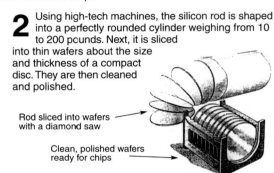

Rod sliced into wafers with a diamond saw

Clean, polished wafers ready for chips

3 Chips in a pattern of circuitry are etched onto the wafer through a series of photographic processes that involve gases and light sensitive material.

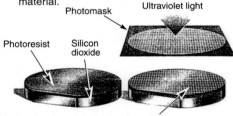

Photomask
Ultraviolet light
Photoresist
Silicon dioxide
Hard resist where light comes through

4 Tiny amounts of chemical impurities, called dopants, are either heated or electrically beamed into the surface of the wafer, giving it positive and negative charges. The location of these charges individualizes one chip from another.

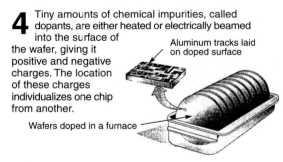

Aluminum tracks laid on doped surface

Wafers doped in a furnace

5 The wafer is cut into separate identical chips. A six-inch wafer, for example, holds about 80 of Intel's Pentium chips.

Cutting chips

Separate chips

6 A speak of dust can ruin a chip, so they are tested with electrical probes. Those that work are attached to a frame and sealed in a plastic case.

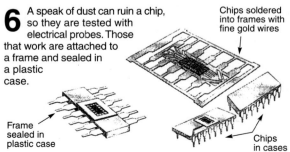

Chips soldered into frames with fine gold wires

Frame sealed in plastic case

Chips in cases

Installation Diagram with Related Parts   Exhibit 4.6 ■

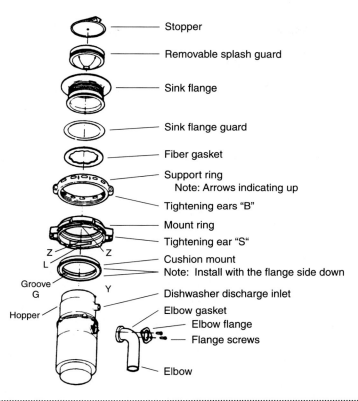

Source: Anaheim Manufacturing, Anaheim, CA: *Food Waste Disposer Owner's Guide* (February 1988), P/N 560C239S01 (phone 800/854-3229), p. 9. Used with permission.

Exploded Drawings

A *line drawing that has been "exploded"* shows the relationships of all the parts to the whole. This graphic is frequently used for complex machines with many parts where the reader is responsible for assembling or maintaining the machine. Exhibit 4.6 is an example of an exploded line drawing by which the reader can identify all of the parts needed to install a food waste disposal system.

Cut-Away Drawings

A *line drawing with part "cut away"* can show the reader the "inside" of the object or process. The line drawing in Exhibit 4.7, also of the food waste disposal, shows part of the object cut away so the new owner can see the "inside" of the unit and understand where to tighten the turntable inside the unit.

■ **Exhibit 4.7 Adjusting Tightness of Turntable**

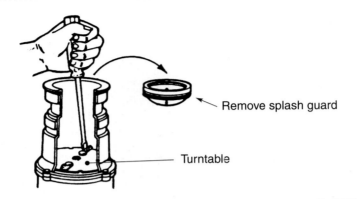

Source: Anaheim Manufacturing, Anaheim, CA: *Food Waste Disposer Owner's Guide* (February 1988), P/N 560C239S01 (phone 800/854-3229), p. 9. Used with permission.

Computer-Generated Graphics

Computer-generated graphics can be quite useful because you can rotate, reverse, or invert drawings (and other graphics) with a few keyboard commands. Computer-generated graphics allow us to analyze objects and processes in very sophisticated ways.

Different software programs can give you a variety of graphic formats to choose from. Authoring programs like AutoCad allow you to create your own graphics, save them, adapt them, and either print them out as needed or "import" them into your word-processed document.

Exhibit 4.8 shows a computer-generated graphic of a fractal, which allows the reader to "see" a mathematical formula in a new way.

HELPFUL HINTS FOR DESIGNING GRAPHICS

The following are some helpful hints that may be useful for designing graphics:

- Select the graphic you want to use by analyzing *where your reader will most need visual reinforcement of key concepts.* This holds true for the object itself or for a process you are describing.

- *Place the graphic as close as possible to the object or concept being discussed.* This helps the reader "see" what he or she is reading about.

- *Introduce and interpret each graphic* to guide your reader's response to the material. Not every reader will respond to information presented graphically in the same way. Your discussion should focus the reader's attention.

The Complete Mandelbrot Set as a 3-D Computer Graphic Exhibit 4.8 ■

Source: Adapted from Morris Firebaugh, *Computer Graphics: Tools for Visualization,* Dubuque, Iowa: W. C. Brown (1993), p. 317.

- *Avoid clutter!* Use one graphic for each idea. Emphasize clarity rather than complexity.
- *Set up the direction of the graphic so that it matches how your reader "reads" text.* Americans tend to "read" pictures just as they read text, from left to right. This means that a process should be shown as the reader will be use it, typically left to right.

Deciding on an Object or Process Exercise 4.6

Prepare the following for discussion with your group.

1. Look at the graphics shown in this last section. Which graphics emphasize **what** the object is (object) and which graphics emphasize **how** the object works (process)?

2. Reread Cavendish's description of the metallic state on page 104. What kind of graphics would help the reader? Where should the graphic be placed? Make a "stick" sketch to go with Cavendish's description. Bring your notes and sketch to your discussion group for review.

3. Think of an ordinary tool you commonly use around the house or on the job. How would you illustrate this tool for a technical description? Would you need both **object** and **process** graphics? Make notes and prepare a "stick" drawing. Share your work with your discussion group. Decide which drawings most clearly show what the object is and how it works.

Turn in a summary of your group work (with rough notes) to your instructor if requested to do so.

Exercise 4.7 What's Wrong With This Graphic?

List areas the writer needs to revise for the following drawing of a handheld fingernail clipper. Bring your notes to your discussion group for review. If you like, you can redraw the illustration. Select the "best" analysis and turn all group work in to your instructor with a brief summary justifying your group's choice.

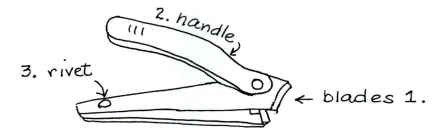

WRITING STRATEGY: USING DOCUMENT DESIGN TO IMPROVE READABILITY

In addition to graphics, the following additional graphic elements will help to make your technical description (and workplace writing in general) easier to read:

- Descriptive headings
- Short lists
- Short paragraphs
- "White" space

Compare the following descriptions of a possible location for a timber harvest, which were written for the general public. Which version is easier to read? Do these four additional graphic elements change the readers focus in the second version?

First Version

Alternate 1. This alternate will harvest timber from two units in the Alsea River analysis area and five units in Lower Fall Creek analysis area. Approximately 10.6 MMBF would be clearcut harvested from 188 acres. This alternate will increase distribution, quality, and quantity of forage of elk. In addition, seeding these sites with a grass/clover mix after burning will draw elk away from private land. Historically, elk grazing has been heavy during winter months, causing landowner complaints. Approximately 1.3 miles of new road construction is required, of which 1.14 miles is system road. Cable logging will occur in the southwest quadrant of Unit A and all of Unit B. Unit C will have a light burn only. This is to limit effects to the soils. This unit will then be seeded for visuals management and elk forage.

Environmental effects to soil and water quality are expected to be low. Visual Quality Objective (VQO) as designated in the Alsea River Unit Plan along Highway 34 is "partial retention" for Unit C and "retention marginal" for Units D and E. This alternative will reduce the amount of contiguous, mature conifer habitat for pine martin to 165 acres, which is below the minimum 250 acres as addressed in the draft EIS for the Forest Plan. Three head wall areas and 3 buffers are planned. Elk forage seeding is planned on 64 acres and release seeding for vegetation management is planned on 23 acres.

Source: U.S. Forest Service Planning Document (1991).

Second Version

Alternate 1—Alsea River and Lower Fall Creek

Size and location of clearcut. Timber would be harvested from two units in the Alsea River analysis area and five units in Lower Fall Creek analysis area. Approximately 10.6 million board feet (MMBF) would be clearcut harvested from 188 acres. Approximately 1.3 miles of new road construction is required, of which 1.14 miles are system roads. Three head wall areas and three buffers are planned.

Alsea River Unit A and B. Cable logging will occur in the southwest quadrant of Unit A and all of Unit B.

Lower Fall Creek Unit C. A light burn is planned to limit effects to the soils. This unit will then be seeded for visuals management and elk forage. Environmental effects to soil and water quality are expected to be low.

Impact on elk forage. This alternate will increase distribution, quality, and quantity of forage of elk. In addition, seeding these sites with a grass/clover mix after burning will draw elk away from private land. Historically, elk grazing has been heavy during winter moths, causing landowner complaints. Elk forage seeding is planned on 64 acres and release seeding for vegetation management is planned on 23 acres.

Negative effects on pine martin. This alternative will reduce the amount of contiguous, mature conifer habitat for pine martin to 165 acres, which is below the minimum 250 acres as addressed in the draft EIS for the Forest Plan.

Compliance to Alsea River Unit Plan. The Visual Quality Objective (VQO) as designated in the Alsea River Unit Plan along Highway 34 is "partial retention" for Unit C and "retention marginal" for Units D and E.

Source: U.S. Forest Service Planning Document (1991).

Graphics, charts, maps, drawings, and photographs are key tools to use in document design and will be covered throughout this book. This section has shown you that using elements of graphic design to organize and present your text (such as descriptive headings, short lists, short paragraphs, and balanced "white" space) can help the reader understand and retain the information more easily.

Exercise 4.8 Reviewing Document Design

Draft answers to the following list of questions about the changes in these two versions of a page from the report on "Alternate 1—Alsea River and Lower Fall Creek."

1. Who is the intended audience? What does this audience want to know?
2. In the one-paragraph layout, what information is most accessible to the reader? What information is most accessible in the second version?
3. What further changes would you recommend for the second version?
4. The report this description came from compared five alternative locations. Which format would make a close analysis of each alternative clearer to the reader?
5. The original report used the one-paragraph layout. What reasons would support such a choice?

Bring your list to your discussion group for review. As a group, select one or two key revision suggestions. Turn in all group work to your instructor with a brief summary justifying why this is important.

■ EDITOR'S CHECKLIST FOR TECHNICAL DESCRIPTION

When you prepare a technical description, you are typically writing for many readers, whether you are preparing the description for a set of instructions, or a manual, or simply working up an operating checklist.

Your readers need to be able to trust you absolutely. There is no margin for error in this kind of informative writing. The larger your audience, the more diverse their backgrounds will be. Some members of your audience may not have good reading skills. This problem makes it even more important that your writing be accurate and easy to read.

Use these questions as an informal checklist to help you during the reviewing and proofreading stages.

Review Content

___ 1. Is all information complete and accurate?
___ 2. Are key terms defined clearly?
___ 3. Are definitions or metaphors provided for key terms where needed?
___ 4. Does this description meet the needs of my readers?
___ 5. Does the introduction preview what's in this document?
___ 6. Does the discussion section include sufficient information? Need more facts?
___ 7. Does the conclusion summarize and interpret, given the writing purpose?
___ 8. Are graphics placed where they clarify key ideas, objects, or processes?

Review Organization and Style

___ 1. Is information presented in a logical order?
___ 2. Are ideas expressed simply, clearly, and concisely?
___ 3. Can any jargon be edited out or defined?

Analyzing Technical Description   Exercise 4.9

1. Find an example of a technical description from an operating manual or set of instructions. Consider content, organization, style, and format, along with guidelines from this chapter. List the strengths and weaknesses of your sample description.

2. Bring your notes and your sample description to class for discussion with your group. Compare your sample descriptions and select one from your group to analyze.

3. As a group, expand on the strengths and weaknesses of this sample. Turn in your group's work to your instructor with a brief memo summarizing what your group learned from this exercise.

Review Document Design

___ 1. Do graphics reinforce or clarify key points in the description?

___ 2. Do headings, lists, and short paragraphs make the description easier to read?

___ 3. Does the document appear balanced in its use of "white" space?

___ 4. Have punctuation, grammar, and spelling errors been corrected?

When you are asking someone to review your draft description, it can be helpful to focus your reader's attention by identifying the two or three questions you are most concerned about. Try this for your next peer review.

WRITING STRATEGY: PREPARING SPECIFICATIONS

Frequently when a company wants to make a major purchase of equipment, it will send out an RFP (request for proposals) to various vendors (sellers of the equipment) to ensure the company gets the best price possible.

Specifications (a detailed list of the criteria for the equipment, its performance, and cost) are usually attached to the RFP. If your company is large, a purchasing office may routinely prepare such specifications and will ask technicians for help. You may also be asked to draft a set of specifications. Specifications are also important when designing or building a variety of projects—from an oil pipeline that cuts across Alaska to the new office building downtown. Specifications help everyone on the project to understand exactly what materials are needed and their exact quantity and dimensions, and to predict the costs of a project.

Because you must write as specifically as possible about exactly what kinds of materials and tolerances are needed, writing specifications is usually considered a very challenging assignment. Specifications that are accepted become a contractual agreement between your company and the vendor, with legal complications if the specifications are not met. A typing error that adds a zero to a measurement or misplaces a decimal point could cost your company hundreds of dollars. For these reasons, specifications are usually proofread very carefully.

Some Suggestions on Preparing Specifications

In drafting a set of specifications, prepare to write very formally, following well-established conventions. Here is an example of the exacting contract language from a set of specifications:

> The Contractor shall provide the goods, services, and materials for the work whatsoever, and those same goods, services, and materials whatsoever shall be subject to inspection and test by the Owner including raw materials, components, processes, intermediate assemblies, and end products at all times, including the place of work and the place of record, and in any event, prior to performance.
>
> Source: Tim Whalen. "A History of Specifications: Technical Writing in Perspective." *Journal of Technical Writing and Communication,* vol. 15, no. 3, pp. 235–236.

All this means is that the buyer (Owner) has the right to inspect the materials and the workmanship of the seller (Contractor) before accepting the product.

Because many specifications are considered *boilerplate* (standard wording that is used over and over again with few changes), you may be able to reuse previously written specifications, adjusting the ranges for new costs or performance expectations. Sometimes you may have to prepare specifications from the absolute beginning. Vendors can be helpful in providing information and often publish specifications as part of their product brochures. However, if you use specifications provided by a vendor, you may discover that only that vendor's product meets your specifications.

Exhibit 4.9 is an example of some specifications that AST Research, Inc., included in its ad directed to a professional audience for its Pentium notebook. As you can see from this set of specifications prepared for a possible buyer, the writer assumes the buyer has a certain technical background.

■ Exhibit 4.9   The Complete Pentium Notebook

Processor:
  120, 90 or 75 MHZ Pentium
  256 KB Level 2 Cache

RAM:
  8 MB expandable
  to 40 MB

Screen:
  11.8" (Diagonal)
  Super VGA 800 x 600

  TFT or 10.4" (Diagonal)
  VGA TFT

Multimedia:
  Sound Blaster 16-bit
  sound system

Hard Drives:
  500 MB, 800 MB
  or 1.2 GB

Battery:
  Lithium Ion, for up to 6 hours

Weight:
  6.2 pounds

Chapter 4: Writing about Tools and Equipment   **117**

A form is often used to complete a set of specifications, whether for routine purchases or to coordinate larger projects that involve more than one type of equipment. The key parts to most specifications include:

- A physical description of the equipment, its capacity, and its tolerances.
- A functional description of the equipment that defines what it can do.
- An operating description that describes how it works and sets performance ranges.
- Cost for the equipment, installation, training, and maintenance.
- Deadlines for designing, testing, and implementing the equipment.

Once you complete the specifications, you will need to have it proofread closely by someone who is knowledgeable about the equipment and how it will be used.

WRITING STRATEGIES: DESCRIBING A TECHNICAL OBJECT

Describing a technical object, whether for specifications, a set of instructions, or a product manual, is a very special kind of writing that requires precision. This ability to write precisely comes with careful observation and experience in using the tool or piece of equipment.

Here is a process you can use to learn about any tool or piece of equipment you're going to use or have to train others to use.

Analyze What the Object Is

1. Read all material available.
2. Notice any notes, cautions, or warnings.
3. Observe the tool closely and draw it.
4. If possible, take the object apart and put it back together, changing your drawing as you gather more information.
5. Notice parts and subparts and label them on your drawing as you work.

Understand How the Object Works

1. Use the object.
2. Observe how you are using it.
3. Draft an outline that shows the key steps in using the object.
4. Ask someone else to use the tool and consider the following closely.
   a. Notice any difficulties the user is having.
   b. Revise your draft outline of the key steps.

118   Part II: Improving Writing Productivity

    c.   Notice operating principles.

    d.   Note limitations (what it can **not** do).

    e.   List the variety of users.

Draft the Technical Description

1. Draft your description that shows **what** and **how.**

2. Include "stick" drawings or rough sketches.

3. Add titles and labels to drawings.

4. Plan document design to include headings, lists, short paragraphs.

5. Check use of white space.

6. Double-check that all illustrations are placed where needed.

7. Reconsider who is using the document and revise.

Once you have completed your description, have someone read and use your final draft if at all possible. This final "pilot test" will help you further refine your description. This is especially useful if your description or instructions will be used by many different people.

WRITING STRATEGY: WRITING PERSUASIVELY

Sometimes when we write about tools or equipment needed for our jobs, we must persuade our readers to repair or replace equipment. If the amount of money is small, a routine memo stating what, why, and by when is usually all that is needed. This kind of persuasive writing gives the reader a reason to say "yes" to your request. Notice how a memo can justify the purchase of a useful but inexpensive combination tool (see Exhibit 4.10).

Notice with this routine request to approve a purchase order, the facts presented make the reader's decision almost automatic.

However, if the equipment is expensive or represents a major change in operations, then your supervisor will want more information and analysis to support the final decision—which may be made at higher levels. You will need to report what the problem is and also include what the options are, the main advantages and/or disadvantages for each option, and perhaps a conclusion with your recommendation. When a company plans to make a major equipment purchase, many people become involved in the decision.

Chapter 4: Writing about Tools and Equipment **119**

A Routine Persuasive Memo   Exhibit 4.10 ∎

WRANGER MANUFACTURING   **MEMO**
". . . Toledo's Best"

April 15, 1997

To: Ron Earhart, Senior Plant Manager

From: Susan Anthony, Shift Supervisor

SUBJECT: Approve PO for $600—Leatherman Tools for Shift Techs

Leatherman tools are justified for technician use by the versatility of the tool and the time it will save. A Leatherman Super Tool is carried in a sheath on the belt. It features needlenose pliers, regular pliers, wire cutters, plain knife blade, serrated knife blade, 9" ruler, can/bottle opener, large screwdriver, medium screwdriver, small screwdriver, Phillips screwdriver, metal/wood saw, file, awl/punch, electrical crimper, and wire stripper.

It will save time on small maintenance operations by allowing technicians to make on-the-spot repairs and adjustments without having to carry a full tool kit for small repairs. Tool boxes are usually stored inside the clean space. The Leatherman tool can be carried everywhere with ease. For most simple repairs, a Leatherman tool can be used simply and easily.

If we estimate that the average technician will save one hour per week looking for tools that this tool will provide and multiply that hour times 15 technicians and times four weeks per month, we can save 60 hours per month spent searching for tools. If the average wage is $15.00/hr, this translates to a savings of $900.00 per month, plus time saved in production down time.

Purchasing Leatherman tools now will save money. The catalog price for a Leatherman Super Tool is $59.99. *DiscountShopper* currently has the same tool for $39.99. There are 15 shift technicians that would like to have them, for a total of $599.85 (from *DiscountShopper*) and a possible savings of $300.00.

Please approve the attached purchase order so I can arrange to have these Leatherman tools purchased at discount. Thank you.

EDITOR'S CHECKLIST FOR WRITING MEMOS   ∎

Memos are a basic communication tool in the workplace. They are used so frequently both to inform and to persuade that your planning, drafting, and revising skills need to be highly developed so that you can produce effective memos efficiently.

Use *Appendix A, Letters and Memos,* to understand the parts of a memo. Use the following editor's checklist to review your draft memo.

Plan Your Memo

____ 1.   Why am I writing this memo? Should I use a telephone call or meeting instead?

____ 2.   What is the problem? Do I have enough information?

____ 3.   Who am I writing this memo for? Are there other readers?

Part II: Improving Writing Productivity

    ___ 4. What are some questions or concerns my readers will have?

    ___ 5. Who else in my work area is involved? What questions or concerns will they have?

Check Content and Organization

    ___ 1. Can my reader tell what this memo is about only by reading the subject line? Does the subject line include topic + purpose + urgency (in five to seven key words)?

    ___ 2. Does my introduction concisely state the problem and purpose of this memo and include a preview of what's in this memo?

    ___ 3. Is each idea supported with the right amount of information or analysis?

    ___ 4. If I present options, is each explored fully and without bias? With findings reported as specifically and concisely as possible?

    ___ 5. Does my conclusion summarize and recommend? Restate reasons for this recommendation? Request action by my reader?

Check Style and Document Design

    ___ 1. Have I expressed my ideas concisely? Are there any places where fewer words will say the same message?

    ___ 2. Have I expressed my ideas clearly and directly, using words my readers will understand and avoiding jargon (words that only technical insiders will know)? Have I edited out any business cliches (phrases that have been used so often they have no meaning)?

    ___ 3. Is my tone courteous—even if my readers may not agree with me?

    ___ 4. Does my overall format follow company guidelines?

    ___ 5. Do I use lists, headings, and short paragraphs to make the information easy to read or analyze?

For important memos, it can be useful to have a co-worker or supervisor review your work and give you feedback on its probable impact. If your memo affects many people, large projects, radical changes, or significant budget amounts, your readers will likely have questions about what you propose, and your co-workers will likely have useful suggestions.

Encourage your discussion group to use these questions to check memos you prepare for this class.

Analyzing a Persuasive Memo   Exercise 4.11

Read through the following memo, making a list of what the writer should change. Use the Checklist for Writing Memos to help you identify problems in four key areas: content, organization, style, and format.

As your instructor directs you, either prepare a memo for your instructor, highlighting the changes you suggest and following memo-writing guidelines yourself, or rewrite the memo so that it is more persuasive, adding or deleting information as needed.

DRAFT MEMO

09-06-96

To: Mort Selivonski, supervisor, plant operations
       cc: Stephen Garcia, president

From: Sara

SUBJECT: PERSUASIVE REQUEST FOR REPLACEMENT FOLLOWING UP ON PREVIOUSLY SUBMITTED SHIFT REPORTS FOR JAN., JUNE, AND SEPT. OF THIS YEAR

It has come to my attention that the blowers for our three operating plants do not work effectively to remove noxious exhaust fumes from the assembly line. Workers have complained bitterly. The union is threatening to file a complaint with OSHA.

I have repeatedly requested that you approve purchasing new blowers. After investigating two optional blowers that could do a better job in my humble opinion, I recommend we purchase the Industrial Fan Model #176353567, costing $25,000 for nice fans for all three plants. (I get a discount because my brother-in-law is a salesman there.)

Please do not hesitate to contact the undersigned if there are any important questions or problems. You know where to reach me on Thursday mornings.

EDITOR'S CHECKLIST FOR WRITING DESCRIPTIONS

Use this checklist to help you prepare technical descriptions.

___ 1. **Title.** Does your title identify very specifically the object you are describing? Include part of model numbers if needed?

___ 2. **Introduction.** Do you introduce the topic, identify the purpose of your technical description, and preview what's in this description? If warnings or cautions are needed, are they integrated throughout the document?

___ 3. **Definition.** Does your introduction include a formal definition (object = class + distinguishing characteristics) to clarify exactly what your object is? Are key terms defined throughout your description?

___ 4. **List of Parts.** Have you included a list of the key parts and subparts? Are the parts organized into categories of parts if they are very complicated?

Part II: Improving Writing Productivity

___ 5. **Illustrations.** Have you included drawings or photographs of the object as a whole? Of key parts? Do your illustrations have clear labels and a clear title? Are they provided wherever needed, for example, to show a key part or operation?

___ 6. **Description.** Does your description include physical characteristics, such as model or part numbers, sizes, shapes, weights, colors, materials, textures, or any other key physical descriptors? Have you included functional or operating principles? A comparison with other tools or machines that the reader may be familiar with? Are all measurements and any calculations accurate?

___ 7. **Conclusion.** Do you end with a conclusion appropriate for your audience and writing purpose? Appropriate for this object?

___ 8. **Writing Style.** Have you written clearly, concisely, and simply to achieve a direct and easy-to-understand style? Is your work free of errors in punctuation, spelling, or grammar?

___ 9. **Format.** Have you used headings, paragraph length, lists, and graphics to make understanding, using, and remembering this information easy for your reader?

■ CONCEPT REVIEW

Can you define the following key concepts from Chapter 4 without referring to the chapter? Write your own definitions in a journal entry or review them out loud.

object and process   expanded definition

mechanism description   stipulative definition

parenthetical definition   specification

formal definition   process analysis

■ WRITING SKILLS REVIEW

To build your skills in editing and proofreading, continue your systematic review of *Appendix C: Writing Skills Review.* Your assignment:

___ Read "Using Apostrophes" in *Appendix C: Writing Skills Review.*

___ Complete Exercises C.24 and C.25 in "Using Apostrophes" and the Summary Exercise on Sentence Combining. Check your answers at the back of *Appendix C.*

___ Prepare a paragraph summarizing your progress for your instructor or use the form your instructor provides to report your progress. Be sure to highlight any areas you need to further review.

SUMMARY EXERCISE

What's Wrong With This Description?

Read the following description that includes a physical, functional, and operational description. What suggestions would you make to improve this document?

Prepare a memo for your instructor that lists revision suggestions. Please organize your suggestions into these four categories: content, organization, style, and format. Chapter checklists may be helpful. Be sure to note where graphics would be most useful and what kind to use.

TITLE: DESCRIPTION OF A COFFEE PERCOLATOR

Audience: Staff of a large restaurant

Purpose: To describe what the percolator is and how it works.

A coffee percolator is an electric appliance used to brew coffee. The percolator is made up of several component parts: the pot, which contains a heating element in its bottom section, the internal parts (taken as one unit) and an electric cord, which connects the heating element with a source of electricity.

The pot itself is usually made of stainless steel and has the capacity to hold approximately two quarts of liquid. The bottom of the pot contains a sealed section housing the heating element. When the heating element is connected to electricity by means of the electric cord, the element becomes hot and heats the water inside the pot.

The viscera of the pot are the vital organs of the percolator. They are usually made of aluminum and are of three parts: a hollow tube about eight inches long, fixed to a parabolic disc at one end, a reservoir which sits atop the tubing, and a cap that fits over the top of the reservoir (see Figure 1).

The reservoir has a piece of tubing in its center larger than the main tubing. The reservoir sits down over the main tubing; ribs on their side of the main tubing prevent it from sliding down all the way to the disc.

When water boils, the parabolic disc, which sits down on top of the heating element, forces water up through the tubing and out over the cap situated on top of the reservoir.

The reservoir contains the coffee grounds. The bottom and cap of the reservoir each have holes in them; when hot water boils up through the tubing and splashes on top of the cap, it drips through the holes and wets the coffee grounds. As the water drips through the reservoir, it picks up the flavor of the coffee grounds. This liquid then drips through the bottom of the reservoir into the holding pot below.

Figure 1: Making Coffee

This entire process takes about 15 minutes. The result is a steaming, aromatic, hot pot of delicious coffee.

124 Part II: Improving Writing Productivity

■ ASSIGNMENTS

Assignment 1: Describe a Technical Object.
Describe a simple object that you can use as a tool. Some students have described a can opener, a dog comb, a wrench, and a stapler. The fewer moving parts your object has, the easier it will be to describe it!

Your audience will be someone from your class who has never used this particular tool. You will need to prepare a one- to two-page memo to your instructor that includes:

- An introduction (problem + purpose + preview).
- A formal definition of the object that includes the term, its class, and its distinguishing characteristics (T = C + DC).
- A description of the object (physical description) and its purpose (functional description).
- A description of how this object works (process description).
- At least one labeled drawing of the object (you can use a simple pencil drawing) that illustrates the object (**what**) or how it works (**how**).
- Headings to separate the different sections.
- A conclusion (summary or analytical conclusion).

After your draft is complete, ask someone from your discussion group to review your draft description. Write out two or three questions you are most concerned about. Revise your description once you have written feedback from your reviewer and turn in the completed description to your instructor.

Assignment 2: Prepare a Set of Specifications.
Pick a simple tool or piece of equipment you have worked on either at home or at work. Assume you are working for a small company that has decided to purchase this equipment, but that you must go out for a bid.

Prepare a set of specifications that includes physical and functional descriptions, performance variances, and cost ranges. You may use a form or prepare a memo to your supervisor (instructor).

Assignment 3: Review a Manual.
Find two or three pages from a manual for a tool or machine you are familiar with. Use the reading guidelines in this chapter to read and review the manual. Do you think this document is easy to read? Why or why not? Make notes on your reading process.

Select one page to revise, following content and document design guidelines from this chapter. Turn in both the original and your revision to your instructor with a brief memo that summarizes the reading and revising process you used and which phases were most difficult. Conclude your memo by suggesting what students might need to know who are beginning a similar assignment.

Chapter 4: Writing about Tools and Equipment   **125**

Assignment 4: Interview a Professional.   Interview a working professional from your field to discover exactly what kinds of tools you will be using, which tools and equipment will be most important in your profession, and why. Include four to five different tools or pieces of equipment. Ask the person you are interviewing exactly what kind of training would best prepare you for the workplace. Summarize your findings in a memo to your instructor.

PROJECT: WHAT HAPPENED AT ALBANY MOTORS?   ■

Divide up the following assignments with your discussion group as your instructor directs you. Each person should draft a response to his or her assignment and bring the completed draft to the discussion group for review. Use the checklists or sample documents provided in this chapter to review the work, make any needed revisions, and turn the completed project in to your instructor.

Background: Your team has just been hired by Albany Motors, a small company with 20 employees that specializes in servicing and reconditioning diesel trucks. Albany Motors wants to buy a photocopy machine.

Prepare Specifications: You have been asked to write a list of specifications for the new photocopy machine. Your new supervisor, Rhonda Johnson, wants you to prepare a draft list of specifications in a memo to her that identifies the physical, functional, performance, and cost ranges for the specifications. Here's some background information.

1.   All 20 employees will have access to the photocopy machine.
2.   The photocopier must fit into a small alcove and must be no larger than 4' x 5' x 3'.
3.   Albany Motors has set aside $8,000 to purchase the equipment.
4.   These features are desired: multiple copies, sorting, stapling, front and back copying.
5.   Color is not necessary.
6.   Some way to count copies per month would be helpful.
7.   Albany Motors is interested in a service or maintenance contract.
8.   The photocopier should be delivered within 30 days of the date of the order.
9.   Some comparative information on lease options would be useful.

Prepare Checklist: Albany Motors has now purchased the photocopy machine, but the manual is too difficult for employees to use. You need to prepare a checklist that new employees can use when they come in to use this new machine. Draft an outline that identifies the key steps in the process of using a photocopy machine. Be as specific as possible. Your assignment is to prepare the outline.

Part II: Improving Writing Productivity

Prepare Persuasive Letter: Your machine has just broken down for the fourth time in two weeks, despite repeated visits by the service person. Write a letter to your vendor, summarizing the problems and requesting a replacement machine.

Prepare Recommendation Memo: Despite a replacement machine at no cost, the photocopy machine has not met the needs of Albany Motors. Write a memo to your supervisor, Rhonda Johnson, summarizing the problems and recommending replacement.

··

■ JOURNAL ENTRY

The First Time I Used a Tool . . .

Describe the first time you used a complicated machine or tool—like driving a car or tractor. Who taught you? What made the experience successful or unsuccessful?

What did you learn about giving people instructions from this experience? About working with new equipment? What factors specifically gave you more confidence? What advice would you have for someone beginning a similar task?

5

Using Forms

Writing on the Floor and in the Field

Chapter PREVIEW

This chapter prepares you to complete and design routine and specialized forms and to write memos. You'll practice several ways to write about information—using lists, headings, and tables. You'll also help your reader understand numbers more easily by translating them into percentages or using a flowchart to analyze key steps in a process. Finally, you'll review forms that can help you keep track of appointments, projects, and schedules.

Several writing strategies will be introduced: *completing forms* so they are clear, concise, and complete; *using numbers and analyzing tables to make your point;* and *organizing information into direct and indirect memos* so that readers can use the information more efficiently. The chapter will conclude with a section on specialized forms, such as flowcharts and project management forms.

Journal ENTRIES

Begin thinking about your experiences in using forms and routine memos as an employee, a student, or a customer. Write a short one-paragraph response to any of the following questions in your class journal or prepare answers for class discussion, as directed by your instructor.

1. What kind of forms have you completed? When did your task seem easier? When did completing forms seem like more work than it was worth?

2. Why do companies use forms?

3. When you think of an employee who works out on the road, or who travels from plant to plant, or who delivers products throughout the state, what kind of communication does that person have with the main office? What kind of forms does that person need to complete?

4. How would you react if your co-worker said, "I hate filling out forms. You won't catch me spending a lot of time with this paperwork."

Chapter READING

Each day we fill out forms and make decisions that are then documented in forms. Because forms are such a pervasive part of our lives, they are nearly invisible, and we may not think about the implications of *how* we fill them out or the impact they have on others.

The Association of Records Managers and Administrators has a code of professional ethics that sets guidelines for the issues we should think about when we work with forms. Read through the Code, which presents these guidelines. Then, read an excerpt from the annotated Code that adds comments to two of the guidelines.

THE CODE OF PROFESSIONAL RESPONSIBILITY

Patrick Cunningham

Preamble

Information and records management is that field within the information profession responsible for managing the creation, use, maintenance, and disposition of records generated in the normal functioning of all types of organizations.

The Association of Records Managers and Administrators (ARMA International) is a not-for-profit organization representing professionals in the field of information and records management. Its primary purpose is the advancement of records and information management through education and professional development.

Purposes of the Code

This code is intended to increase the awareness of ethical issues among information and records management practitioners and to guide them in reflection, decision making, and action in two broad areas of ethical concern: society and the profession.

I: The Social Principles

Because of their responsibilities to society, information and records managers:

1. Support the free flow and oppose censorship of publicly available information as a necessary condition for an informed and educated society.

2. Support the creation, maintenance, and use of accurate information and support the development of information management systems which place the highest priority on accuracy and integrity.

3. Condemn and resist the unethical or immoral use or concealment of information.

4. Affirm that the collection, maintenance, distribution, and use of information about individuals is a privilege in trust: the right to privacy of all individuals must be both promoted and upheld.

5. Support compliance with statutory and regulatory laws related to recorded information.

II: The Professional Principles

Because of their responsibilities to their employers or clients as well as to their profession, information and records managers:

1. Pursue appropriate educational requirements for professional practice, including a program of ongoing education and certification.

2. Accurately represent their education, competencies, certifications, and experience to superiors, clients, co-workers, and colleagues in the profession.

3. Serve the client or employer at the highest level of professional competence.

4. Recognize illegal or unethical situations and inform the client or employer of possible adverse implications.

5. Avoid personal interest or improper gain at the expense of clients, employers, or co-workers.

6. Maintain the confidentiality of privileged information.

7. Enrich the profession by sharing knowledge and experience; encourage public discussion of the profession's values, services, and skills.

8. Are actively committed to recruiting individuals to the profession on the basis of competence and educational qualifications without discrimination.

The Annotated Code of Professional Responsibility
[These following excerpts of Guidelines 3 and 4 are part of the annotated version of the Code. These additional comments further explain key issues related to the guidelines from ARMA.]

3. Serve the client or employer at the highest level of professional competence.

Using effective information and records management principles and practices, the professional provides service at the highest level of competence. One factor differentiating a professional from other employees of an organization is that a professional is able to separate professional responsibility and judgment from personal feelings and loyalty. This serves the employer's or client's best long-term interests. Anything less demeans the practitioner and, by extension, the profession.

4. Recognize illegal or unethical situations and inform the client or employer of possible adverse implications.

The knowledge and values of information professionals uniquely qualify them to recognize the ingredients of ethically complex issues related to information and records management. The information and records manager pursues a reflective morality, not one limited by custom, tradition, or the moral terrain of a specific work environment. The professional has a responsibility to inform the employer or client that a given decision, action, policy, or procedure may have negative implications. The information and records manager may decide to dis-associate from a client or employer who continues to pursue such a course.

Source: Internet citation: doctype html public: /hp/hq/armahome.html, ARMA Home Page (Association of Records Managers and Administrators). Used by permission.

132 Part II: Improving Writing Productivity

Analyzing THE READING

To build your reading skills, use the following questions to analyze the content and to develop your reaction to this code of ethics. Write a one-page entry in your class journal or prepare your answers for class discussion.

Reading for Content

1. Who is the audience for this code of ethics? Could these guidelines influence anyone who prepares forms or documents? In what ways? How widely should these guidelines be distributed?

2. How would you describe this writer's style? How would you paraphrase the main idea of these guidelines? Could you rewrite one of the guidelines in your own words? How does your writing differ from the way it is written here? Does the writing style affect the usefulness of the document?

3. What are the three most important guidelines in this code? Why are these three the most important?

Reading for Reaction

1. When you are working with routine forms at work, how do you think the privacy of customers, co-workers, and the company should be protected?

2. How does Guideline 3 define professional responsibility and personal responsibility? What distinction is the writer making between professional responsibility (which includes loyalty) and personal responsibility (which involves moral behavior)? What do you think of this distinction?

3. Can you think of any workplace situations that would challenge these guidelines? List one or two situations and suggest how the problem might be resolved.

4. Have you ever been in a situation where you felt you were being asked to make workplace decisions that seemed unethical? Describe the situation (it's not necessary to name the company or the people involved) and how the situation was resolved. What did you learn from this situation?

5. The comments for Guideline 4 explain the need to "disassociate from a client or employer" if the employer consistently doesn't follow these guidelines. What does this mean? What job protection does an employee have if he or she disagrees with how something is done? At what point would you resign rather than carry out a task your employer wanted you to do?

WRITING WITH FORMS

Sometimes it seems as if life is filled with more forms than you ever wanted: telephone message forms, application forms for financial aid or bank loans, insurance forms, rental agreements, tax forms, etc. Whether forms are simple or

Chapter 5: Using Forms **133**

complex, we all have to fill them out. Recall the forms you filled out when you applied for a job. Think of the forms you have completed as a customer to get a refund or to replace or repair a product.

The workplace is no exception. People in every area in a company either fill out or design their own forms. The main purpose behind these forms is to improve overall efficiency of the department and the company. Even though our overall goal is to improve efficiency, sometimes we resist filling out forms. Sometimes filling out forms creates so much paper that we can't find what we need when we need it, or the form doesn't have the right information.

The focus of this section is to help you understand the diversity of forms in the workplace and how you can best use them for both simple, routine tasks and more complex tasks. We'll focus on two key areas where forms are often used.

- *From the floor* means that employees work on the shop floor where products are assembled, sometimes called operations, "on the line" or the assembly line. Here, the key need is to ensure high-quality production. Forms may be used to document work in progress, to report equipment malfunctions or to record the rate of inventory usage. In health care settings, the floor can mean the operating room, the wards, or the laboratory. Forms are used in these areas to ensure high-quality "production."

- *From the field* means that employees work independently outside of company facilities. These workers may include sales staff, police officers, or home health care providers. Forms are used from the field to document what was done, where it was done, and why it was done.

Supervisors scrutinize these kinds of forms carefully to ensure that production quotas and quality standards are met.

WRITING STRATEGIES: WORKING WITH FORMS

As many different forms cross your desk, you may ask, "How can I use the information I'm given in the best way?" On any one day, you may fill out or read test results on new equipment or procedures, training manuals, trip reports, vacation requests, payment authorizations, inventory reports, routine plant inspections, flowcharts of operations, and accident or incident reports. Most of the time, you are working with preprinted forms; other times, you will need to design your own form for a special situation. Frequently, forms are used

- *To process routine information quickly* without wasting time on inventing a new form or looking at a different form each time a request is processed.

- *To document key information for routine operations systematically* without having to look at many different forms saying the same thing.

- *To simplify complex tasks* so that all steps are followed.

134   Part II: Improving Writing Productivity

Forms are used to allow readers to quickly *respond to routine requests*. The key question to ask when filling out routine requests is: Have I included all the information my reader needs to act?

On the invoice from Inkstone (see Exhibit 5.1), employees simply (and efficiently) filled out the amount due. A similar fill-in-the-blank order form from Printing Services (Exhibit 5.2) has detailed information about the jobs to be done as well as the customer (billing information), the date the order was received, and the price of the job. If something went wrong with this routine order, the customer or the supervisor could track down who took this order by checking the initials at the bottom of the form.

■ Exhibit 5.1   **Example of an Invoice**

INVOICE

DATE: *January 3, 1996*

CUSTOMER:

Mirror Press

FROM:
INKSTONE COPY CENTER
213 Sumner Street
Newton Centre, MA 02159

MONTH OF *December '95*

PREVIOUS CHARGES *$158.31*

PAYMENTS *$158.31*

CURRENT CHARGES *$132.70*

BALANCE NOW DUE *$132.70*

THANK YOU

*pd 1/15/96
#3026*

Chapter 5: Using Forms 135

Example of a Fill-in-the-Blank Form   Exhibit 5.2 ∎

PRINTING SERVICES ORDER FORM

JOB TITLE ___Transparency_____

REQUESTED BY ___Camp_____ EXT ___XXX_____

DEPARTMENT ___English_____

DELIVER TO _____ (after 3 p.m.) or ☐ BOOKSTORE WAREHOUSE

BILL TO ___(ML) G_____ _____ _____ _____ _____

Printing Services stamp: JUL 18 4 14 PM '96 — PRINTING SERVICES — 000138

	CODE	or	FUND	ORG	ACCT	PROG			
MULTIPLE BILLING	_____		_____	_____	_____	_____	/ _____	= $ _____
	_____		_____	_____	_____	_____	/ _____	= $ _____
	_____		_____	_____	_____	_____	/ _____	= $ _____
	_____		_____	_____	_____	_____	/ _____	= $ _____
	CODE	or	FUND	ORG	ACCT	PROG	# OF COPIES	TOTALS

PRE-PRESS

PRODUCTION/FORMAT INFO.	SCANNING	OUTPUT	
☐ Keyboarding	☐ Line Art ☐ Halftone ☐ Contone	☐ Laser ☐ IR Paper ☐ Film	
☐ Layout/Pasteup	_____ dpi _____ lpi	☐ 8.5 X 11 ☐ 8.5 X 14 ☐ 11 X 17	
☐ Hard Copy ☐ MAC ☐ IBM Comp ☐ Output ☐ Convert to DOS		Other _____ x _____	
File Name _____	☐ To Disk, save as…	☐ Positive ☐ Negative ☐ Halftone	
Software _____	☐ tiff ☐ eps ☐ pict ☐ ____	☐ 1200 dpi ☐ 2400 dpi ____ lpi	
Fonts _____	Size ____ x ____ or ____ %	☐ Crop Marks ☐ Reg Marks ☐ K Outs	
_____	Adjust _____	☐ Spot Colors ☐ Process Colors	$

PRESS

PRINT	ORIGINAL	PAPER	
____ # of 1-sided pages	☐ New ☐ Re-order (No Changes)	☐ Color _____	1.20
____ # of 2-sided pages…	☐ Revised (Changes)	☐ Bond ☐ Text ☐ Index ☐ Cover	
☐ Top-Top ☐ Top-Bottom	☒ 8.5 X 11 ☐ 8.5 X 14 ☐ 11 X 17	☐ NCR… _____ Parts	
____ Total Copies or Sets	☐ Other ____ x ____	☐ Other _____	$

BINDERY

☐ Collate	☐ 3-Hole Punch	☐ Laminate	
☐ Staple	☐ Spiral bind	☐ Number, starting _____	
☐ Fold (circle one)	☐ Round Corner	☐ Cut _____ x _____	
	☐ Saddle Stitch	☐ Pad… ☐ ____ per pad ☐ NCR	
⌐ ∟ Z ↖	☐ Shrink Wrap… ☐ w/ sticker ☐ w/o sticker	☐ Other _____	$

SUPPLES/INSTRUCTIONS

_____

_____

_____

_____ $

_____

_____

_____

_____

PRINT DATE _____	
DEL'D BY CS	1.20
DATE 7/18/96	
REC'D BY Zpp	$

White Copy - Printing Services   Canary Copy - Billing   Pink Copy - Origination

The same principles are used in the following computerized billing from Pitney Bowes (Exhibit 5.3).

■ Exhibit 5.3   Example of a Computerized Bill

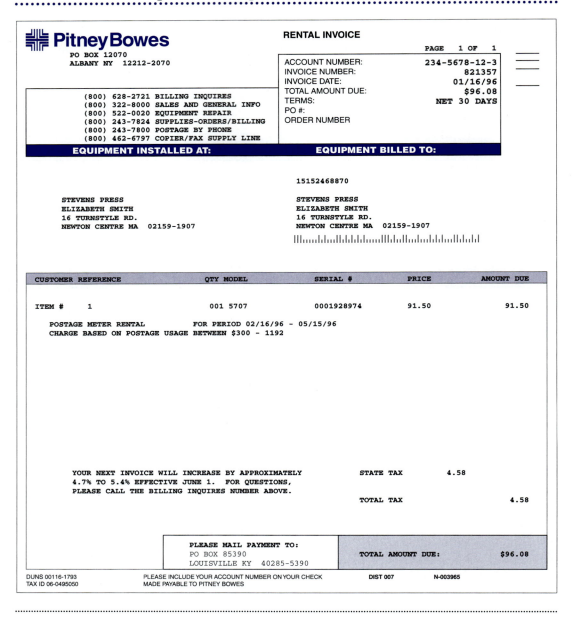

Standardized forms are also used *to document routine operations.* One standardized form is a log, commonly used to record key events about a shift or the performance of a particular machine. A log works as a kind of diary, with entries organized by dates. For example, a ship's log is used to record a ship's speed and progress and to highlight daily events. You might complete a shift log that notes equipment breakdowns during your shift so that the workers on the next shift will know what happened. You might add to a service log or machine log to record repairs or maintenance for a particular machine.

The key question to ask when completing such routine forms is: Has all necessary information been recorded completely and concisely so that it makes sense to the next reader?

If you are new on the job, consider reviewing forms with your supervisor so you can understand "best" practices and avoid behaviors that don't meet your employer's goals for efficient production.

To avoid information overload, we also use forms to *simplify complex tasks.* The following form (Exhibit 5.4) was designed to help one shift understand exactly what happened during the previous 12-hour shift. Notice how the wording reminds the writer of key information that should be included.

If you were preparing this form to pass along your evaluation to the next shift, you would need to answer some difficult questions:

- What is the quality of the completed work?

- Does this work meet production quotas?

- Were deadlines met?

- Do I need additional materials or equipment?

- Have I reported any problems accurately and honestly?

- Does the reader have all needed information?

Companies also use standardized forms to certify that they have met certain legal requirements. For example, the Certificate of Origin for the North American Free Trade Agreement (see Exhibit 5.5) must be completed along with other shipping documents before it is possible for one North American company to buy or sell goods from another North American company.

Forms that make up a patient chart are another example of forms that document key information so it can be used appropriately. Charting involves understanding the situation, gathering and recording information, and then interpreting and applying that information for patient care. Exhibit 5.6 shows an initial assessment form used in core charting developed by nurses at St. Joseph's Hospital in Hamilton, Ontario.

The desire to work efficiently and accurately keeps people using forms and designing new ones!

138 Part II: Improving Writing Productivity

■ Exhibit 5.4 **Example of Form for a 12-Hour Shift**

INSPECTION PASSDOWN

SHIFT   A   B   C   D   DATE: _____

LINE 1: EQUIPMENT UP/DOWN

Time	**Problem**	**Action Taken**	**Current Status**	**Technician**

Follow-up?
Impact on productivity or goals?
Critical materials issue?

LINE 2: EQUIPMENT UP/DOWN

Time	**Problem**	**Action Taken**	**Current Status**	**Technician**

Follow-up?
Impact on productivity or goals?
Critical materials issue?

LAMINATOR: UP/DOWN

Time	**Problem**	**Action Taken**	**Current Status**	**Technician**

Follow-up?
Impact on productivity or goals?
Critical materials issue?

ALIGNER: UP/DOWN

Time	**Problem**	**Action Taken**	**Current Status**	**Technician**

Follow-up?
Impact on productivity or goals?
Critical materials issue?

OTHER COMMENTS:	**SUMMARY:**
Process changes?	LINE 1: EQUIPMENT UP/DOWN
	LINE 2: EQUIPMENT UP/DOWN
	LAMINATOR: UP/DOWN
Reporting Technician:	ALIGNER: UP/DOWN

Chapter 5: Using Forms **139**

Example of a Standardized Form   Exhibit 5.5 ■

**NORTH AMERICAN FREE TRADE AGREEMENT
CERTIFICATE OF ORIGIN**

1. EXPORTER NAME AND ADDRESS	2. BLANKET PERIOD (DD/MM/YY) FROM:

TO:

TAX IDENTIFICATION NUMBER:

3. PRODUCER NAME AND ADDRESS	4. IMPORTER NAME AND ADDRESS

TAX IDENTIFICATION NUMBER:

TAX IDENTIFICATION NUMBER:

5. DESCRIPTION OF GOOD(S)	6. CLASSIFICATION NUMBER	7. PREFERENCE CRITERION	8. PRODUCER	9. NET COST	10. COUNTRY OF ORIGIN

I certify that:

The information on this document is true and accurate and I assume the responsibility for proving such representations. I understand that I am liable for any false statements or material omissions made on or in connection with this document.

I agree to maintain, and present upon request, documentation necessary to support this certificate, and to inform, in writing, all persons to whom the certificate was given of any changes that could affect the accuracy or validity of this certificate.

The goods originated in the territory of one or more of the parties, and comply with the origin requirements specified for those goods in the North American Free Trade Agreement, and unless specifically exempted in Article 411 or annex 401, there has been no further production or any other operation outside the territories of the parties.

This certificate consists of one page, including all attachments.

11.	11a. Authorized Signature	11b. Company
	11c. Name	11d. Title
	11e. Date	11f. Telephone number (voice) (facsimile)

140   Part II: Improving Writing Productivity

■ Exhibit 5.6   **Example of a Patient's Chart**

Initial Assessment				
General	Physical	*immobile and severe headache*		
Appearance	Skin	☐ Clear ☐ Dry ☑ Intact		
	Hygiene	*good*		
Cardiopulminary		*tachycardia and tachypnea*		
Mobility	*unable to stand or walk*		Aided by	*wheelchair*
Vision	*good*		Aided by	
Hearing	*good*		Aided by	
Speech	Language	*English*		
	Impediment	*none*	Aided by	
Appetite, diet	*fair, regular*			
	Dentures	☐ Upper ☐ Lower ☐ Partial ☑ None		
Urination	*good - voids freely*		Aided by	*urinal*
Bowel routine	*daily*		Aided by	*laxatives*
Sleep and rest	*poor, secondary to pain*		Aided by	*analgesics*
Pain and discomfort	*severe pain @ temporal area*		Aided by	*Percocet*
Reproduction and sexuality	*normal*			
Mental status (orientation, memory)	*oriented X3*			
Emotional status (mood, attitude)	*agitated*			
Religion	*Catholic*			
Nationality	*Irish*			
Lifestyle habits				
	☑ alcohol *occasional beer*		☐ drugs *none*	
	☐ tobacco *none*		☐ other	
Occupation	*construction worker*	**Hobbies**	*sports*	
Patient's concerns for discharge	*Pt. states that he's concerned about when he will be able to return to work. States that if he requires surgery no one will be home to help him because his wife works.*			
Signature	*Pam Watts, RN*			
Date and time	*1/8/95 0200*			

Source: Edith McMahon et al. (eds.), "Using Core Charting," *Mastering Documentation*. Springhouse, Pennsylvania: Springhouse Corporation (1995), p. 92. Used with permission of *Mastering Documentation,* copyright 1995, Springhouse Corporation. All rights reserved.

Chapter 5: Using Forms 141

Designing a Routine Request   Exercise 5.1

Prepare a form that students can use at your college to request a grade change. The persons approving the grade change will be the instructor and the Registrar. If you like, design the computer screen that students will use, instead of a paper form. Your goal is to design a form that is easy to fill out, easy to process, and easy to interpret.

Bring your draft to your discussion group for review. As a group, discuss the strengths and weaknesses of the forms from your group. Decide which form is best and why, summarizing your work for your instructor.

If your group designed paper forms, ask one more question: Which of these forms could be most easily converted to a computer screen? Why? Turn in a group memo that summarizes your discussion in either list or paragraph format and attach all draft forms.

Although we may use routine forms as checklists to remind us about what to do or what may have happened, such forms were developed to make sure that information crucial to the company's success is documented.

Because forms are used so widely, sometimes how we work (the processes we use) changes, but forms do not. Keep alert to such changes and look for ways to simplify forms so they can document key events, actions, and needs.

As companies automate, expect routine forms to show up on computers. For example, reports filed by home health care providers visiting patients in their home are increasingly completed on laptop computers. How we fill out forms may change, but we will continue to fill out standardized forms to help our co-workers improve productivity.

PREPARING FORMS FOR AN AUDIENCE

When you are preparing forms that will be used elsewhere in the company, you are preparing information for different audiences—your co-workers, your supervisors, and perhaps your supervisor's supervisor.

Because your primary goal is to ensure that your work meets deadlines, performance standards, and production quotas, you want to spend as little time as

Looking at Forms   Exercise 5.2

Collect four or five forms from your current employer or a friend's employer. Use the checklist on completing forms to analyze why these forms are being used. Prepare notes that identify the audience, the purpose and any strengths or weaknesses for each form.

Bring your forms and the analysis for each form to your discussion group. Compare forms and analyses. Decide which forms are the most useful. Summarize the results of your group discussion in a short memo for your instructor and share it with the class.

HOW DO I FILL OUT ROUTINE FORMS?

Most routine forms have either boxes to check or blank lines to fill in. Although forms are designed to save time, sometimes workers are not sure what information should be recorded, for example, when recording the highlights of a 12-hour shift.

Why is the form needed? Use the following questions to help you decide what information goes onto the form.

- **What** kind of information is needed? **How** often?
- **Why** is this information being collected?
- **How** is this information being recorded? Is it concise yet complete?
- **Who** uses this form? Why and when?
- **Who** else reads this form? Why?
- **Who** monitors the records? When? Why?
- **How** is the information stored? **Where**?
- **Why** am I completing this form?

What can I do to improve how I complete forms? Use the following suggestions to evaluate how you complete forms.

- Consider readability: Print neatly!
- Fill out every part of the form. Write N/A (for *not applicable*) on any blank line.
- Write clearly and concisely: What happened, where, when, and why?
- Use numbers (actual output or performance data) so the reader can easily tell how what happened affected productivity. If you are completing a shift log, document your shift's performance in meeting its goals. What the next shift needs most is information that will help it meet its production goals.
- Double-check names of people, places, and equipment; numbers, dates, and key information for correctness and completeness so that spot-checking or follow-up is easier.
- Develop a system for proofreading numbers.
- Avoid putting personal comments on forms. Shift logs, for example, that casually note down time and close with "Have a nice day" are not informative enough to be useful.

  Example: It has been observed that three of five Juggernaut processors failed. Hope your shift goes better.

  Revised: 3 Juggernaut processors (Stations 3, 5, and 7) failed setup, holding up production for 30 minutes. Ron Clark repaired ignition problems.

- Have someone else double-check your work, especially if the form is complex.

possible filling out forms. So the forms you *do use* must be easy to use and easy to read. They must also be accurate, on time, and complete.

Additionally, how completely, clearly, and accurately the information is presented on a form can sometimes have legal implications. Patients' charts and incident reports may end up as part of the evidence in a liability case.

Many companies also use a variety of forms to monitor, document, and control changes in production. You can understand the importance of some seemingly unimportant forms by asking:

- Why is this information being collected?
- What are the key activities of my company? Who are our customers and why?
- How does this information help meet customers' needs or company's goals?
- What impact does this information have on "the bottom line" (the company's profits)?
- Which forms help my supervisor understand the impact of my work on "the bottom line"? (the company's profits)
- Could this information have legal implications?
- Have I included all the information my reader needs in order to act?

Company operations (the ability to produce products or services) can be affected by many different factors, from changes in customer demand, to shifts in the level of sales, to the availability of materials and workers. A small winery in California nearly went out of business because it did not have a second supplier for bottle corks—and their only supplier in Italy was unable to ship goods on time because of a strike by workers. Managers and technicians reading routine monthly reports did not anticipate this problem.

Nurses must document exactly what happens during a patient's care, considering treatments, procedures, unexpected incidents, and changes in the patient's condition. Nurses tend to put a patient's care first but must never neglect the detailed paperwork involved in that care. From a jury's point of view, "an incomplete chart suggests incomplete nursing care." [1]

Part of your job is giving information to your supervisor so crises can be anticipated and avoided. Regardless of the type of form you are completing or designing for others to use, your forms need to present information that is complete, concise, accurate, and timely.

---

[1] Edith McMahon et al. (eds.), *Mastering Documentation*. Springhouse, Pennsylvania: Springhouse Corporation (1994), p. 297.

Exercise 5.3 Discovering Where Companies Use Forms

Consider the main functional areas of a particular company you are interested in (it can be a major employer or a small company). Identify three or four major areas like personnel, operations, sales, planning, public relations, or maintenance.

What kinds of standard forms do you think each area might need to use? List as many different types of forms as you can think of. Use an asterisk to identify the forms you think are most important. Bring your lists to class.

Appoint a recorder from the members of your small discussion group to take notes of your discussion. Then compare your lists. After your group has discussed the lists, try to come up with two or three categories that you could use to organize these standard forms. You might consider where the form is being used or what the form is being used for.

Then, working as a group, summarize your findings in a memo to your instructor. Your memo should answer the following questions:

- What do you think of the information your group discussed? Were there any surprises?
- Did one area appear more important to a company's success than others? Does one area seem to produce more forms than other areas?
- Do you think the importance of forms changes, depending on the kind of company? Or which area forms are used in?
- What does your group think are the strengths and weaknesses of forms in general? What do you think?

Turn in your group's discussion notes and lists to your instructor.

WRITING STRATEGIES: REPORTING RAW DATA

If you are providing raw data (that is, the information itself—the basic numbers that show a patient's temperature, the number of arrests within a certain period, how many engines were rebuilt, etc.), in addition to being complete, accurate, and timely, the raw data must be organized so that it can be easily read and understood. You can help your reader by organizing and presenting raw data visually—in lists, tables, or charts.

Using Lists to Highlight and Clarify Information

Lists are often used to highlight key information in memos and reports, to help design forms, and to organize checklists and instructions. Which of the announcements in Exhibit 5.7 is easier to read?

Chapter 5: Using Forms **145**

Sample Announcements   Exhibit 5.7 ■

Sample A

ANNOUNCEMENT TO ALL STAFF

All staff will need to attend a safety awareness workshop held in the Clarkson Training Center. Please sign up for one of the following times before September ends: Monday, September 7, at 3:00 PM; Tuesday, September 15, at 8:00 AM; Wednesday, September 22, at 5:30 PM; and Thursday, September 29, at 9:00 PM . Schedule one of these times through your supervisor.

Sample B

ANNOUNCEMENT TO ALL STAFF

All staff will need to attend a safety awareness workshop held in the Clarkson Training Center. Please **sign up for one** of the following times before September ends:

Monday	Sept 7	3:00 PM
Tuesday	Sept 15	8:00 AM
Wednesday	Sept 22	5:30 PM
Thursday	Sept 29	9:00 PM

Schedule one of these times through your supervisor.

Using Headings to Direct the Reader's Attention

The more complicated or extensive the information is you are presenting, the more useful headings may be. Headings give a descriptive title for the information provided. Look at the two boxes in Exhibit 5.8. Which is easier to read?

Sample Lists   Exhibit 5.8 ■

Sample A

Inventory use for January	$ 48,937
Inventory use for February	$ 69,756
Inventory use for March	$ 74,293
FIRST QUARTER INVENTORY	$192,986

Sample B

Inventory use by month:

January	$ 48,937
February	$ 69,756
March	$ 74,293
FIRST QUARTER INVENTORY	$192,986

Part II: Improving Writing Productivity

You may need to report raw data over a specific period. For example, you may report on the inventory used daily, weekly, or monthly, or you may summarize the kinds of machine malfunctions and lost production time by the month or by the quarter in a special form or in a simple list that appears right in your memo as shown above.

Using Tables to Present Data

Tables are formal lists of information, in which the data are organized into categories and columns so that readers can quickly analyze and compare what is being presented.

Tables are useful because they can summarize complex information in a very concise and visually appealing way. Although tables are used frequently in formal reports, they are also useful in summarizing raw data in memos, letters, and routine short reports.

How should information be organized on a table? Several different patterns are commonly used. You can organize *chronologically* (by date or time period), or *functionally* by categories (such as products or locations). You can also put *the most important information first*. Your writing purpose and your audience's needs should help you decide what order of information to use.

Notice how information on the length of employment and education level is presented in a simple table format for registered nurses (see Exhibit 5.9). This information is organized by category.

The challenge for the reader is in trying to decide how this information might be useful. This particular table shows the number of years of training that hospital nurses have. For example 39% of nurses hold two-year degrees. The table also shows that about 46% of all nursing staff (whether full-time or part-time) have worked five years or longer at the same hospital. Therefore, a hospital administrator can anticipate a 20% to 50% turnover at any time. Your job as a writer is to decide what the implications are for your particular situation and your particular reader.

Notice that descriptive titles, such as the very formal title shown with Exhibit 5.9, can be useful in telling readers the **topic** and **purpose** of this table. Many readers look at a table (or other graphics) without reading the text that comes with it. Thus, a title that shows **what** (subject) and **why** (purpose) helps the reader to understand the information quickly.

Introducing and Interpreting Tables

Should your table just speak for itself—without an introduction, discussion, or conclusion? Not all readers have the same ability to translate raw data into a meaningful conclusion. Readers can compare complex data more easily if tables are presented with:

Hospital Registered Nursing Personnel Summary, 1992   Exhibit 5.9

Characteristic	
Length of Employment at Hospital	
Full-time: Less than 1 year	15.5%*
1 year to less than 2 years	14.3
2 years to less than 5 years	25.0
5 years or more	45.2
Part-time: Less than 1 year	13.1
1 year to less than 2 years	13.1
2 years to less than 5 years	25.2
5 years or more	48.7
Educational Attainment	
Associate degree in nursing	39.4
Nursing diploma	27.7
Baccalaureate degree in nursing	21.0
Master's degree in nursing or higher	2.0

*Based on a sample of 5,417 U.S. short-term acute care hospitals.
Source: U.S. Dept. of Commerce. *Statistical Abstract of the United States, 1995,* 115th ed. Washington, D.C.: U.S. Government Printing Office (1995), p. 124. Used by permission.

- An **introduction** that clarifies the purpose of the table.
- A **discussion** section that interprets the most important information in the table.
- A **conclusion** that summarizes and interprets the table.

Sometimes your introduction and discussion can be quite concise, but this additional analysis is very useful to your reader. The more complex the information (and the more important the relationships between the data), the more important it is to give your reader some kind of summary statement of your findings when you use tables or other graphics.

This next exercise should help you to understand how your readers can benefit from your interpretation of the data being presented. The more complex the information, the more necessary it is to add your analysis.

Analyzing a Table   Exercise 5.4

Exhibit 5.10, an excerpt from the *Statistical Abstract of the United States,* allows the reader to compare the years of schooling different groups have completed by marital status, gender, and employment.

Read that data and list three or four implications of these facts. Were you surprised by any relationships between different sets of data? Do you think there are any significant findings from this data? For example, which group faces the greatest risk of divorce? What does this information mean to someone who wants to drop out of high school? Is job security linked to how much education someone has?

Bring your list of implications to class for discussion with your group. As you work in your group, use an asterisk to highlight the most significant findings.

148 Part II: Improving Writing Productivity

■ Exhibit 5.10 Years of School Completed, by Selected Characteristic: 1994

[For persons 25 years old and over. As of March. Based on Current Population Survey.

| Characteristic | Population (1,000) | Percent of Population | | | | | |
		Not a high school graduate	High school graduate	With some college, but no degree	With an associate's degree*	With a bachelor's degree	With an advanced degree
Total persons	164,512	19.1	34.4	17.4	7.0	14.7	7.5
Age:							
25 to 34 years old	41,946	13.6	34.5	19.9	8.5	18.2	5.2
35 to 44 years old	41,527	11.3	33.2	19.4	9.1	17.9	9.2
45 to 54 years old	29,522	14.9	34.0	17.9	7.1	15.2	11.0
55 to 64 years old	20,737	24.4	37.2	14.8	4.7	11.1	7.7
65 to 74 years old	18,087	32.2	36.6	13.5	4.0	8.5	5.1
75 years old or over	12,692	45.2	30.6	10.8	2.5	7.0	3.9
Sex: Male	78,539	19.0	32.3	17.3	6.3	15.9	9.2
Female	85,973	19.3	36.2	17.4	7.5	13.7	5.9
Race: White	139,760	18.0	34.5	17.5	7.1	15.1	7.9
Black	18,103	27.1	36.2	17.5	6.3	9.5	3.4
Other	6,648	21.0	26.0	14.3	6.3	22.0	10.4
Hispanic origin: Hispanic	13,714	46.7	26.2	13.3	4.7	6.2	2.9
Non-Hispanic	150,798	16.6	35.1	17.7	7.2	15.5	7.9
Region: Northeast	33,797	17.3	37.8	13.4	6.6	15.6	9.3
Midwest	38,427	17.3	37.8	17.0	7.2	14.1	6.7
South	57,025	22.4	33.4	17.4	6.5	13.7	6.6
West	35,262	17.4	28.9	21.5	7.9	16.3	8.0
Marital status:							
Never married	24,026	17.8	31.3	17.2	7.0	19.1	7.8
Married spouse present	103,987	16.2	34.8	17.4	7.3	15.8	8.5
Married spouse absent	6,208	29.0	35.4	17.3	5.8	8.5	3.9
Separated	4,404	28.1	37.4	18.6	5.5	7.0	3.3
Widowed	13,273	42.0	34.0	11.6	3.2	6.4	2.8
Divorced	17,017	17.0	34.7	21.9	8.5	10.9	6.1
Civilian labor force status:							
Employed	102,325	10.7	33.7	19.2	8.5	18.3	9.6
Unemployed	6,268	24.1	38.7	17.4	6.3	9.8	3.7
Not in the labor force	55,151	34.5	35.1	13.8	4.2	8.7	3.8

*Includes vocational degrees.

Source: U.S. Dept. of Commerce. *Statistical Abstract of the United States,* 1995, 115th ed. Washington, D.C.: U.S. Government Printing Office (1995), p. 158.

Interpreting and Summarizing Raw Data

When you are organizing much information into tables for your readers, you may need to prepare a summary of the data. Simply showing the raw numbers may not clearly show the relationships or significance of any specific numbers.

Chapter 5: Using Forms **149**

Notice how the following summary of revenue and expenses for Pacificorp's electric operations (Exhibit 5.11) shows changes over six years, ending the summary with a percentage to compare the last two years and a second percentage to show the rate of annual growth, compounded over five years.

You can read this table in two ways: *across each line* to compare categories by each year, and *down each column* to compare categories within one year.

To analyze raw numbers, like these revenue figures appearing on Pacificorp's financial statement, you may need to translate the totals into percentages.

Calculating the Growth Percentage.
Do you want to find out how much the company grew during one year? You can calculate the growth percentage by subtracting the base year amount from the following year amount. Then you divide this number by the base year amount.

Looking again at Pacificorp's figures, to calculate the *growth percentage* in total revenues from 1993 to 1994, first subtract the base year ($2,506.9 million in 1993) from the latest year ($2,647.8 million in 1994). You then divide $140.9 million by the base year of $2,506.9 million. The answer is 0.056, or rounded up, to 6% growth from 1993 to 1994. If you are an investor, you can decide if that is a reasonable growth rate.

Summary of Revenues and Expenses for Pacificorp's   Exhibit 5.11 ■
Electric Operations, 1989–1994

Electric Operations								Millions of Dollars/for the year
	1994	1993	1992	1991	1990	1989	1994 to 1993 Percentage Caparison	5-Year Compound Annual Growth
Revenues								
Residential	$ 724.9	$ 698.9	$ 649.8	$ 663.8	$ 646.4	$ 646.4	4%	2%
Commercial	570.4	543.9	526.9	517.4	509.0	517.3	5	2
Industrial	726.3	696.2	695.6	674.9	673.8	670.6	4	2
Other	30.7	29.8	29.9	34.2	34.3	38.2	3	(4)
Retail	2,052.3	1,968.8	1,902.2	1,890.3	1,863.7	1,872.5	4	2
Wholesale - firm	456.2	422.5	356.5	264.7	209.9	190.3	8	19
Wholesale - nonfirm	76.5	77.3	71.3	59.9	78.4	79.0	(1)	(1)
Wholesale	532.7	499.8	427.8	324.6	288.3	269.3	7	15
Other	62.8	38.3	32.4	36.9	32.5	33.9	64	13
Total	$2,647.8	$2,506.9	$2,362.4	$2,251.8	$2,184.5	$2,175.7	6	4

Source: Pacificorp annual report, 1995. Used by permission.

Part II: Improving Writing Productivity

Calculating a Breakdown Comparison.
Do you want to understand *the breakdown* or what percentage of revenues came from the different categories of revenue—residential, commercial, industrial, and other? You can calculate the percentage of each category by dividing each individual number by the total.

For example, to calculate what percentage of revenues came from each category, divide $724.90 by $2,052.30 (the total for retail revenues). The resulting number .353 becomes 35%. Which is the most significant contributor to retail revenues? Which category showed the most growth between 1993 and 1994? You may be able to guess, but calculating percentages for your reader makes the analysis more efficient. Which of the following tables is easier to read?

Revenues (in millions)	**1994**		**1993**
Residential	$ 724.9		$ 698.9
Commercial	570.4		543.9
Industrial	726.3		696.2
Other	30.7		29.8
Total Retail	$2,052.3		$1,968.8

Revenues (in millions)	**1994**		**1993**
Residential	$ 724.9	35%	$ 698.9
Commercial	70.4	28%	543.9
Industrial	726.3	35%	696.2
Other	30.7	15%	29.8
Total Retail	$2,052.3	100%	$1,968.8

Does the breakdown remain the same for both years? Calculate the breakdown in 1993 to see any changes.

To analyze such complex figures would require more than one calculation, especially when the breakdown figures show little change in sources of revenues between 1993 and 1994. The analyst could conclude that all categories of revenue increased between 1993 and 1994, but without calculating the percentages, do you know which category grew most?

These two percentages can be used to translate raw numbers into relationships that are easier for your readers to understand. Once you have the percentages, you can choose a variety of other commonly used charts to show the information visually.

Why is interpreting and summarizing raw data important to you as a writing skill? Most readers understand what percentages mean more quickly than they understand the relationships between different pieces of raw data. Many readers are intimidated by numbers. You can help your readers understand raw numbers by translating them into simpler formats and then by interpreting the data.

Using Percentages for Analysis   Exercise 5.5

1. Working alone or with a partner, read over the table of raw data on numbers of visits to emergency rooms (see Exhibit 5–12). Your goal is to find out if a specific group is using the emergency room more than other groups to determine if there should be any changes in emergency room staffing or procedures.

2. Calculate percentages to check relationships between different age groups (reading down the table) and to check trends within each specific different age group (reading across the table). After you have translated the data in the table to percentages, what does this information mean? List several implications or questions to share with your discussion group.

3. Share your findings with your discussion group. Considering the audience for this information and your purpose (to inform? to persuade?), what recommendations could be made? Summarize your discussion in a short memo to your instructor and hand in all notes and homework.

Reporting on Performance

When you are reporting on your team's ability to meet production quotas, you are also evaluating the performance of people in your work team. To do so fairly involves setting performance standards, communicating them clearly to all team members, and, where needed, requesting training or other support so each team member can be fully productive. Such information needs to be presented honestly and tactfully.

Hospital Emergency Room Visits: 1993   Exhibit 5.12 ■

	\multicolumn{4}{c}{Number of Visits (1,000)}
Characteristic	Total	Urgent*	Non-urgent	Injury-related
All visits	92,814	42,048	50,766	37,712
Age				
Under 15 years old	23,285	9,176	14,108	9,114
16 to 24 years old	15,583	6,291	9,292	7,644
25 to 44 years old	28,064	11,526	16,537	12,563
45 to 64 years old	12,822	6,488	6,334	4,625
65 to 74 years old	5,496	3,523	1,973	1,511
75 years old and over	7,565	5,044	2,521	2,257

*Patient requires immediate attention.

Source: U.S. Department of Commerce. *Statistical Abstract of the United States,* 1995, 115th ed. Washington, D.C.: U.S. Government Printing Office (1995), p. 129. Used by permission.

Reporting on performance involves:

- Analyzing and setting performance standards.
- Understanding and assigning tasks.
- Collecting data on performance or productivity.
- Reporting on production quotas, timeliness of production, and quality of performance or production.

If you have "bad news" to report, it may be better to talk with your supervisor informally before preparing a written report.

Work logs, shift reports, and progress reports are commonly used to evaluate the effectiveness of a work team. With a work log or shift log, you simply record the main events that occurred during your shift or on your project by completing a daily log. These logs can then be used to compile weekly or monthly progress reports for project managers or supervisors. The format for this kind of report varies widely for different companies and different software programs.

Exercise 5.6 What's Wrong with This Monthly Report?

1. Read this monthly report of machine malfunctions. What changes would you suggest to the writer and why? Please consider content, organization, style, format, and the way in which both raw data and summary information are presented.

2. If your instructor asks you to, revise the memo, adding your interpretation of the raw data. You may need to invent missing information. Review your findings with your discussion group. Then, turn in your completed report (with the group's comments) to your instructor.

Pilot Steel Manufacturing

DATE: July 19, 1996

TO: Roger T. Rogers, Plant Supervisor

FROM: Your name

SUBJECT: MONTHLY REPORT OF MACHINE MALFUNCTIONS

Here is this month's report on machine malfunctions. I think we need some safety and preventative maintenance training. What do you think?

PLANT	#1	#2	#3	#4
Compressors	4	3	7	0
Boilers	1	3	5	1
Hydraulic Lifts	2	3	4	1
Circuit Breakers	4	5	6	3

Chapter 5: Using Forms   **153**

WRITING STRATEGY: WRITING COHERENTLY

Sometimes workplace writers don't know which ideas should come first or why. The goal is to write so your reader moves logically from one idea to the next in a way that best meets your writing purpose, which can be informative or persuasive.

Most workplace readers expect information to be presented logically. If you write 1-2-3, your reader will expect 4-5-6 to follow. This means selecting the organization pattern that best fits the information you present.

The easiest way to ensure logical development of your ideas in a memo format is to follow an introduction, discussion, conclusion pattern for every memo you write.

Sometimes readers only want information. In this case, you will need an introduction to introduce the topic and purpose of your memo and then a discussion section listing the information. Even with this kind of **informative writing**, you can still conclude with a summary statement.

At other times, readers will request your analysis or recommendation. For this more **persuasive writing**, you will need discussion and conclusion sections that interpret the data more fully and present a recommendation.

Informative writing usually follows a *direct* order or pattern. The main idea comes first, followed by supporting facts. See the following example.

> We need to replace Acme Paper because they have consistently given us the wrong paper weight, their shipments have been late, and their billings have been inflated.

If you are writing to a favorable audience (an audience that agrees with you), the **direct** pattern gives your reader no surprises. Your audience reads the most important information first, and then works through supporting details if needed.

Persuasive writing generally uses an *indirect* order or pattern, with reasons leading the reader to your conclusion. Note the following example.

> Because of their late shipments, inflated billings, and consistently wrong paper weights, we need to replace Acme Paper.

If you are writing to a hostile audience (an audience that does not agree with you) or a neutral audience (an audience that is undecided), the **indirect** pattern may be more effective. If you use the indirect pattern, the reader has a chance to understand your concise summary of the reasons and facts that have led you to make your recommendation.

Regardless of the order or pattern you select, direct or indirect, memos should still start with an introduction that states the problem, the purpose of the memo, and previews what's contained in the memo.

Remember the work done in this chapter on analyzing and interpreting numbers. Recommending a purchase, for example, would require persuasive writing

Exercise 5.7 Using a Direct or an Indirect Pattern

Identify which order (*direct* or *indirect*) would probably work best for each of the following reports, letters, and memos—and explain why. Which documents would require supporting numerical data for greater credibility with the reader?

1. A notice of staff layoffs from the company president to all staff.
2. A request for a new telephone system for the company to the chief operations officer and president.
3. A response to a customer complaint letter.
4. A report on monthly sales to the plant managers.
5. An annual report that goes to all shareholders.
6. A request for supplies to the purchasing department.

Come to class prepared to discuss which of these would be most difficult to write.

---

and would most likely rely on some number "crunching." Using numbers is part of your persuasive writing strategy. Even in routine memos, you may have to decide which is best: direct or indirect order.

In addition to deciding whether to use a direct or indirect order to organize how you present information, you may need to use one of several other commonly used organizing patterns.

Being familiar with these different writing patterns will help you decide which pattern is best for a specific form or memo.

We can use the following *descriptive patterns* to describe tools and mechanisms:

- Definition (What is X?)
- Spatial (How is X organized, by direction or location?)
- Chronological (What happened when? By date?)
- Process analysis (How does X work? Step-by-step?)
- Instruction (How should X be done?)

We can use the following *analytical patterns* to write more complex documents, such as proposals or recommendations.

- Analysis (What's wrong?)
- Comparison and contrast (How is X different from or similar to Y?)
- Cause and effect (What has caused X? What are the effects of X? Of doing X?)
- Problem + solution (What's wrong and what should be done about it?)
- Recommendation (Which is best and why?)

Sometimes more than one pattern is used to organize your writing.

Deciding on an Organizational Pattern — Exercise 5.8

Read the following passages and identify the organizing pattern used for each.

1. When you enter the bank, you will see the new accounts desk to your left. Immediately in the center of the bank is the teller platform separated from the main lobby by counters. You can reach the vault by taking the stairs that are on your right.

2. If you want to earn a promotion, you need to anticipate the needs of your customers, your co-workers, and your supervisors. Think about your department's and your employer's goals. Ask how you can improve your performance.

3. If you were hired within the last three months, please complete the enclosed pretax earnings form and make an appointment to review it with the personnel department before January 15th. If you are renewing your FSA contribution from a previous year, you may either complete the form and forward it to the personnel department or call to make an appointment if you have any questions.

4. As a sales clerk, I worked 37 hours every week and punched a time clock four times daily. Now, as co-owner of the hardware store, my work hours are much longer, rarely less than 50 hours per week.

WRITING STRATEGY: WRITING CONCISELY

Sometimes workplace writers use too many words or include too much information when they fill out a form or write a memo. Other writers may have exactly the opposite problem. Their writing is so pruned down that it becomes cryptic, almost a code. With this kind of writing, the reader must work very hard to understand what has been written.

This section will help you develop a concise style without becoming so cryptic that your readers can't easily get the main message.

Usually writers cut out wordiness in the **revision** stage. You can ask the following questions to help focus your revision:

- Where does my rough draft seem wordy? Underline these areas.
- How can I write this sentence using fewer words? Go back over underlined areas and revise by cutting out extra words or restating the ideas more simply.
- Have I used formatting techniques (headings, lists, and short paragraphs) to present ideas concisely?

After you have completed your revision, ask:

- Does my draft meet my writing purpose and reader's needs?
- Have I left out any key ideas or information? If yes, put them back in as brief a form as possible.
- Have I used a friendly yet professional tone?

Frequently when we are writing a rough draft, we use more words than we need to express our ideas. That's because during the **drafting** stage, it's most important to get all your ideas and supporting information down on paper—without stopping to criticize or edit the draft.

156   Part II: Improving Writing Productivity

Exercise 5.9   **Revising for Conciseness**

1. Read the following two incident reports. What are the key problems explained in both reports? Which report best meets company needs? Why? Does Incident Report 1 adequately document routine operations?

2. Revise Incident Report 2 for conciseness. The "best" report will be concise without sacrificing key information.

Some hints for getting started:

- Underline key information.
- Decide what is most important.
- Decide in what order the information should be presented.
- Restate in your own words, using lists, headings, and short paragraphs (if needed).
- Revise for conciseness.

Bring your revised incident report to your discussion group for review. You may want to check Appendix A: Quick Format: Memos and Letters, for memo-writing guidelines.

3. Discuss the revision process you used and have your group select the "best" revised incident report. As a group, summarize why this revision works well, and turn in your summary with working papers to your instructor.

Incident Report 1

September 20, 1996

MEMO TO: Regina Jackson-Huerta

FROM: Your name

SUBJECT: NEW PAPER SUPPLIER

We need to find a new photocopy supplier before the end of the month. What should we do next? Acme Paper has been terminated.

Incident Report 2

September 20, 1996

MEMO TO: Regina Jackson-Huerta

FROM: Your name

SUBJECT: PROBLEMS WITH PHOTOCOPY PAPER LED TO MY RECOMMENDATIONS THAT WE FIRE ACME PAPER AS INADEQUATE

Last Thursday I terminated Acme Paper as our sole supplier for photocopy paper. As you know, Acme has been our sole supplier for the last four years; however, we have been having a lot of problems with them about the quality of the paper they provide. Also, they have been late. Further, they have consistently avoided returning our phone calls, have been rude to our workers when routine requests have been made and they have made mistakes in their billing. I just got tired of dealing with all these problems. The last straw happened on Wednesday. When our photocopy machine report came in, I found very high breakdowns with a report from the service repair person that the wrong weight of paper was being used at all our plants. We have a standing order for 10% bond, but we were given 4% bond. I checked with two other photocopy supply companies to find that on average this has cost us 20% over what it should have cost us. If I annualize our photocopy paper costs, we could have saved over $3,000 per year, and we probably would have improvements in service as well as quality of product. If it would not be too inconvenient, please give me the go-ahead to find a new supplier as our current supplies will run out by the end of the month.

Thank you for your interest and support. If you have any questions, please do not hesitate to contact the undersigned.

During the **revising** stage, try to check content and organization in the first phase of editing. You can then work on style and format in the final phase of editing. Checking for wordiness is an important part of revising your style. With practice, you can develop a style that is complete yet concise.

WRITING FROM THE FIELD

Police officers, home health care providers, security guards, and sales staff are only a few types of workers who routinely work in the field, away from the main office and without direct supervision. These people must make independent decisions that follow the policies and procedures of their employer. Their success often depends on good written and oral communications.

Forms developed for use in the field often include many of the elements that were previously discussed:

- Fill-in-the blank and checklist sections to make sure basic information is acquired.

- Memo sections to supplement basic information with additional details or a narrative of an incident.

- Graphic chart areas that allow the writer to diagram the incident.

In the case of reports written by health care workers or police officers, the data collected or statistical information may be used by courts, lawyers, the police, and insurance companies, as well as state and federal governments. Therefore, the information on these routine forms must be absolutely accurate, clear, and concise.

Exhibit 5.13 illustrates how to use a typical graphic record. Exhibit 5.14 is a sample patient care flow sheet on which routine interventions are documented. Which of the two reports is easier to read? Which would be most useful to the medical staff? Why?

Health care workers and police officers have a special obligation to observe events fairly and report them objectively, even in the case of very traumatic events. Using routine forms can help the writer report the kind of information needed, but years of training in the field are also needed to develop professional objectivity—and the ability to closely observe "the facts."

Reporting from the field may change from setting aside time back in the main office to complete reports to filling in the blanks on a computer screen right at the site as different industries begin using mobile computers. These records would be instantly linked to a database in the main office. In the case of police officers, these links could give immediate background on suspects; for health workers, medical case histories could be available. We are just beginning to understand how changes in the way in which we enter and store data electronically can result in more efficient use of such data by employers.

We can anticipate change in how we report information as we gain access to more and more information. One challenge for managers will be deciding how best to use the increasing volume of information that is available. As recorders of information, our challenge will be reporting the information most efficiently, accurately, and objectively.

■ Exhibit 5.13 **Graphic Record**

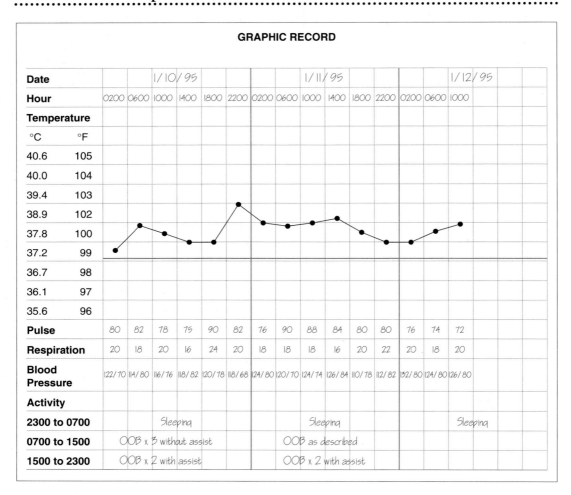

Source: Edith McMahon et al. (eds.), "Using a Graphic Record," *Mastering Documentation*. Springhouse, Pennsylvania: Springhouse Corporation (1995), p. 123. Used with permission of *Mastering Documentation,* copyright 1995, Springhouse Corporation. All rights reserved.

Chapter 5: Using Forms   **159**

Patient Care Flow Sheet   Exhibit 5.14 ■

PATIENT CARE FLOW SHEET

Date 1/22/95	2300 – 0700	0700 – 1500	1500 – 2300
RESPIRATORY			
Breath sounds	Clear 2330 PW	Crackles LLL 0800 SR	Clear 1600 MLF
Treatments/results	————	Nebulizer 0830 SR	————
Cough/results	∅ PW	Mod. amt. tenocious yellow mucus 0900 SR	∅
O2 therapy	Nasal cannula @ 2 L/min PW	Nasal cannula @ 2 L/min PW	Nasal cannula @ 2 L/min PW
CARDIAC			
Chest pain	∅ PW	∅ SR	∅ MLF
Heart sounds	Normal S1 and S2 PW	Normal S1 and S2 SR	Normal S1 and S2 MLF
Telemetry	N/A	N/A	N/A
PAIN			
Type and location	Ⓛ flank 0400 PW	Ⓛ flank 1000 SR	Ⓛ flank 1600 MLF
Intervention	meperidine 0415 PW	Repositioned and meperidine 1010 SR	meperidine 1615 MLF
Pt. response	Improved from #9 to #3 in 1/2 hr PW	Improved from #8 to #2 in 45 min SR	complete relief in 1 hr MLF
NUTRITION			
Type	————	Regular SR	Regular MLF
Toleration %	————	90% SR	80% MLF
Supplement	————	1 can Ensure SR	————
ELIMINATION			
Stool appearance	∅ PW	∅ SR	∅ soft dark brown MLF
Enema	N/A	N/A	N/A
Results	————	————	————
Bowel sounds	Present all quadrants 2330 PW	Present all quadrants 0800 SR	Hyperactive all quadrants 1600 MLF

Source: Edith McMahon et al. (eds.), "Using a Flow Chart," *Mastering Documentation.* Springhouse, Pennsylvania: Springhouse Corporation (1995), p. 125. Used with permission of *Mastering Documentation,* copyright 1995, Springhouse Corporation. All rights reserved.

WRITING STRATEGY: USING SPECIALIZED FORMS

Many different forms can be designed by your employer, as we have seen in this chapter. Some specialized forms can be used to support individual planning, task analysis, and project management.

Exercise 5.10 **Reporting from the Field**

1. You have just observed a minor traffic accident involving two vehicles. One of the drivers was injured enough to require an ambulance.

Write a brief and objective narrative of the incident, describing the sequence of events in chronological order exactly as they occurred. Work hard to use *absolute precision* in your narrative. Use a location you are familiar with and invent any additional details that may be needed to clearly describe what happened. Your narrative will be attached to the police report of the incident, so please use the diagram and narrative section of the typical report shown on the next page.

2. Bring your draft accident report to your discussion group. Appoint a recorder to take notes as you review all the accident reports, analyzing which meet the needs of the officer who must use this report. Consider content, organization, style, and format—with particular attention to clarity, conciseness, and coherence. Summarize your discussion in a memo to your instructor, attaching all accident reports from your group.

Forms for Tracking a Schedule

Notice how the Do Sheet (see Exhibit 5.15) helps an individual worker to identify tasks that have high priorities and to monitor continuing tasks. This form also provides a record of the work in progress on a certain date.

These kinds of "time management" forms and systems are very useful for people who are working on many tasks and projects at the same time.

Forms for Analyzing Tasks

When workers coordinate a major project, they need to know **who** is going to do **what** by **when** for all phases of the project.

Usually this kind of task analysis occurs during the proposal phase, when the company is deciding whether or not to take on this major project. The key question asked at this stage is: Can we do this job with current staff, current resources, and current facilities at a competitive cost?

To answer this question requires analyzing and costing out the individual tasks that workers will need to carry out to complete the project. Many times in the early years of a company, these estimates are inaccurate. This is the reason for keeping such close cost records on any project. Such cost "histories" will enable the company to improve its ability to project costs for future jobs.

One form that can be used to analyze work flows and identify key tasks is the **flowchart**.

*Organizational flowchart*s show how something is organized, with the primary emphasis on the relationship between the parts. Exhibit 5.16 shows the main divisions of a police department and their relationship to each other.

Chapter 5: Using Forms 161

POLICE TRAFFIC ACCIDENT REPORT

PAGE ___ OF ___

LOCAL CASE NUMBER	ACCIDENT DATE	ACCIDENT TIME	ROAD ON WHICH ACCIDENT OCCURRED		MILEPOST	151 HF
	DAY OF WEEK		INTERSECTING ROAD	☐ WITHIN ☐ NEAR _____ ☐ FEET ☐ MILES ☐ NORTH ☐ SOUTH ☐ EAST ☐ WEST		LOCAT
	TIME POLICE NOTIFIED	TIME POLICE ARRIVED	CITY/TOWN	☐ WITHIN ☐ NEAR _____ ☐ FEET ☐ MILES ☐ NORTH ☐ SOUTH ☐ EAST ☐ WEST		WEATH
	TIME EMS NOTIFIED	TIME EMS ARRIVED	COUNTY	DISTRIBUTION		LIGHT

☐ PROPERTY DAMAGE   ☐ INJURY   ☐ FATAL (TTY SENT)   ☐ HAZARDOUS MATERIALS

☐ PUBLIC PROPERTY DAMAGE   ☐ HIT AND RUN   ☐ PHOTOS TAKEN   ☐ TRUCK JACKKNIFED

DMV ID NO.	OM 20

UNIT ☐ MOTOR VEHICLE ☐ PROPERTY ☐ PEDESTRIAN ☐ OTHER _____	ACTION TAKEN	SRF TV

NAME (LAST, FIRST, MIDDLE)	LOCAL ID NO.	SEX	RACE	DATE OF BIRTH	SAF OC

ADDRESS (CITY, STATE, ZIP CODE)	PHONE	☐ HOME ☐ MESSAGE ☐ WORK	TCD TY

DRIVER LICENSE NUMBER	STATE	CLASS	INSURANCE COMPANY - - POLICY NUMBER	VEHICLE DAMAGE ☐ 18 OVERTURN ☐ 19 UNDERCAR ☐ 00 NONE ☐ 99 UNKNOWN	TCD CC

VEHICLE PLATE NUMBER	STATE	CLASS	COLOR	USE ARROW TO SHOW FIRST IMPACT	RD CHA
YEAR	MAKE		MODEL/MOTORCYCLE CC'S	STYLE	RD FLO
REGISTERED OWNER NAME AND ADDRESS					NO. LM

02   03   04   05   06   07
01   15   16   17   08
14   13   12   11   10   09

DRIVER TAKEN TO	BY	VEHICLE TAKEN TO	BY	FIRE ☐ YES ☐ NO

PED TYPE	PED ACT	PED VIS	DESIG SP	STATD SP	VEH MOV	TR CONFIG	TRL TYPE	ALC INVL	BAC TEST	LIC VIOL	LOCATION	EQUIPMENT	EJECTION	INJURY	CARE
12	13	14			15	16	17	18	19	20	21	22	23	24	

▼ DIAGRAM AND/OR NARRATIVE ▼

USE ARROW TO INDICATE NORTH

162 Part II: Improving Writing Productivity

■ Exhibit 5.15   Do Sheet

DATE :			REMINDER :		
■ DO SHEET ■					
✳ = DO TODAY					
URGENT	**Priorities**	**TO SEE OR CALL**			
		AWAITING DEVELOPMENT			
		What		**Who**	**When**
CONTINUING ATTENTION					
Keep on top of your job – Do it now • Don't fear failure • Be decisive					

Organizational Structure of a Police Operations Bureau (representative model)   Exhibit 5.16 ■

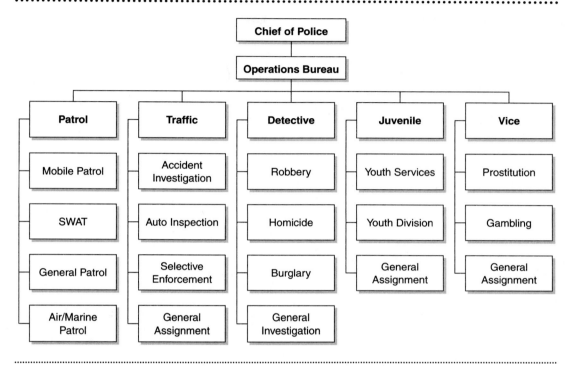

Source: Russell L. Bintliff, *Police Procedural: A Writer's Guide to the Police and How They Work.* Cincinnati, Ohio: Writer's Digest Books (1993), p. 16. Copyright 1993 by Russell L. Bintliff. Used with permission of Writer's Digest Books, a division of F&W Publications, Inc., Cincinnati, Ohio.

Process flowcharts also show how something happens. Exhibit 5.17 shows the complex relationships among different parts of the criminal justice system.

Flowcharts can help the reader to visualize the most important steps in a process or, as shown in Exhibit 5.17, different steps in a cycle. The flowchart can also help the writer to think through key steps in a project or to identify key parts in an organization.

Using a flowchart to map out *what* happens *where* and *when* can help you to understand all the parts of a process or project. You might prepare a flowchart to help you analyze the steps taken in a particular task with the purpose of improving efficiency, or understanding operations, or finding out where and why the work is slowing down (often called a "bottleneck").

One major task of a manager is to constantly look for ways to improve efficiency in processing products or services for customers. If you, as a technical

164 Part II: Improving Writing Productivity

■ Exhibit 5.17   **Typical Flow of Events in the Justice Process**

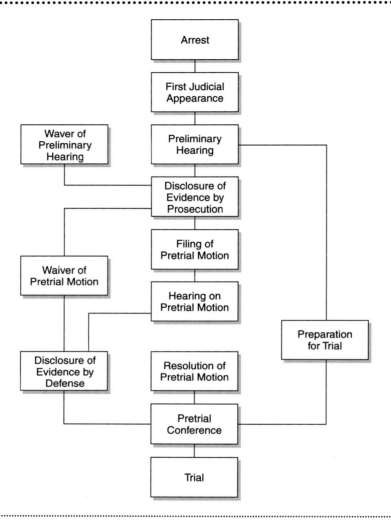

Source: Russell L. Bintliff, *Police Procedural: A Writer's Guide to the Police and How They Work*. Cincinnati, Ohio: Writer's Digest Books (1993), p. 232. Copyright 1993 by Russell L. Bintliff. Used with permission of Writer's Digest Books, a division of F&W Publications, Inc., Cincinnati, Ohio.

specialist, are already thinking about ways to save your company money, you have increased your value to your employer.

If many people are involved and the task or process is complex, errors can occur that can cost your company significant amounts of money. Use a flowchart to look for ways to improve productivity!

Drafting a Flowchart   Exercise 5.11

1. Make a list of the steps that occur in a task or process you are familiar with. For example, you might list how to prepare a patient to ride in an ambulance, or how to clip a cat's nails, or how to apply for a job. You could list the steps a person in your field would follow to fill a customer's order. Then analyze your list. Is it complete?

2. Organize your draft list into a simple flowchart. After you have completed your flowchart, share it with your discussion group. Analyze which is easier to read—the list or the flowchart. Why? Does the flowchart help you understand the task being described? Is the flowchart complete? Turn in discussion notes and final flowcharts to your instructor.

- Does the flowchart describe every step?
- Are there bottlenecks (places where the work seems to slow down)?
- Are there duplicated steps?
- Are there places where steps could be simplified?
- How can the process or task be used to catch and correct errors?

Forms for Managing Projects

If you are responsible for managing a larger project, you will need a more complicated kind of time management form—one that will allow you to monitor the work done by several team members, to track task completion and deadlines, and perhaps to report on the costs of the project.

Your company may already have a project management form similar to this one used by a national engineering firm (Exhibit 5.18). Read over this form and consider these questions:

- What is the most important information on this form?
- How will the project manager use this form?
- How will the project administrator use this form?
- Before the form can be used, what are some key actions that must be taken?

How Are Planning Forms Used in the Workplace?   Exercise 5.12

Interview a professional in your field to discover how this person manages his or her time. Ask if the employer provides any kind of standardized forms to help employees manage time or support planning. Don't be surprised if the employer doesn't have a systematic way of approaching planning. Ask for sample forms if they are available.

Summarize and analyze the information you gather, reporting it to your discussion group and submitting it to your instructor in a memo with forms attached.

PROJECT _____ MONTH _____ YEAR _____ PAGE ____ OF ____

NAME	REG.	DISC.	TASK	1	2	3	4	5	6	7	8	9	10	11	12	13	14	15	16	17	18	19	20	21	22	23	24	25	26	27	28	29	30	31	TOTAL DAYS	DAILY COST @TARGET	TOTAL COST @TARGET

PROJECT MANAGER:	NOTES:	TOTAL LABOR @ TARGET
PROJECT ADMINISTRATOR:		LABOR @ _____ (Proj. Mulitplier)
SENIOR CONSULTANT:		SERVICE CENTER EXPENSES
		TRAVEL
PROJECT START DATE:		OUTSIDE SERVICE
TARGET COMPLETION DATE:		OTHER
WORK PLAN COMPLETION DATE:		TOTAL EXPENSE
		CONTINGENCY
PROJECT NO.	CLIENT:	TOTAL FEE

Chapter 5: Using Forms **167**

In the last few years, forms and calendars have been developed into a system to help individual workers take control of their valuable time. Some companies use "families" or "systems" of such work plan forms to make sure that all levels of work are properly managed, that individual workers are accountable for tasks and deadlines, and that costs remain under control.

WHAT'S COMING NEXT?

This chapter has helped you to consider forms as an important way to document work. Your writing and communication strategies will need to focus on being clear, concise, complete, accurate, and unbiased. Your careful completion of forms will support many other kinds of audiences—some immediate and some unanticipated.

This chapter has also introduced the idea of organizing information so that it meets the needs of the reader. You'll use the direct pattern most often to simply present information that the audience needs. You'll use the indirect pattern if you are writing or speaking with a more persuasive intent. Most of the time, reasonable people will agree with your suggestions if they understand what is being proposed, why it is being proposed, and what the benefits will be. This often means preparing the reader for your suggestions before you present them.

In Chapter 6, Communicating with E-Mail, you'll continue to practice clear, concise, and accurate writing to solve problems, learn how e-mail works, and develop a process that will help you to systematically deal with information that is delivered by a variety of changing technologies. Your goals will be to produce work that meets very high standards and to improve your turnaround time—in spite of increasingly complex tasks, processes, and information.

CONCEPT REVIEW ■

Can you define the following key concepts from Chapter 5 without referring to the chapter? Write your own definitions in a journal entry or review them out loud.

raw numbers	assembly line
shift log	passdown form
production quota	productivity
time management	troubleshooting
tracking	percentage analysis
direct order	indirect order

Part II: Improving Writing Productivity

flowchart

table

preprinted forms

standardized forms

coherence

clarity

conciseness

incident report

■ WRITING SKILLS REVIEW

To build your skills in editing and proofreading, continue your systematic review of *Appendix C: Writing Skills Review.* Your assignment:

___ Read "Revising for Correct Pronouns" in *Appendix C.*

___ Complete Exercises C.26 through C.30 in "Revising for Correct Pronouns." Check your answers at the back of *Appendix C.*

___ Prepare a paragraph summarizing your progress for your instructor or use the form your instructor provides to report your progress. Be sure to highlight any areas you need to further review.

■ SUMMARY EXERCISE

What's Wrong with This Flowchart?

The flowchart in Exhibit 5.19 shows the decisions that a salesperson can make to improve customer purchases. Working alone or with a partner, analyze this flowchart to see if you agree with its content, organization, style, and format. List your editing suggestions and review them with your discussion group. Turn in your notes if requested to do so by your instructor.

■ ASSIGNMENTS

Assignment 1: Design and Use a Machine Log. You have been asked to design and complete a machine log for your company's new photocopy machine, which will be kept with the machine. This machine log will be used by all three shifts and will be approved by the plant manager.

1. Design a machine log to be used to report key incidents that will be used by all three shifts. Decide what kinds of information should be collected and in what order they should appear on your shift log. Leave room for the supervisor's approval. Once you have your draft machine log ready, share it with your discussion group, making any revisions you think necessary.

Flowchart Exhibit 5.19 ■

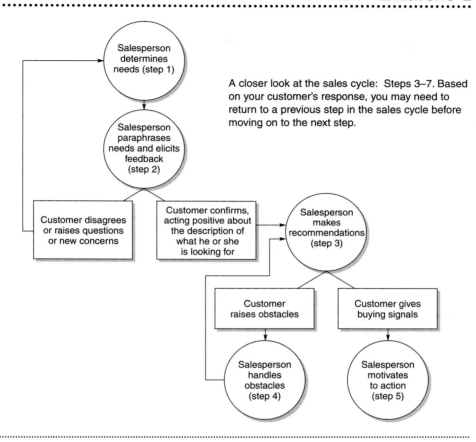

Source: Steven Trooboff et al., "Understanding the Sales Cycle," *Travel Sales and Customer Service,* ICTA (1995), p. 69.

2. Fill out your shift log using the information that follows. Be ready to answer your supervisor's questions about any information you may choose not to include in your shift log. Turn in your completed shift log to your instructor.

Here is what happened on the graveyard shift. When you first got to work at 9:00 PM, you found the machine out of paper. You refilled the paper on your routine rounds. At 10:30 PM, you replaced the toner. At 11:45 PM, the electricity to the plant went out for 17 minutes. At 1:12 AM, you pulled jammed paper out of the drum of the machine. At 3:00 AM you found the machine jammed again. This time, you could not fix the jam. You put an out-of-order sign on the machine and alerted the next shift to call a repair person. It is now 5:30 AM and you need to prepare your shift log before going home.

Part II: Improving Writing Productivity

3. Bring your completed shift log to your discussion group for review. As a group, discuss the strengths and weaknesses of the content of the forms from your group. Turn in a summary memo of your discussion to your instructor with the completed shift logs.

Assignment 2: Complete a Field Report.

Visit a local company or school workplace, observing closely what tasks are done by workers in a particular department. For example, you might observe the campus bookstore or cafeteria, or you might visit a local manufacturing company, nursing home, or library.

Prepare a memo that introduces where you went, who you saw, and what you observed. Include a flowchart that shows how the tasks of the department fit together in meeting the purpose of this department. Did you observe any bottlenecks or breakdowns in service? In a memo to your instructor, summarize your findings, concluding with any suggestions for improving the work flow.

Assignment 3: Prepare a Work Plan.

1. Using any of the ideas in this chapter, design your own time management forms by the day, week, or month. As you put together your planning sheet, consider the classes you are taking, the due dates for major assignments, the tasks and time it will take to complete these assignments, your job commitments, or other personal commitments.

2. Review your time management and planning forms with your discussion group, making any needed changes. Track your performance daily for the next two weeks, completing your form as fully as possible. Make notes about parts of this form that work well for you and why.

3. At the end of the two weeks, review your completed time management forms and write an analytical summary of how the forms worked for you, considering advantages or disadvantages. Do you have any recommendations? Turn in your summary with your completed planning sheets to your instructor.

■ JOURNAL ENTRY

In this journal entry, you will analyze a persuasive situation. Think of a time at work or at school when you needed to persuade your supervisor, a colleague, or a friend to do something he or she did not want to do.

First, describe the situation and problem fully; then, list the strategies you used to persuade your audience. Note down verbal and written strategies. Be as specific as possible.

Next, write an analysis of the situation and what happened. Were you successful? What would you change if the situation were to occur again? Consider the timing of your actions, level of information you provided your audience, and the reasons why you were persuasive or not persuasive.

6

Communicating with E-Mail

Chapter PREVIEW

This chapter introduces you to how workers write and respond to e-mail. E-mail demands an instant response, so you'll practice writing and revising on the spot for clarity, accuracy, and conciseness in your writing.

Part of this chapter will help you understand the strengths and weaknesses of communicating via computer. You'll also work on developing a method you can use to systematically process large amounts of information that you receive via computer. E-mail is only one of several important communication tools that will change rapidly over the next several years.

Journal ENTRIES

Begin by thinking about what kind of privacy you have when you write e-mail, who monitors materials sent by computer, and who has access to the information may affect how you use communication technology in your workplace. Write a short one-paragraph answer to any of the following questions in your class journal, or prepare answers for class discussion as directed by your instructor.

1. How do people use the computer to do banking, pay for their groceries, register for school, and file their tax returns? In how many different ways are computers being used in your field? How recent are these changes?

2. When do you use e-mail instead of a phone call or letter? When is e-mail more efficient?

3. How do you redesign information from a memo to fit a computer screen that's about half a page and still have it readable?

4. If your supervisor sees a co-worker reading your e-mail, what should he or she do?

Chapter READING

Jeanne Walker Harvey suggests that the guidelines discussed in this next article can help you and your employer prevent legal problems related to e-mail.

SET YOUR RULES ON E-MAIL

Jeanne Walker Harvey

Just as computers have replaced typewriters, electronic mail—or e-mail—is rapidly taking the place of printed memos in companies that rely on computers.

Despite its convenience, e-mail could cause legal problems for unsuspecting employers. It sometimes pits an employee's right to privacy against an employer's business interests. Employees typically develop a sense that e-mail messages are private and that they can write anything they wish and no one will ever see it—or have a right to see it—except the recipient.

A potential problem employers should be concerned about is defamatory or discriminatory statements that employees may make about one another on the e-mail system. Employers also must protect trade secrets from dissemination through e-mail to unauthorized recipients inside or outside the company.

The legal conflicts that can arise over e-mail represent a burgeoning and unsettled area for the law. For example, company officials who read "private" e-mail messages and use the contents as grounds for dismissal might become the target of wrongful-termination and breach-of-privacy claims. On the other hand, an employer might sue after a search of a former employee's computer files shows evidence that he had e-mailed confidential material to the company that subsequently hired him.

Whether the employee or the employer prevails in lawsuits like these depends on the facts in the cases and on the law in the jurisdiction. But certain factors may help tip the scale in favor of the business owner. Chief among them is having a written e-mail policy.

Here are guidelines for writing one:

• Emphasize that the e-mail system is the property of the employer and is intended solely for carrying out company business. The policy should state that all messages transmitted via e-mail will be treated as business messages.

Any employee who sends a personal message on the system should be aware that such a message will be viewed as a business message and not a personal, confidential message of the employee.

• Make it clear that company managers can rightfully enter the e-mail system and review, copy, or delete any messages, and disclose such messages to others.

Often employees believe their private password entitles them to privacy of their e-mail messages. Your policy should state that the use of passwords to gain access to e-mail is for the protection of the company, not the employee. The company has the right to enter the system, and the employee should not assume that messages are confidential even though a private password is used.

Because the law regarding e-mail and the right of privacy is still developing, employers should search e-mail systems only if they have a compelling business reason to do so and only after consulting an attorney.

Chapter 6: Communicating with E-Mail   **175**

• Incorporate the e-mail policy into the company's overall confidentiality policy designed to protect trade secrets. Employees need to be reminded that information transmitted via e-mail may be confidential information, and reasonable efforts must be made to protect its security.

The policy should state that your company's confidential information should not be forwarded via e-mail outside the company or even to employees within the company unless such recipients are authorized to receive such information.

To protect the company's proprietary information, the policy should tell employees they should not leave e-mail messages on their screens when they leave their desks. Employees should change their passwords frequently to avoid unauthorized access. Also, employees must not copy and send by e-mail any information, including software, that is protected by copyright laws.

• Advise e-mail users to keep their messages businesslike and refrain from using the system for gossip and personal messages. E-mail should not be profane, vulgar, defamatory, or harassing.

• Ask employees to sign an acknowledgment of receipt of the written e-mail policy. The employee is then put on notice that he or she is to comply with the stated policy and that the company may access the employee's messages.

The acknowledgment should state that the company has the right to modify or revoke the policy upon notice to the employee.

Before adopting a written e-mail policy, an employer should consult an attorney; statutes and case law may vary among jurisdictions. Of course, a written policy cannot be a perfect shield against oncoming legal difficulties, but it may prevent some problems and enable a company to carry on its business effectively with the aid of its e-mail system.

Source: Jeanne Walker Harvey, "Set Your E-Mail Rules," *Nation's Business* (August 1995), p. 48. Excerpted by permission, *Nation's Business,* August 1995. Copyright 1995, U.S. Chamber of Commerce.

Analyzing THE READING

Reading for Content

1. How and where does Harvey narrow the topic of e-mail in general to the topic of an e-mail policy? Can you find a sentence that introduces the main idea? Does this article have a conclusion? Where is it? Does the conclusion connect back to the main idea? Why or why not?

2. Who is Harvey's audience and what is her purpose? Do you think readers will be convinced by Harvey's discussion? Why or why not?

3. How would you describe Harvey's style?

176 Part II: Improving Writing Productivity

Reading for Reaction

1. What is your reaction to this article? Were there any surprises about the legal or ethical issues that managers (and workers) need to consider when using e-mail?

2. Do you agree with all of the guidelines that are presented? Which guideline do you think is most important and should be followed by all employees? Why? Which guideline do you think is the least important. Why?

3. What skills do you think you would need to use e-mail effectively? What kinds of communications technology are being used in your field?

WRITING WITH E-MAIL

"Thank you for calling. I'll get that price information to you right away!" Your company has a chance to land a major order, but how will you get that information to your potential client "right away"? Will you send your pricing information by regular mail, sometimes called "snail mail" because a document takes days to arrive instead of minutes? Deliver it personally? Send it by fax? Will you use "voice mail" or an electronic mail system? And how will you know which system is best for this particular customer?

You can count on communications technology changing rapidly in your future. For example, larger companies are finding good cost savings and boosts in productivity when they switch from telephone systems staffed with people to voice-mail systems that automatically take care of telephone messages through a system of menus. The menu might sound like: "Press '1' to place an order, press '2' for customer service," and so on.

Some people want to talk to a "real" person and don't like this new technology. However, for many of us, such communication technology saves time; new technologies are changing the tools we use to send voice and written messages to our customers, our co-workers, and our supervisors.

You can use this chapter to review one of the most important new communication technologies: e-mail (electronic mail). The chapter can also help you to develop your writing style so that you can use e-mail effectively and efficiently. The most important writing strategies for e-mail are to write clearly, concisely, correctly, and courteously.

What Is E-Mail and How Does It Work?

E-mail is a software program that can be used to link your computer to every other co-worker's computer in a department, a plant, or an entire company; it can also link different companies. Once computers are linked together in a network, workers can then "mail" messages electronically to each other.

Chapter 6: Communicating with E-Mail   **177**

E-Mail Split Screen   Exhibit 6.1 ■

```
WP Mail - SEC:CAMPB    Thursday, May 19, 1994 11:30 am
┌In Box─────────────────────────────────────────Envelopes:  13─┐
  Cynthia Yee      05/04 10:39    STUDY ON WRITING FOR ENGLISH/DEV ENG -Re
  Dorie Nelson     05/07 10:41    Transfer $$ for ENG classes
  Arthur Bervin    05/12 13:04    Alternate WR 121 topics
  Arthur Bervin    05/13 09:49    Learning Center Report
  Dorie Nelson     05/16 10:45    TECHWR TEACHER FALL TERM - Reply -Reply
  Paul Snyder      05/17 17:01    P/T C/O LIB. MATERIALS
  Barney Cazort    05/17 21:27    email practice
  Dorie Nelson     05/18 17:25    TECHWR TEACHER FALL TERM - Reply -Reply
 •Cindy Epps       05/19 10:12    Transfer $$ for ENG Classes           CC
┌Out Box────────────────────────────────────────Envelopes:  35─┐
  Mail message     watsone, CHESTE  CARL PERKINS $$$ 93-94
  Reply            CCS:Chestep      CARL PERKINS $$$ 93-94 -Reply -Reply
  Mail message     smithv           COMPUTER STUFF
  Mail message     fosterg          BOOK ORDERS
  Reply            ETHERIA          Advising on Friday, June 17 -Reply
  Mail message     Bervina          ENG DEPT MEETING & WR121 FINAL
  Mail message     yeec             STUDY ON WRITING FOR ENGLISH/DEV ENG
  Mail message     yeec             INTERNET AD HOC

Tab Out Box; F1 Undelete; Shift-F1 Setup; F2 Search; F3 Help; F7 Exit;
1 Read; 2 Del 3; Save; 4 Info; 5 Group; 6 Mail Msg; 7 Phone Msg; 8 Folders: 1
```

Most of the time, you will use e-mail to send short, work-related memos to your co-workers. The computer delivers the e-mail message almost instantly, whether you are writing to 1 person or 30 people.

Some companies use computer menus to simplify your communication choices. You might choose **e-mail** as one option to send or find information. You might also choose **fax** or **voice mail.** You also might use the Internet to find information about a new technology by connecting to a computer outside your company, perhaps a government agency or a university library.

Most e-mail programs have some sort of "in box" for incoming messages and an "out box" for your outgoing messages. Exhibit 6.1 shows a "split screen" with the in box at the top of the screen and the out box at the bottom of the screen.

You can sometimes decide what to do with a message (read it, forward it to someone else, delete it, or respond to it), just by reading the subject line. Many routine messages may just be read and deleted. Others require some action. You may need to write a response or forward the message to someone else. You could also print out a "hard copy" of any message you received.

The bottom half of the e-mail index screen shows an out box, which lists all the outgoing messages this worker has sent. You could print these messages, file them electronically onto a disk, or delete them.

Part II: Improving Writing Productivity

■ Exhibit 6.2   **Routine E-Mail Message**

```
Cindy Epps       05/19  10:12       Transfer $$ for ENG Classes
┌─ Items in Current Envelope ───────────────────────(Tab next item)─┐
│ MESSAGE                                                            │
From:      Cindy Epps (EPPSC)
To:        NELSOND
Date:      Thursday, May 19, 1994 10:13 am
Subject: Transfer $$ for ENG Classes
Budget numbers to use to transfer the '94-'95 PT $ for replacements/ENG207-8-9:
40000-4430-16010-44434
***Would you please send me a copy of the transfer documentation for my
records. . . . THANK YOU!
CC:        CAMPB

Ctrl-F4 Move to Folder; Alt-F7 Launch Application;
1 Next (Read); 2 Delete; 3 Save; 4 Info; 5 Previous; 6 Forward; 7 Reply;
```

Exhibit 6.2 is an example of a routine message confirming approval of a purchase. Note that this e-mail message is easy to act upon, and it is easy to read. It includes the necessary information about account numbers so the reader knows exactly what to do. The task can be completed without further communication. Notice also that this e-mail message includes both upper- and lower-case letters. Messages in all capital letters are hard to read!

As employees and managers become more comfortable with using e-mail, more information finds its way onto the computer screen. Two examples of Pacificorp's routine e-mails, a daily "News Hotline" and an excerpt from a press release to all employees, are shown in Exhibits 6.3 and 6.4.

■ Exhibit 6.3   **Daily News Hotline**

```
PACIFICORP NEWS HOTLINE

TUESDAY, JANUARY 9, 1996

   DUE TO EXTREME WINTER WEATHER CONDITIONS ON THE EAST COAST—WHERE PACIFICORP'S
K PLUSLINE IS LOCATED — THE K PLUS BENEFIT SERVICE REPRESENTATIVES WERE NOT AVAILABLE
TO ANSWER CALLS YESTERDAY. DEPENDING ON STORM CONDITIONS, THEY MAY NOT BE AVAILABLE
TODAY.

   OTHERWISE, THE K PLUSLINE FUNCTIONED NORMALLY ON MONDAY. ALL REGULAR TRANSACTIONS
THAT CAN BE ENTERED THROUGH THE AUTOMATED PHONE SYSTEM—LIKE ACCOUNT BALANCES, TRANS-
FERRING BETWEEN FUNDS AND LOAN APPLICATIONS—WERE NOT SUSPENDED. HOWEVER, DUE TO THE FACT
THAT THE FUND PRICES WERE NOT AVAILABLE FROM ALL MUTUAL FUND MANAGERS, ALL INVESTMENT
FUND TRANSFER REQUESTS WHICH WOULD NORMALLY BE PROCESSED BASED ON MONDAY'S FUND PRICES
WILL BE PROCESSED BASED ON TUESDAY'S PRICES.

LOGOFF — PF2     PAGE FORWARD — PF8     LIST OF OTHER DAYS - PF5
```

Source: Pacificorp e-mail, 1996. Used by permission.

Australia Utility Purchase — Exhibit 6.4

Nov. 15, 1995

To: Officers, AVPs, Directors, Managers & Supervisors
From: Corporate Communications

RE: Australia utility purchase

Please share the following information with your staff and others who may not receive e-mail messages.

Pacificorp announced tonight that the company successfully bid for Powercor, a local electricity distribution company in the State of Victoria, Australia. Below is the news release distributed to the media.

Additional information has been faxed to field offices and delivered to Portland and Salt Lake City headquarters locations. Please check the Hotline screen on the IMS system of the mainframe for further details as well.

A telephone conference call will be scheduled later this week for all employees. The time of the call and instructions for joining the teleconference are coming via separate e-mail, or can be found on the hotline.

Thank you

News Release:
FOR IMMEDIATE RELEASE
November 15, 1995

PacifiCorp (NYSE, PPW) announced today that its wholly owned subsidiaries, Pacificorp Holdings, Inc., and Pacificorp Australia Holdings Pty Ltd., have agreed to purchase Powercor, an electric utility in southeast Australia, for approximately $1.6 billion.

Source: Pacificorp e-mail, 1996. Used by permission.

Most workers use e-mail to solve routine problems. When you write an e-mail message, you'll need to include *who, what, where, how,* and *why*—so that your reader can act on your e-mail without returning to you or someone else for further clarification.

Understanding E-Mail — Exercise 6.1

Based on the experiences you have had with e-mail, write a brief response to each of the following questions, discussing your answers with your discussion group. Try to come to some consensus about the value of e-mail in your group and summarize your findings for the class.

- What kinds of communications do you think e-mail should be used for?
- How would e-mail support day-to-day work? Would e-mail be useful in an emergency? Give an example of two or three such emergencies.
- Consider a typical company in your field. How would e-mail be used and why?
- How does an industry's or a company's way of getting the work done influence how e-mail is used?
- What other questions should be asked about e-mail?

180 Part II: Improving Writing Productivity

WORKING WITH E-MAIL DAY TO DAY

Why do people like to use e-mail? The biggest strength of an e-mail system is that workers can send and receive short memo-like messages within seconds. Thus, workers don't have to play telephone tag or wait for a return telephone call to get a critical piece of information, and with some programs, it's easy to get a printout of the e-mail message.

You'll see all kinds of e-mail messages. Many e-mail messages are requests for information—and usually your response is needed right away! The following are some other types of e-mail messages:

- Routine requests or reminders about meetings or key deadlines.

- General announcements of companywide information or policies.

- Personal messages (some companies limit this type of message).

- Rough drafts that require your response.

- Networked messages to work groups (the same message is sent to the entire work team at the same time).

Good communication means staying in touch and staying on top of all the information that you need to carry out your job.

You can develop your own system for dealing with information overload by asking the following questions when you look at the index for your inbox:

- **What** is this e-mail about specifically?

- **How** urgent is this e-mail?

- **What priority** does it have over other work?

- **Who** sent this e-mail? **What** are their deadlines?

- **What** must I do to respond? **How much time** will it take?

You will see the same kind of problems with e-mail that you encounter with any other kind of communication. However, when mistakes are made via e-mail, they can occur companywide! Some problems are related to the person sending the e-mail, other problems can be linked to the person receiving the e-mail, and still other problems are related to the medium of e-mail itself.

Problems Related to Receivers

Suppose someone at your employer grows organic apples and wants to sell them to co-workers at a deeply discounted price. Unless your employer has a policy restricting such personal advertisements, your co-worker can, with the flick of

MANAGING E-MAIL

How do we manage information overload? One successful way is to just use a systematic way to respond to your e-mail.

- Set aside a specific time to read incoming mail.
- Use the index to preview your e-mail. If you see junk mail, delete it!
- Skim read your e-mail with a sense of your priorities in mind.
- Decide whether you want to act now, act today, or delete a message.

What will you do with those e-mails you send and receive? Should you delete them? Print them? Save them? Set up a folder system for them? How often will you "clean out" your in box and out box by deleting, saving, or printing "old" incoming messages? Keep in mind that your e-mail messages take up space on the company network of computers.

Try to set aside a specific time, either every day or about once a week, to clean up your e-mail messages. As you gain experience in working with e-mail (and with the specific program your employer uses), you'll develop more skills in using it, and there will be more sophisticated e-mail programs to use.

a command key, "broadcast" this personal ad to everyone in the company via e-mail, increasing the number of messages waiting for your action.

What's in our future? Will video junk mail that beams out TV-like commercials invade the workplace? Your strategy should be to develop a policy that will make e-mail at your company an efficient resource.

Many workers report being overwhelmed by the sheer volume of messages they receive, whether by voice mail, fax, or e-mail. Many employees receive 30-60 messages a day. We can expect the volume of incoming messages to increase. We will need to manage and respond to each message, whether important or trivial. Learning how to delete nonessential e-mail messages without wasting time or without accidentally deleting important e-mail messages is a valuable skill. E-mail programs that help us to screen, preview, or filter our incoming mail—without excluding important information—will be in demand.

Most workers check their e-mail regularly throughout the day simply by turning on their computers and then working through the new messages, that is, answering, filing, deleting, and acting. However, not all workers check their e-mail frequently. A few workers may feel overwhelmed by the volume of messages they receive. To avoid seeing all this new work, some workers may not even turn on their computers.

If you are sending critical information by e-mail *and* you are not sure the person will be receiving your e-mail, take the extra time to call your co-worker on the phone. Leave a message that asks the person to check e-mail, especially if your co-worker needs to take action by a specific time or day.

Many people think of e-mail as a way to reduce the flow of paper. However, files saved electronically can get lost. Most programs allow you to print out paper copies of important messages or "download" (save) them on a disk.

Your challenge as a workplace writer is to deliver your work within tighter deadlines, in spite of increasingly complex tasks, processes, and information. You can achieve this balance by recognizing that technologies will change and by developing your own systematic way of working with whatever new technologies evolve.

Problems Related to Senders

Because you may find it easy to respond to e-mail "instantly," your error rate may go up. The second after you press the SEND button, you may discover you didn't include an important piece of information!

Exercise 6.2 Developing a Process for Reading E-Mail

Today is September 14th. You have just logged on to your computer after being away for one week, and you find 75 e-mail messages. Describe how you would process these messages efficiently. The following excerpt of your index for your in box shows a few of your messages:

```
In Box                             Envelopes: 75
Shelly Chu      09/04 08:00   Giorgio order
Paula Smith     09/04 08:04   Payroll authorization
Peter Mosau     09/04 08:11   Request for backorder
R O'Donnell     09/04 08:15   Vacation request form—reply req
Rita Sanchez    09/04 08:22   Computer services usage report
R O'Donnell     09/04 08:27   Welcome to new employees
Sam Saunders    09/04 08:44   Refrigerator for sale (PAID AD)
Peter Mosau     09/04 09:00   Approve Printer Purchase Order
Shelley Chu     09/04 09:00   Giorgio order
Rita Sanchez    09/04 09:00   Computer downtime scheduled TODAY
```

Answer the following questions before coming to class:

- Which is the clearest subject line? Which subject line is most difficult to understand?
- Which messages have top priority? Which can be deleted without reading them? How did you know?
- How many different types of messages are included in this excerpt? Write down at least four types of messages.
- If you are a new employee, which messages are the most important? If you are working for the purchasing department, which messages are most important?

After you have finished your analysis, work with a partner to draft a list of suggestions for using e-mail, which you could give to a new employee of this company. Use any experience you have had with e-mail to help you write these suggestions. Turn your finished work in to your instructor after sharing it with the class or your discussion group.

Sending out two or three follow-up messages to correct your first e-mail can result in more clutter and even more e-mail messages as people help you correct the problem. If your receiver has already acted on your message, the misinformation spreads further throughout the organization, costing more time. Your strategy should be to plan your e-mail messages carefully and double-check them for correctness and completeness.

You may send out an angry response to an e-mail message because you are upset about a situation. You read and you react instantly. The instant, angry response is so common that it's called "flaming" your reader. However, such messages, like any letter, memo, or report sent out with negative information, can also affect your co-workers and supervisors negatively. Your strategy should be to think about the emotional impact of your e-mail before sending it. If your e-mail system has a feature that allows you to pull back "unopened" mail, use it when you need to!

Writing an E-Mail Message   Exercise 6.3

You've just worked 20 hours of overtime in three days to make sure an important order is completed. Your immediate supervisor is grateful, but you are tired and you want to go home on time tonight. You check your e-mail messages one last time and find the following message.

```
To: All Personnel Working Overtime This Week
From: Paul Rogers, Accounting
Subject: Overtime Overruns

It is inexcusable that the overtime budget has escalated this month.
Bring all project records and time sheets to a meeting this afternoon
at 4:45 PM to justify your overtime. No excuses! Confirm your attendance
by return e-mail.
```

1. Draft your e-mail response to this message. Bring it to class to share with your discussion group.

2. As a group, first list the criteria a good e-mail message should meet. Remember to consider content, organization, style, and format—as well as audience and purpose.

3. Select which draft e-mail (and strategy) meets these criteria *and* solves both the reader's and the writer's problem.

4. Share your group's selection and findings with the class and turn in your group work with rough drafts to your instructor.

Problems Related to E-Mail Technology

Problems you may find related to e-mail technology tend to fall into two categories: technical or ethical.

Technical Problems. We can expect some technical problems to gradually subside through improved software. For example, some e-mail programs automatically ring a bell every time an incoming message comes in on your computer. No

184 Part II: Improving Writing Productivity

matter what else you may be doing, the bell interrupts your work. Although not all e-mail messages are urgent, when the bell rings, you will react as if you were receiving an important message. Your strategy should be to disable the bell!

Another technical limitation occurs if all your co-workers are not "wired" into the e-mail system. How do you quickly know who's on and who's not? Your strategy should be to learn to use the features that will help you to deliver mail to different groups of readers. Some e-mail programs allow you to set up your own personal "groups" of employees so you can send e-mail to several people at once just by typing in a code name. For those not on the e-mail system, make sure these readers get paper copy by mail or fax.

Another important technical limitation can occur when the power goes off. How do you retrieve that critical background information when you can't even turn your computer on? If important information is stored without a printout (hard copy), and there's a power outage, your key information may be inaccessible for minutes, hours, or days. Sometimes when there are outages, messages can be lost in "cyberspace." Your strategy should be to print out important messages.

Ethical Issues. Ethical issues will take longer to resolve, and they may involve quite a bit of discussion inside and outside your company. For example, how private is your e-mail? Co-workers can be surprised to discover that their employers can monitor all incoming and outgoing e-mail messages. Since e-mail tends to be less formal, writers sometimes cross over the line from professional to personal writing. A "funny" joke sent to a good friend will not seem so funny if it's accidentally sent to all employees.

Privacy and security are very real issues, especially in larger companies with hundreds of employees and perhaps global networks for their communications systems. Proposed laws may require companies to inform their employees if their e-mail is being monitored under guidelines passed by Congress with the Privacy for Consumers and Workers Act (1993).

Another issue is who "owns" the e-mail message? You as the sender? You as the receiver? Perhaps the employer owns the message, since it was created and sent through company facilities. These definitions will be decided in the courts, and careful attention will be paid to laws defining copyright, privacy, and employee/employer rights of ownership.

WHAT WRITING SKILLS ARE NEEDED TO USE E-MAIL?

Even though you will need some keyboarding skills to enter text and numbers, and you will need to learn how to use the e-mail system itself, writing "good" e-mail messages means using the same writing guidelines you have seen throughout this book: **conciseness, clarity, correctness,** and **courteousness.** These are all essential to your writing effective e-mail messages.

Drafting an E-mail Policy   Exercise 6.4

Your employer has had an e-mail system for approximately two years. Everyone really likes the quicker access to information, but people are starting to complain about the volume of e-mail.

1. You (and your work team) have been asked to draft five e-mail guidelines for your company that will be policy into the next decade. Work with a partner to first pick a company or industry you are familiar with. Then, draft a list of five possible guidelines that will set company policy for e-mail use.

2. Exchange your guidelines with one other team, and then discuss whether the guidelines reflect the company's "culture." List the strengths and weaknesses of these guidelines and return them to the other team.

3. Once you have your team guidelines back, review needed changes. Write a brief response to the peer review, and, if asked to do so by your instructor, revise your list into a final recommendation memo following the Editor's Checklist for Writing Memos (in Appendix A).

When you write concisely, sometimes the "tone" of the message seems abrupt. Over time, such messages can affect the "social glue" that holds work teams together. Therefore, some workplace writers make sure they talk to their co-workers informally every day. Others sparingly use *emoticons,* little faces made with characters to add expression to their e-mail. Some common ones are :) happiness, :(sadness, and :0 surprise.

One of the questions to ask yourself when you look at your screen right before you press the SEND button should be: Will my message seem too abrupt or cold? When you write and send an e-mail message, your message arrives almost instantly. You usually do it now and do it fast! However, your strategy should be to use a writing process that helps you to **plan, organize, write,** and **double-check** your e-mail message *before* you press that SEND key. Use the Editor's Checklist that follows.

EDITOR'S CHECKLIST FOR WRITING E-MAIL MESSAGES

Plan:

Think about both your purpose and your audience's purpose before you begin writing.

- What information do they need?
- What order do they need this information in?
- Are there any deadlines or special requirements?

Ask yourself why you're using e-mail. Would a phone call be faster? Would a letter or memo document the problem or action more permanently?

Organize:

Put the main idea first. What key information must be included?

- Use a "key word" outline to check the order and completeness of your e-mail.

186 Part II: Improving Writing Productivity

Write:

Keep it short! Is your e-mail as concise as possible? Is it also clear, courteous, complete, and correct?

- Solve the reader's problem. If follow-up action is needed, state it! Is there a benefit to your reader? If you can tell this to the reader concisely, do it!
- Be specific. Include dates, times, and deadlines. Name people, places, projects, and goals.

Check:

- **Content:** Is your information complete and accurate for this audience and this purpose?
- **Style:** Is your tone courteous? Concise? Clear?
- **Format:** Have you used the half-page size most effectively? Is your information easy to read? Did you use a list to highlight important or statistical information?
- **Proofreading:** Did you proofread punctuation, spelling, grammar, and for typos?

Because some e-mail systems do not allow employees to look back at a message when they are writing, important information may be left out. Always write so that your e-mail can be read on its own—without supporting documents. Finally:

- Think of e-mail as public. Do you want everyone to read your e-mail?

WHAT'S COMING NEXT?

Although this chapter has focused on e-mail as one key communication tool, other technologies are being developed and linked together to improve communications.

While some may argue that nothing will replace face-to-face talking, many of us will use fax machines, voice-mail systems, video phones, bulletin boards, video conferences, and bar-coding systems to track all kinds of work in progress. Some of our co-workers may be "telecommuters," people who work at home but who use the computer to stay "connected" to the workplace.

We can successfully adapt to these new technologies by keeping current with what's happening in our field, by talking to people who are currently working with these new technologies, and by experimenting with new techniques as they become available.

■ CONCEPT REVIEW

Can you define the following key concepts from Chapter 6 without referring to the chapter? Write your own definitions in a journal entry or review them out loud.

e-mail   computer bulletin board

e-mail address   modem

Chapter 6: Communicating with E-Mail   **187**

in box and out box   telecommuting

voice mail   junk mail

WRITING SKILLS REVIEW ■

To build your skills in editing and proofreading, continue your systematic review of
Appendix C: Writing Skills Review. Your assignment:

__ Read "Revising for Effective Verb Use" in Appendix C.

__ Complete Exercises C.31 through C.36 in "Revising for Effective Verb Use." Also
complete the "Summary Exercise on Verb Review," and check your answers at the
back of Appendix C.

__ Prepare a paragraph summarizing your progress for your instructor or use the form
your instructor provides to report your progress. Be sure to highlight any areas you
need to further review.

SUMMARY EXERCISE ■

What's Wrong with This E-Mail?

Review the chapter reading, consider the editor's checklist for writing e-mail messages,
and then complete the following steps, working alone or with a partner.

1. Read the following e-mail message and underline the key problems the writer wants
solved. Then, list the revisions you want the writer to make before sending this e-mail.
Use the checklist on writing e-mail messages to make sure you have checked all aspects
of this message!

2. Review your editing suggestions with your discussion group. If your instructor
requests a revision, prepare a new version of this e-mail that will solve the problem without alienating either Caroline or Cheryl.

```
To: Cheryl Caplan, Operating Shift Supervisor
From: Jackie Hernandez, Personnel Supervisor
Subject: Reprimand for Caroline Smith

I am requiring that you immediately meet with Ms. Smith to reprimand her
about her inadequate completion of forms and adherence to working hours. She
definitely needs to improve her use of the time clock and her attitude. On
September 5, 6, and 15, she did not complete the hourly check-in sheet at
all. It looks like she worked 4.5 hours, 9.0 hours, 8.4 hours and 6.0 hours,
respectively. On September 9, 13, and 22, she did not sign her hours worked
report. When I brought this to her attention she told me to talk with her
supervisor because she was only doing this the way she had been told. I
assumed this meant you. You should know how to complete the forms properly.
How do I know what to pay her if her correct hours are not reported and
validated??? Your signature was missing also. Please report your satisfactory
resolution of this serious problem to the undersigned.
```

Part II: Improving Writing Productivity

ASSIGNMENTS

Assignment 1: Research the Impact of Electronic Mail.

Working alone or with another person, find two or three articles about the effects electronic mail has on companies. Look for articles that explain the skills needed for employees to successfully use e-mail, improve productivity, or show how e-mail systems are being used by different kinds of employers. Summarize each article and draw conclusions from the information you have gathered. Present the results of your research in a memo to the class.

Assignment 2: Research Future Communications Skills.

Working alone or with another person, find two or three articles about trends in communications technologies. Look for articles that explain the next level of services or products. Use Craig Crossman's article, "Try Tracking Your Parcels," reprinted here, to get started. Then, working with a partner, write a memo that justifies converting to an e-mail system for a busy hospital, a growing mid-sized electronics firm, or a company of your choice. Include the following in your memo:

- Specify the benefits and weaknesses of such a system. Emphasize what is best for the audience you select.
- What problems would such a system solve or create? Point out some safeguards that will need to be taken.
- List performance criteria a new system must meet. Consider initial cost, maintenance, safety, security, and any other factors you think are important for this industry.

Suggestions on getting started: Work with another person to draft an outline, draft questions that will be answered in each section of the outline, and then draft the answers to each question. After you are satisfied with your rough draft, share it with another group for feedback and revise, turning the final copy in to your instructor.

TRY TRACKING YOUR PARCELS

Craig Crossman

Remember when fax machines were a novelty, a few short years ago? Now they're everywhere. You can even call a restaurant, and they'll be happy to fax you a take-out menu. And you can then fax your order to the restaurant!

It's smart business to utilize communications tools to enhance your company's performance. I recently saw a business card with the usual name, address, phone and fax number. What made me smile was that it also listed the company's computer bulletin board phone number. That made me realize computer communications are becoming as mainstream as faxes. Having a modem is as necessary as having a printer on most business and personal computers. A company that can

Chapter 6: Communicating with E-Mail **189**

make its services available to those millions of computers and modems could open up another market.

Federal Express has made the savvy business decision to do just that. It's offering free Windows, DOS, and Macintosh software that allows your computer to access FedEx's package-tracking system. Not only is the software free, the service is free, too.

FedEx Tracking Software lets you see the disposition of any package. You fill in details such as your name and FedEx account number. The program automatically dials an 800 number and registers you with the Federal Express system. Now, you're ready to see how your packages are doing. Simply enter a package's air bill number (you may track up to 14 packages at one time) and press the "track" button.

In a few moments, you see a summary of your packages. Clicking on the "detail" button displays information on each package, one at a time, along with details of the important steps in its delivery. Details include delivery priority, where and when it was picked up, the origin location, what hub processed it, when it was put on the van, delivery date and time, even who signed for the package. You can print everything out, or save this information to disk.

Granted, you could phone a Federal Express customer-service representative who would give you the same information, although my experience is that they usually won't offer details about the path of a package.

To order the software, call (800) 817-8300.

Source: Craig Crossman, "Try Tracking Your Parcels," *The Miami Herald* (January 23, 1994), p. C5. Reprinted with permission of *The Miami Herald*.

JOURNAL ENTRY ■

In this journal entry, you will explore communications style. Make a rough draft list of several instances when you received "bad" news and several instances when you received "good" news—in a letter or memo, face-to-face, or by e-mail. Consider news from friends, family, supervisors, and co-workers.

Pick two of these incidents to write about more fully.

- What happened exactly?
- What was your reaction—at first and then later, on reflection. Did you need to go back to the "sender" for more information?
- What kind of follow-up did you do, once you had the news?
- Exactly how did you get the additional information you needed?
- Was your relationship with the sender changed because of the medium he or she used to send information to you?
- Was the information easier to understand when you were talking face-to-face? Why?

Conclude by summarizing a few ways that communication by letter, memo, or computer could be improved by using techniques we use in face-to-face communication.

BUILDING COMMUNICATION SKILLS

Part III will help you to refine your writing skills using a variety of formats—from informative memos and letters to instructions to persuasive letters and proposals.

You will develop your writing process (planning, writing, and revising), and you will practice using criteria to evaluate your work (thoughtful content, logical organization, a clear and concise writing style, and a readable format). You will also review strategies for successfully participating in and leading meetings.

7

Writing for Co-Workers

Informing and Instructing

Chapter PREVIEW

The chapter will help you with communication skills that are used every day in the workplace: planning, giving, and following up on instructions.

Some information comes from our supervisors or senior managers as they explain the company's policies and procedures—why and how to follow routine administrative guidelines, like whether or not you can accept a gift from a vendor (an example of a policy) or what forms you need to fill out to qualify for additional insurance (an example of a routine procedure). You may be asked to write routine memos about company policies or procedures.

But not all information on how we carry out our jobs is written down. This chapter will clarify how to write informative memos that explain policies and procedures, how to prepare formal instructions on how to carry out a task, and how to prepare checklists to make sure all steps in a task are completed.

Journal ENTRIES

Begin by thinking about how instructions have been given to you and how you give instructions to others. Write a short one-paragraph response to any of to following questions in your class journal or prepare answers for class discussion, as directed by your instructor. .

1. What qualities do you admire in someone who gives you "good" instructions?
2. What strategies do you use to get others to follow your directions? What do you do if your co-workers or friends do not want to carry out the task?
3. What strategies would you use to give instructions to a co-worker who could not read? Who had difficulty understanding English?
4. When should instructions be written? When should they be spoken?
5. After a task has been carried out, what kind of feedback would help the worker carry out the task correctly the next time?
6. Have you had any experiences in working with teams to plan, carry out, and then evaluate tasks on the job? How did this work? What were some strengths of this approach? Were there any weaknesses?

Chapter READING

In this next article, Janine S. Pouliot describes how some companies have changed their training to make sure employees understand key health and safety information.

A VISUAL APPROACH TO EMPLOYEE SAFETY

Janine S. Pouliot

Many of the 38 employees at the Pettit National Ice Center in Milwaukee routinely handle hazardous chemicals and solvents to create smooth-as-glass ice for such skaters as Olympic gold-medalists Bonnie Blair and Dan Jansen.

For safety's sake, it's essential that the workers know about toxic fumes and respirators involved in the ammonia-based ice-making system.

But Bill Greinke, executive director of the Olympic training center, wasn't happy with the standard written safety materials about the system. Experience showed a highly technical message often doesn't get through to workers who need it.

So when it came time to institute his own program, Greinke decided to do things differently. Instead of relying solely on written safety manuals, he also used videos.

"The absolutely best way to educate the work force about safety is in the medium in which they're most at home," say Glenn Gronitz, president of Quality Safety Management, a Milwaukee consulting firm and designer of Greinke's program. "Television and videos are what people know."

Alternative safety-training methods are cropping up in many businesses in the form of board games, training by consensus, and hands-on sessions, as well as videos.

Most businesses still rely on the difficult-to-read Material Safety Data Sheets (MSDS) to comply with federally required safety training, Greinke points out. "But there's no standard format," he says. "They're prepared by the manufacturers of hazardous products, so each MSDS is entirely different. You really have to hunt to find the basic information."

Finding alternatives to written safety materials is becoming increasingly necessary for companies today. About 90 million Americans demonstrate low competency in basic reading, math, and reasoning, according to a 1993 study by the U.S. Department of Education. Data sheets such as the MSDS generally are written at a high-school or college level.

In addition, a recent Census Bureau report indicates the number of U.S. residents for whom English is a foreign language is nearly 32 million. Yet common safety-training programs are typically geared to an English-speaking audience.

There is a gap in communications in the workplace for safety training, according to Elizabeth Szudy, co-author with Michele Gonzalez Arroyo of *The Right To Understand: Linking Literacy to Health and Safety Training.* As part of a study conducted by the Labor Occupational Health Program, affiliated with the School of Public Health at the University of California at Berkeley, Szudy and Arroyo, safety trainers in the program, assessed current safety materials aimed at today's work force.

The authors concluded that most information on health and safety is ineffective because of the way it is written and presented. Tiny print, complex wording, and few illustrations often leave many workers confused.

"Information should be targeted to a wide audience with varying language skill levels," according to Szudy. "Employers should just assume that some percentage of their employees are grappling with the written word."

Here are 10 techniques recommended by experts and business people for conveying safety information without relying strictly on written materials:

Use visual aids.   Gronitz, of Quality Safety Management, used generic safety-training videos at the Pettit National Ice Center to introduce the subject of safety. This approach comes from experience: "For 20 years, I've worked with supervisors who can't read or write."

He also uses widely available standard safety videos and customizes them to meet the particular firm's needs. "The more important the point is, the more often we repeat it," he says.

Foster Participatory Training.   When Mark TenBrink, environmental manager of Micro Metallics, in San Jose, Calif., conducted safety training, he used easy-to-follow participatory methods to sidestep language and literacy problems. Micro Metallics determines the recycling value of scrap metal in used manufacturing and electronic equipment.

The firm's workers were divided into groups of two to four and then were given a large sheet of paper with the outline of the facility on it. "They were asked to indicate a fire hazard with a red X, or the location of poisons with a blue X, and so on," TenBrink says. A small X symbolized a minor danger; a large X meant a more serious hazard.

From their daily encounters with potential hazards, employees already knew where to place many—but not all—of the marks. Different employees had different levels of knowledge about the location of potential hazards. The exercise helped identify gaps in knowledge.

Play a Game.   The familiar activity of a board game also served to explain safety concepts at Micro Metallics. The center of the board displayed the outline of a human body depicting internal organs. Players moved around the board by selecting cards that asked about the effects of various chemicals on organs. A volunteer comfortable with reading aloud recited the questions to the team.

The object of this kind of friendly, noncompetitive play is to encourage workers to take part without feeling intimidated, Szudy says.

Turn Tests Into Learning Sessions.   "Certain [federal] regulations dictate that you test after a training course," says Dana Zanone, environmental-affairs manager of Myers Container Corp., in Richmond, Calif. "How you test is left up to you."

Zanone transforms the exams into an opportunity to reinforce safety messages covered during training. First, the material is presented in English and Spanish. A review is conducted immediately, and "we literally give them the answers," Zanone says. Then he gives the test as a collective exercise. "We read the question aloud, and the group comes to consensus on the answer," Zanone says. "That way I'm assured everyone understands."

Use a Translator.   A translator is essential in reaching a multilingual work force. Robert Borovicka, plant manager at Plastonics, a plastic-coatings business in Hartford, Conn., recruits an employee fluent in Spanish and English to provide translations. Of Borovicka's 25 employees, more than half use English as a second language, and several speak no English.

Offer the Real Thing.   Quality Safety Management's Gronitz supplements presentations with individualized hands-on training. When showing employees how to use a respirator around hazardous chemicals, he asks each worker to try on the apparatus. He then sprays vegetable oil into the air. If employees can smell or taste the mist, they know they're not using the gear properly.

Show Workers You Care.   When Cary Grobstein was vice president of sales administration at Cardinal Color, Inc., a paint company in Paterson, N.J., he wanted his workers to know he wouldn't require them to do anything he wouldn't do himself. Se he poured the nontoxic chemicals that make up paint onto his hands to prove they were safe to handle. But he quickly explained that they were hazardous if ingested.

"It's important that workers understand management is sensitive to their concerns about handling chemicals," he says. Grobstein, who now owns LBL Sales, a chemical brokerage firm also in Paterson, often works beside employees to demonstrate how to handle chemicals.

Offer Positive Reinforcement.   Roger Sheaffer, owner of Sheaffer, Allan and Hoyme Safety Consultants Inc., in Addison, Texas, says successful accident-prevention programs require upbeat reinforcement. "Many of these workers are doing hard manual labor and are unaccustomed to receiving positive feedback," he says. And he frequently compliments employees for following safety procedures demonstrated in training.

Train in Context.   Rather than suddenly springing safety training on workers, Laurie Kellogg, health and safety specialist for the International Ladies' Garment Workers' Union in New York, recommends placing the discussion in a comprehensive framework.

Kellogg, who conducts training for entrepreneurial businesses with cross-cultural, multilingual work forces, suggests letting people know ahead of time what will happen. She begins her sessions with a clear introduction of the topic

Chapter 7: Writing for Co-Workers   **197**

and why it's important to each employee. "They're not going to take you seriously if you just throw information at them," she emphasizes.

Create a Sense of "Ownership" in Safety.   Greinke of the Milwaukee skating center gave an employee safety team authority to resolve hazards in the work-place—not just bring them to management's attention. "I oversee what they do, but they take the initiative and have their own small budget," he says. "Because they're directly involved, they're more responsive to the solution."

For example, the team recognized a concrete and metal stairway posed a danger to workers with wet shoes. They reacted by adding a sandpaper strip to the edge of each step. "We were looking into purchasing expensive treads, but the employees' solution was just as effective," Greinke says.

Clearly, there are many ways for a company to convey information about hazards and to cultivate employee commitment to safety. It's also a safe bet that relying solely on written materials won't get the job done.

"Offer a range of approaches, and let the employees select what's best for them," Szudy advises. "It may be a little extra work, but the payoff is the absolute assurance that you're getting your message across."

Source: Janine Pouliot, "A Visual Approach to Employee Safety," *Nation's Business* (February 1995), pp. 25–26. Excerpted by permission, *Nation's Business,* February 1995. Copyright 1995, U.S. Chamber of Commerce.

Analyzing THE READING

To build your reading skills, use the following questions to analyze the article and to develop your reaction to it. Write a one-page entry in your class journal or prepare your answers for class discussion.

Reading for Content

1. Why are companies becoming more involved in safety training their employees?

2. What kind of instructions are needed to protect the safety of workers? When? How often?

3. How is safety training currently being delivered to employees—by most companies and by a few innovative companies? Which ways do you think will help employees most? Which ways do you think may be more costly in terms of time or staff?

Reading for Reaction

1. Pouliot says that a standard technical manual "often doesn't get through to workers who need it" (25). Why do you think workers resist written instruction? What strategies could you use to overcome this resistance?

2. Why are workers confused by most current safety-training materials? What strategies could you use as a writer to improve written training materials?

Part III: Building Communication Skills

3. Several techniques are presented that companies could use to improve their training. Which two do you think would be most successful in a small company? Why? Which two would be the least successful in a small company? Why?

WRITING FOR CO-WORKERS

When we write for our co-workers, most of the time we are giving them information or directions so they can do a better job.

Your co-workers are most likely to speak the same language and to understand the day-to-day complexities of your job. They are also most likely to work very closely with you on larger projects.

The same guidelines that lead to good writing for your customers or your supervisors also apply when you write for your co-workers. Your co-workers will appreciate:

- **Thoughtful content** that is relevant to the problem being worked on.

- **Logical organization** that gives the reader information in the order she or he needs it.

- A **writing style** that presents information clearly, concisely, courteously, and completely—for this particular reader.

- A **document design** that uses both *layout* (how information is placed on the page) and *format* (how paragraphs, headings, and lists are used) to make the information easy to understand, easy to use, and easy to remember.

When co-workers are your primary audience, the **timeliness** of your work is also crucial. Nearly everyone has worked with someone who can't be trusted to complete work on time or who works right to the deadline, turning in work at the last minute that needs additional revision. You can work around this problem by building extra days into your work schedule, but these days can be expensive.

What motivates people to produce work at the highest quality and on time? The quality of your own work, your positive attitudes, and your generous support and encouragement of your co-workers can set an example for your co-workers and help build a productive work team.

Although you will find yourself writing persuasive memos and reports for your co-workers, this chapter emphasizes three of the most common kinds of writing for co-workers:

- **Informative memos** used by all levels of workers to solve problems.

- **Instructions** used by workers to learn new tasks or skills.

- **Checklists** used by knowledgeable workers for routine and complex tasks.

Analyzing a Task Exercise 7.1

Think about an important job you recently completed—either at work or at home. Your purpose in writing this description is to help others understand the decisions you made to work effectively on a difficult or challenging job and (if applicable) what impact this had on your co-workers or your employer. Read through the whole exercise first before you begin.

1. Answer the following two sets of questions as you prepare to write a description of this task and your analysis of what made your performance successful. Jot down key words to help you in writing your later draft. As you make notes in response to these questions, be specific and concise.

Describe the Job

- What was the job?
- Were other people involved? Was expensive equipment involved?
- What kind of direction or instructions were given to you?
- What was your knowledge of the task?
- Did you use a checklist or instructions?
- Was there a deadline or time limit to the task?
- Did someone have to "approve" your final work?
- Did you have any problems in successfully completing the job?

Analyze the Impact of This Job on You and Others

- How important was this task?
- How much responsibility was involved?
- What made the job successful?
- If there were problems, what factors contributed to the problems? How might they have been solved?
- As you think back over the job now, what could have improved your performance?

2. Review your answers to the previous questions. Now prepare a rough draft that summarizes your experience and share the draft with your discussion group.

3. Use the following criteria to give peer feedback to each member in your discussion group. Make any needed revisions and turn in your completed description to your instructor.

Focus questions for peer review: As you read and review the rough drafts within your group, remember to highlight the writing strengths and make suggestions on how the draft could be improved.

1. **Thoughtful content:** Can you easily tell what the task was? Were there problems? If so, were the solutions clearly presented? Do you need any additional information or analysis? Is the writing purpose clear? Are main ideas supported with enough facts, examples, or other specific description?

2. **Logical organization:** Is there an introduction, a discussion section, and a conclusion? Is there some sort of time order (then and now)? Did the writer use transitions to move smoothly between description and analysis?

3. **Writing style:** Is the information presented clearly? Concisely? Completely enough for you to understand it? Do you feel the tone is courteous?

4. **Document design:** Should changes be made in paragraph length, headings, or lists to make the information easier to understand, analyze, or remember? Does the writer change to a new paragraph each time a new topic is introduced?

200 Part III: Building Communication Skills

WRITING STRATEGY: WRITING MEMOS THAT INFORM

Often, you will need to prepare memos that tell people how to solve a problem or what to do in a particular situation. Sometimes you'll need to write memos that explain company policies or procedures, or you'll need to document that you have completed a task.

You'll need to present the information using a proper format, but the information must also be presented so that it is easy to understand. Your audience needs to use the information you are presenting, which highlights the most important writing challenge: How do I organize and present this information so that it is absolutely clear to all readers?

An effective memo that gives instructions will give the reader **what** information is needed, **when** it is needed, using a **style** and **document design** that is easy to understand and remember.

Exhibit 7.1 is an example of a patient's chart that documents exactly what was done with and for this particular patient. Notice how the chart shows who, what, when, and where. The reader can also see why the actions were taken. This chart guides the work that the health care team is providing to this particular patient, and the chart could be used later by the institution either to analyze patterns in care given or to legally validate treatment.

Ask these questions when you are writing to give information on policies or procedures or documenting tasks:

- **Who** is my reader?

- **Why** does my reader need this information?

- **What kind of information** does my reader need to understand what I have done or to complete the task?

- **Will the reader** need additional resources?

- **How, where,** and **when** will my reader use this information?

- **What further steps** should be taken to make sure the task has been completed correctly?

Informative memos are usually routine and follow a very specific organizational pattern. Read the two examples of informative memos presented in Exhibits 7.2 and 7.3. Notice how the writer uses a direct, "no nonsense" style and a very straightforward organizational pattern to present the instructions. As you read, consider these questions:

- Can you find an introduction?

- How much information is given in the discussion section?

Chapter 7: Writing for Co-Workers   **201**

Patient Progress Chart   Exhibit 7.1 ∎

Date	Time	Notes
11/22/94	1600	P#1: High risk for ineffective breathing pattern related to possible smoke inhalation.
		IP#1: Assessed respiratory rate and breath sounds q 1 hr. to R/O pulmonary edema and bronchospasm. Taught pt. how to perform deep-breathing and coughing exercises and taught use of incentive spirometer. O_2 applied @ 2 L/min via nasal cannula. EP#1: Pt. maintains patent airway and normal RR and depth. Pt. understands the importance of performing deep-breathing and coughing exercises q 1 hr. Pt. has normal ABG levels. ———— *Deborah Ryan, RN*
11/22/94	1600	P#2: High risk for decreased CO R/T reduced stroke volume as a result of fluid loss through burns. IP#2: Teach pt. to report any restlessness, diaphoresis, or light-headedness, which may indicate shock. Evaluate VS and hemodynamic readings at least q 2 hr. Monitor urine output q 1 hr. Monitor ABG levels. Provide and monitor I.V. therapy.
		EP#2: Pt. maintains normal VS and stable hemodynamic status. ABGs WNL. Pt. has adequate urine output. Pt. verbalizes signs and symptoms of shock. Pt. receiving adequate replacement through I.V. therapy. ———— *Deborah Ryan, RN*
11/22/94	1630	P#3: Pain related to second-degree burns over 20% of body.
		IP#3: Assess pain q 2 hr and medicate q 3 to 4 hr with morphine, as ordered.
		EP#3: Pt. reports a decrease in pain rating from 8 to 2 on a scale of 1 to 10, with 10 being the worst pain imaginable. *Deborah Ryan, RN*

Source: Edith McMahon et al. (eds), "Using the PIE Format," *Mastering Documentation.* Springhouse Pennsylvania: Springhouse Corporation (1995), p. 77. Used with permission from *Mastering Documentation,* copyright 1995, Springhouse Corporation. All rights reserved.

- Does the writer use a conclusion?
- What impact does the writer's tone have on the reader?
- How does the memo persuade the reader to take action?
- What format strategies are used to make this memo easier to read?

Notice that in the first memo on new tax guidelines, the audience is told what to do next, not asked. This arbitrary tone is often used to just get the job done.

Part III: Building Communication Skills

■ Exhibit 7.2 **Memo about Procedure**

ABC INTEROFFICE MEMO

To: Distribution List
From: Shirley Kyte
Date: March 22, 1996
Subject: IRS RULING—TRAVEL REIMBURSEMENT

On March 1, 1996, the IRS ruled (Opinion #345-QR) that all travel reimbursements must be included in the employee's tax forms as nontaxable compensation.

Please act immediately to add such reimbursements to the records of all employees to be included in the end-of-year reporting.

ert

Distribution:
A. Abrams
K. Chin
L. Jabbar
B. Kahookaulana

■ Exhibit 7.3 **Memo That Provides Information**

MEMORANDUM

To: All Employees
From: Brenda Middlecoff, Human Resources Director
Date: January 19, 1996
Subject: Promoting Opportunities for Women Abroad

The Women's Bureau of the U.S. Department of Labor develops policies that promote the welfare of working women worldwide. These international activities are conducted primarily through the Organization for Economic Cooperation and Development (OECD) and the United Nations.

If you want more information about the resources and expertise of this organization, brochures are available in the human resources department.

urs

In the second memo, the writer provides background information. Which of these two memos is most likely to be more useful?

Notice that both of these memos are very short! Not one extra word is used.

UNDERSTANDING THE PARTS TO AN INFORMATIVE MEMO

You will find three basic parts to any informative memo: an introduction, a discussion, and a conclusion.

The **introduction** to an informative memo is very important, because it gives the reader a reason or context for the information that follows. Ask the following questions any time you prepare an introduction:

1. What is the **purpose** of this memo? Can my reader identify the purpose from either the subject line or the first paragraph?

2. What is the **problem?** Can my reader understand the extent of the problem by reading the first or second paragraph of the memo?

3. Does my reader need any **background** on the problem?

4. What's in this memo? Have I included a **preview** of the main kinds of information in this memo?

The **discussion** section or sections either summarize or present the information needed. Sometimes lists or tables are used to present the information clearly. Sometimes supplementary sheets of information are attached to the memo to give the reader further information. If you have organized your information into different categories, you may use more than one section (a section typically starts with a heading and includes two or three paragraphs).

The key concerns you will have in preparing the discussion section are:

*   Does this information meet my reader's needs? Does it help to solve the problem?

*   Is this information correct? Is it complete?

*   Is this information presented as clearly and concisely as possible?

*   If requested, have I analyzed the information so the implications are clear?

*   Does my reader need any **background** information on the problem?

The **conclusion** section is read carefully by most readers. For this reason it is a very important part of any informative memo; however, sometimes writers will not include a conclusion, because they believe providing the information itself is enough to meet the reader's needs.

Part III: Building Communication Skills

Conclusions can highlight key information, make connections that may not be clear from the information itself, or state any action the writer wants the reader to take. Ask the following questions when you write the conclusion to an informative memo:

- Does my conclusion restate the purpose and the main findings of this information?

- If I request action, have I done so courteously? Have I included a deadline for the needed action?

- Have I thanked anyone who provided special services or facilities?

- If my reader has questions, does he or she know how to reach me easily?

- Have I buried my main points and conclusions in too many words or workplace cliches?

Although a good informative memo will typically include an introduction, a discussion section, and a conclusion, if you are writing informally (for example, to someone you know very well or your employer is a small company) or the information is very routine, you may skip using one or more of these parts of a memo.

■ EDITOR'S CHECKLIST FOR WRITING INFORMATIVE MEMOS

Use this checklist to help focus your revision before you complete your informative memo. Ask someone in your discussion group to act as your peer editor and to answer the following questions about your rough draft.

Encourage your peer editor either to set aside about 20 minutes to work through the list with you, discussing each question, or to write out sentence-length responses to each question. Check Appendix A: Letters and Memos for additional guidelines on writing memos.

Check audience and purpose:

___ 1. Who is the audience for this information? Is there more than one audience?

___ 2. Why do they need this information?

___ 3. Will my audience accept or resist this information?

___ 4. How exactly will this information be used?

___ 5. What do I want my audience to do with this information?

Check content and organization:

___ 1. Can the reader tell the topic and the purpose of this memo from the subject line and the introduction?

___ 2. Does the reader need more background information?

___ 3. If tasks are given, are they presented in the order they should be completed?

___ 4. Is the information presented in an order that makes sense to the reader?

___ 5. Does the conclusion emphasize any action the reader needs to take?

___ 6. Does this memo say where the reader can get help or additional resources?

Check style and document design:

___ 1. Is the information presented as clearly and concisely as possible?

___ 2. Is the tone of the memo courteous and helpful?

___ 3. Is the vocabulary easy to understand for all readers?

Check document design:

___ 1. Do headings help the reader by highlighting key steps or concepts?

___ 2. Are all the paragraphs short and easy to read?

___ 3. Are lists used to highlight key information?

___ 4. Does the document follow standard memo format?

Allow time for editing and proofreading:

___ 1. Has this memo been edited for correct spelling, punctuation, and grammar?

___ 2. Has the memo been proofread for typing errors?

What's Wrong With This Memo?   Exercise 7.2

1. Read and analyze the following informative memo, using guidelines from this chapter. Make a list of suggestions for the writer, considering content, organization, style, and format.

2. Review your list with your discussion group, making any revisions or additions needed. Turn in your final analysis to your instructor.

MEMO TO: All employees

FROM: Franklin Perry

SUBJECT: USE OF TIME CLOCK

I am sorry to announce that beginning next Monday, February 26th, all employees must punch in and punch out whenever entering or leaving the workplace.

It has come to my attention that many of you are lying on your time cards, and I cannot tolerate this type of cheating; therefore, I simply have no recourse but to institute a time clock system. The time clock will be located adjacent to the entrance to each work area. Because this is a place of business, any person excessively late for work or late returning from lunch is dangerously testing his or her future employment. If you have any questions, please do not hesitate to contact the undersigned. Thank you. I am sure the new system will be accepted by all.

WRITING STRATEGY: PREPARING EFFECTIVE INSTRUCTIONS

Preparing effective instructions involves understanding the task to be done, knowing your audience needs, and presenting the information so that it is easy for this particular audience to use.

Why are instructions needed? In most workplaces there is a high turnover of workers and constantly changing equipment or processes (or even performance criteria). Workers must be flexible as they constantly adapt to new people and new tools, equipment, and processes. They also must adapt to changing company markets or policies.

Most instructions are given verbally, but the more complex the task, the more likely it is that both the supervisor and the worker will want written instructions.

Workplace tasks often need to be done exactly the same way—no matter who is doing them or where or when the tasks are being done. You may be working under hazardous conditions, using expensive machinery, or starting a very complex task that will be completed by another worker.

If you are responsible for giving direction to another employee or a group of employees, you may need to use written instructions for the following reasons:

- To ensure everyone understands the exact skills, tools, and steps needed to complete some process.
- To set clear performance standards.
- To give workers something they can later use as a reference to help them do the task correctly.
- To document the quality and the amount of instruction that you gave to your co-workers.

What kinds of instructions are given? Some instructions stay with the equipment. Most of us have worked with a photocopy machine with a small instruction booklet attached to it. The booklet stays with the machine so that when problems occur, the person (who may not be completely trained in using the photocopy machine) has a chance to solve the problem.

Exercise 7.3 Thinking about Instructions

Have you ever been asked to carry out a task without clearly understanding "why" or "how"? Was this a stressful situation? How did you handle it? How does this experience affect how you give instructions? Use a list format to record your reactions, then share them with the class or your discussion group. As a group, list four suggestions on how to handle similar problems. Turn in your suggestions to your instructor.

Chapter 7: Writing for Co-Workers 207

SELECTING A FORMAT FOR INSTRUCTIONS

Instructions may come to us in a narrative form, as part of a memo, in a formal instruction format, or as a checklist.

Instructions can be routine, part of each day's work, or special, one-time-only guidelines. Other instructions may be published quarterly or once a year. A common annual set of instructions that most of us have had experience with is the set of forms sent from the Internal Revenue Service to help us file our tax returns.

How do we know what format works best for what types of instructions? This chapter will present three types of formats for instructions:

- A **narrative** style, usually in paragraph form.

- A **checklist** style, usually in outline format.

- A **combined** narrative and checklist style.

PREPARING INSTRUCTIONS IN A NARRATIVE STYLE

The following set of instructions is presented in the narrative style; it was followed in the early 1950s to serve barbeque beef to 75,000 people.

Construct four underground brick pits, each 8 feet wide, 15 feet long, and 8 feet deep, covering the bottoms with large rocks to hold the heat. Build a fire in each of the pits and let them burn from 8 to 10 hours, so there is a solid bed of charcoal. Cover charcoal with sheet iron and place a layer of wet burlap sacks on the sheet iron.

Meanwhile, cut 40,000 pounds of prime steer beef (forequarter only) into 25-pound chunks, tie in cheesecloth and saturate cloth with barbeque sauce. Place the meat in prepared pits on the wet burlap sacks. Cover pits with boards, then with canvas and finally with about 18 inches of sand to hold in the heat. Leave meat in the pits, undisturbed, for 14 to 15 hours.[1]

Although this recipe includes some specifics, much information is missing. It relies on an experienced reader. The narrative acts as a reminder of what to do at key points, but someone who is working through this for the first time would probably prefer a checklist with more details!

Written instructions can look very much like a manual for a product or a piece of equipment. Both give readers the information they need to carry out a task, using a specialized format to make the job easier.

---

[1]Ruth Berolzheimer, ed. *The Prudence Penny Binding of the United States Regional Cookbook.* Chicago: Consolidated Book Publishers (1957), p. 613.

Part III: Building Communication Skills

■ Exhibit 7.4 **Excerpt from a Consumer Product Manual**

A Little Help from a Friend   Your NELSON RAINSHOWER® sprinkler is going to perform wondrous feats by the square feet for you, so you want to be good to it in return. Here are some tips for the care and feeding of your NELSON RAINSHOWER® sprinkler.

Keep the Darn Thing Clean   The nozzle and spray tube will be kept fresh as a daisy if you remove the plug and flush the thing out with water. There's a handy little device in the end that can be taken out to clean the nozzle holes.

Possum Position   Store your NELSON RAINSHOWER® by hanging it with the coupling end down so all the water will drain out of it. If there's not water, it can't freeze, right?

A *product manual* can be used to help consumers use their new product safely and effectively. Product manuals also can help to resell the consumer more products. Notice how humor is used in the consumer-oriented narrative reprinted in Exhibit 7.4.

Instructions for a general audience may appear in a recipe on the back of a cereal box. On the job, however, instructions usually take a memo form or checklist format for more experienced workers. You may also need to clarify the performance standards for your co-workers, because often these are not included with the instructions. The performance standards may specify:

- The actual amount of the product or service.

- The expected quality of the product or service.

- The acceptable rate of errors or "spoilage" of materials or poor service.

- The acceptable range of time to complete the task.

Some employers set "zero defects" as their quality standard. This means no errors will be tolerated. Some workplaces are organized around "total quality management." This means that every worker is responsible for the highest quality possible. Either one of these conditions can create stress for the new worker who is unsure of what he or she can do.

No matter how carefully we may direct someone to do a task, we may leave out important information or make inaccurate assumptions about what our audience either knows or needs to know. For these reasons, it is very important to listen carefully and clarify even the hint of a question—especially those from a new worker.

In fact, poorly defined tasks are a major cause of workplace stress, and sometimes newer or inexperienced workers are afraid to ask questions—right when they need the information most. You can minimize confusion by making sure your instructions show exactly *what* to do, *how* to do it, and *how* to get help if help is needed!

Clarifying Performance Expectations — Exercise 7.4

Over the next few days, the next time you are asked to do something, listen carefully to the request, then, as soon as possible, write down the spoken and unspoken performance expectations your supervisor (or friend, or teacher) has of you. What were your own performance expectations? Try to record two or three of these requests. Observe others and record any particularly useful instructions you hear being given.

Share your observations with your discussion group. Additionally, appoint a recorder for your discussion group and then discuss the following:

Before Carrying Out the Instructions:

- How are instructions usually given?
- What do people do when they give instructions?
- How are performance expectations communicated?
- Do people tend to ask questions to clarify instructions?

After Completing the Instructions:

- What happens after the task is completed?
- How is performance actually evaluated?
- Is feedback given to people after the task is completed?
- Exactly how is feedback given? Is it positive or negative?
- How does this feedback affect the person immediately?
- How does feedback affect future performance?

Look over the notes recorded during your group's discussion. Did analyzing performance expectations in this way uncover any unexpected information. Turn in a summary of your work and your draft responses to your instructor, polishing your findings into a short memo if your instructor requests this.

UNDERSTANDING THE PARTS OF INSTRUCTIONS

Most of us tend to think that the step-by-step part of instruction is really all that is needed when we prepare a set of instructions. As the following list shows, much more is involved, especially when the task is new to the worker. Also, the more complex the task, the more important it is for the person carrying out the task to understand *why* an action must be done a certain way.

Not every task or equipment breakdown can be anticipated with a set of instructions. People need to be informed enough so they can make their own judgments about what to do next—and when to call in outside help. In some

Part III: Building Communication Skills

manufacturing settings that run 24-hour shifts, for example, technicians must be called from home to solve serious technical problems. How will the worker know when to make that call?

Here are the key parts to a set of instructions. As you read this list, think about a task you recently learned and ask yourself which part(s) were useful.

At the Beginning (Introduction):

- A **clear title** lets the reader know exactly what these instructions will cover.

- An **introduction** tells the reader the purpose of these instructions and previews what's in these instructions.

- A **definition** or **theory of operation** gives the reader a sense of *what* this produce or process is.

- A **description** of the product or process expands the definition and gives the reader a sense of *how* this works and why it is useful.

- **Cautions** or **warnings** alert the reader to any potential hazard in materials or processes. (Put cautions and warnings *before* the steps they affect).

In the Middle (Body):

- **Step-by-step instructions** take the reader through all the steps needed to complete the process or assemble the product (for example, Stage 1, Stage 2, or Phase 1, Phase 2, etc.).

- **Graphics** (typically line drawings or photographs) show exactly what the product or process is made up of or exactly what happens in a particular step.

At the Ending (Conclusion):

- A **troubleshooting section** tells the reader how to identify and solve some common problems; it may list where to go for *help.*

- A **conclusion** may re-sell the reader on using or purchasing the product or process or it may list local suppliers or additional attachments, products, equipment, or services.

Notice how these parts of a formal set of instructions were used in Exhibit 7.5, "How to Make Perfect Toast," which was prepared for a class exercise by Kelly, Kevin, and Teresa.

PUTTING TOGETHER DOCUMENT DESIGN FOR INSTRUCTIONS

How do you know you have good instructions? You might correctly respond, "I know a good set of instructions when I see it!" That's because the design of the document (how the instructions are laid out on paper) can make or break well-written instructions.

How to Make Perfect Toast Exhibit 7.5

The process of making perfect toast no longer has to be a "trial-and-error" task. With the use of a conventional toaster and these instructions, the user will transform plain sliced bread into a tasty breakfast or snack item. For your convenience, these instructions have a troubleshooting guide to help assist you in solving the common problems that occur when making toast. Although making toast is very safe, it is recommended that small children have adult supervision.

Materials Needed

Sliced Bread

Toaster   Electrical Outlet

Knife

Condiments

Set Up

1. Find an area of a counter or table that is uncluttered and close to an electrical outlet.
2. Gather needed materials and condiments

WARNING

HOT!

H_2O + Electricity = Electrocution

212   Part III: Building Communication Skills

■ Exhibit 7.5   (concluded)

Carrying Out Instructions
Step 1.
Plug toaster cord into wall outlet.
Step 2.
Select desired darkness setting. **Note:** It is recommended that first-time users put selector to center position.
Step 3.
Put bread slices into toaster.
Step 4.
Depress control.
Note: This lowers the bread and activates the heating element.
Step 5.
Wait for toaster to eject toasted toast. **Note:** If toaster is in good working order, it will eject bread within approximately two minutes.
Step 6.
Toast should be evenly toasted and ready to garnish with desired condiments.
Note: If toast is not evenly toasted, too light, or too dark, then consult the section titled "Troubleshooting."
Step 7.
If plain toast is desired, then skip this step and consume.
If condiments (toppings) are desired, then evenly spread desired condiments on the face-up surface of the toast using a butter knife. **Note:** It is recommended that the first topping be butter or margarine.

Clean Up
1. Unplug toaster from outlet and put it away. **WARNING** Toaster may be hot and cause serious burns.
2. Put all condiments away, and wipe work area with damp cloth.

Troubleshooting

Symptom	Correction
Untoasted, undertoasted	• Check that toaster is plugged into outlet. See Step 1.
	• If toaster was plugged in and operating, then the darkness control should be set to a darker setting. See Step 2.
Unevenly toasted	• Check if bread slice size is too large for compartment.
	• Turn toast over and insert untoasted side first. Restart. **Note:** this may provide unsatisfactory results due to excess toasting.
Overly toasted or burnt	• Adjust darkness control to a lighter setting. Restart at Step 2 with a new slice of bread.
Bread sticking out of toaster	• Check that toast handle is depressed. See Step 4.
	• Check that bread slice is the correct size for the toaster.
Toast stuck—heating element on	• Unplug toaster and remove bread. **WARNING:** The toaster may be very hot, so care must be taken not to touch it.

This step-by-step guide contains all the information needed to prepare appetizing toast. A troubleshooting section is available to solve problems that may occur while making toast.

Source: Kelly Dungan, Kevin Mockel, and Teresa Trueba, class project, April 24, 1992. Used with permission.

Your purpose in designing instruction is to make the instructions easy to read, easy to understand, easy to remember, and easy to use. You already have been using several key elements of document design in preparing memos and forms. They are:

- Headings

- Lists

- Short paragraphs

We'll now use these techniques in preparing a set of instructions, adding:

- Graphics

- Notes, cautions, and warnings

Headings

Headings are key words that describe the subject being discussed next. Notice how clearly headings are used in sample instructions in this chapter to *divide* the discussion into manageable sections.

Headings can divide a set of instructions into phases (or stages) and, most usefully, can divide the step-by-step portion of your instructions into manageable tasks. The most common phases are setting up, carrying out, and cleaning up.

Headings also predict what the next section will be about. Just like a subject line for a memo, headings need to be descriptive, clear, and concise.

The following are three types of headings you can use.

Functional Headings.   These show the *function* or task that this section carries out. *Introduction, List of Materials,* and *Bibliography* are examples of functional headings.

Topic or Subject Headings.   These show *what* the section is about. *Aquaculture, Emerson Plant Operations,* and *Troubleshooting* are examples of subject headings.

Descriptive Headings.   These include both a *topic* and a *purpose.* They usually show *what* the section is about and *why* it is important. *Setting Up a Profitable Aquaculture System; Recommend Expansion of Emerson Plant Operations to Detroit;* and *Debugging Your New Computer* are all examples of descriptive headings that show both *what* and *why.*

How do you decide which of these types of headings are best for your writing? You can consider your audience, the information, and your writing purpose.

Part III: Building Communication Skills

Writers use *parallel form* in headings for instructions. Notice how headings are used in the sample sets of instructions in this chapter. Headings are a very powerful organizing tool, because they:

- Start with the same kinds of words.
- Show where major phases or stages begin and end in the instructions.
- Highlight the major steps in the instructions.

Parallel form is a kind of logical structure that can be used to hold any document together and make the information easier to read and easier to remember. We'll see how parallel form works in this next discussion on lists.

Lists

Lists can be used to highlight key information because they separate out the information rather than compressing it into a paragraph format.

Which information would you find easier to memorize?

Treatment of Fractures

To treat fractures, the person administering first aid must restore the fracture fragments to their normal anatomical position in a process called reduction. Then, whether as emergency first aid or actual treatment, the health worker must maintain the reduction in place until healing occurs (immobilization). Finally, the health worker promotes regaining of the normal function and strength of the affected part through rehabilitation.

Treatment of Fractures

1. Restore fracture fragments to their normal anatomical position (reduction).
2. Maintain reduction in place until healing occurs (immobilization).
3. Promote regaining of normal function and strength of the affected part (rehabilitation).[2]

Perhaps this is why students use lists or outline chapters for very difficult subjects to emphasize key information. You can use this powerful tool to organize the most important information for your reader.

Psychologist G. A. Miller became very well-known for his work on the limits of memory. Apparently, our brains cannot process more than four to seven items at a time (plus or minus two). If we are tired, our ability to remember drops to the lower end of this range. Very complex tasks obviously involve more than four to seven different pieces of information.

---

[2] Lillian Sholtis Brunner and Doris Smith Suddarth, *Textbook of Medical-Surgical Nursing,* 6th ed. Philadelphia: J. B. Lippincott Company (1988), p. 1577.

For these reasons, watch the length of your lists of steps carefully—especially when more than seven steps are involved. Read the lists in Exhibits 7.6 and 7.7. Which list would be easier to memorize?

The list in Exhibit 7.7 has information clustered around key word headings that reinforce the task to be completed. Each supporting list has no more than seven items. That list also uses parallel form for both headings and subheadings.

Notice again how each heading in Exhibit 7.7 starts with an *-ing verb* (getting, eating). Each item in the supporting list starts with a *command* or an *action verb* (get, find, do, eat, make, etc.). This is *parallel form.*

Simple List   Exhibit 7.6 ■

Turn off alarm	Prepare eggs and toast
Get up	Eat breakfast
Shower	Read paper
Brush teeth	Do dishes
Bring in paper	Get briefcase
Feed cat	Find train token
Water plants	Get umbrella
Make orange juice	

Clustered List with Headings   Exhibit 7.7 ■

GETTING READY FOR THE DAY
    Turn off alarm
    Get up
    Shower
    Brush teeth
    Bring in paper
    Feed cat
    Water plants

EATING BREAKFAST
    Make orange juice
    Prepare eggs and toast
    Eat breakfast
    Read paper
    Do dishes

PREPARING TO LEAVE
    Get briefcase
    Find train token
    Get umbrella

With parallel form, you are setting up "categories" that "presort" the information for the reader. This presorting makes it easier for the reader to memorize the material.

Parallel form is especially important in writing instructions. If you begin each direction with the command form of the verb, you will emphasize the action the reader needs to take. Starting with *action verbs* (also called *commands*) helps readers understand exactly what action is next.

Which of the following is easier to read?

Unit 1. Stitch front to back at shoulder seams, matching small dots. Allow ⅜" seam allowance. Stay stitch neck.

Unit 1. Stitch front to back at shoulder seams, matching small dots. Allow ⅜" seam allowance. To stay stitch, machine stitch through single thickness of fabric to prevent stretching of bias or curved edges on seam line or ⅛" from seam line within the seam allowance. (adapted from Simplicity Pattern #7768, 1976)

Either one of these sets of directions could work well, depending on the knowledge of the reader. The first version assumes an experienced reader; the second version assumes a novice reader. Clear instructions depend on the writer selecting words precisely, arranging them for greatest clarity, and then presenting them as concisely as possible.

A final suggestion inspired by Dr. Miller: Try to avoid overloading your reader with too much information. Put only *one* task in each step in your instructions.

Exercise 7.5 Using Parallel Form

Prepare a list that tells how to complete a favorite task. Develop your outline of the step-by-step instructions so that it includes at least three stages or phases and at least three action steps under each major category.

After you have drafted your step-by-step sections, use parallel form and descriptive headings. "Cluster" ideas into sections of main ideas and supporting ideas. Indent five spaces to show supporting steps. Bring your draft list to your discussion group for review.

Notes, Cautions, and Warnings

You will find three kinds of warnings commonly used in many different fields—*notes, cautions,* and *warnings.*

- **Notes** give the user some additional information that will make the task easier or more efficient.

- **Cautions** alert the user to some minor hazard. If the suggestion is not followed, the user may experience a slight shock, a minor injury, or damage to the equipment.

- **WARNINGS** must jump off the page! Warnings command the reader's immediate attention! If a warning is not heeded by the user, serious injury or death may result.

Notice how international notes, cautionary notes, and warnings have been used in two pages from a set of instructions for a lawn mower (see Exhibit 7.8).

These notes, cautions, and warnings have used internationally accepted symbols; however, you can use other graphic techniques to highlight warnings, such as <u>underlining</u>, *italics,* **bold,** or CAPITAL LETTERS. You can also vary your font size or put your warning into a simple box.

Example of Notes, Cautions, and Warnings   Exhibit 7.8 ■

IMPORTANT: This unit is shipped without gasoline or oil. After assembly, service engine with gasoline and oil as instructed in the separate engine manual packed with your unit.

➤ NOTE: Reference to right- or left-hand side of the mower is observed from the operating position. Refer to parts identification on page 4 (not shown) for location of parts when assembling the mower.

ASSEMBLY INSTRUCTIONS

Tools Required for Assembly
(1) Pair of Pliers
(1) 1/2" Wrench*
(1) 5/16" Wrench or Nutdriver*
*Or one 6" Adjustable Wrench

UNPACKING

1. Remove the lawn mower from the carton by opening the top flaps and lifting the unit out. Be careful of the staples. Make certain all parts and literature have been removed from the carton before the carton is discarded.
2. Disconnect and ground the spark plug wire against the engine. Check beneath the deck for any cardboard packaging. Remove if present.
3. Stretch out all control cables and place on the floor. Be careful not to bend or kink the cables at any time during assembly.
4. Remove page four from this manual and lay the contents of the hardware pack on the illustration for identification.

ATTACHING THE LOWER HANDLE (Hardware A)

1. Attach the lower handle by placing the bottom holes in the lower handle over the weld pins on the handle mounting brackets extending through the rear of the deck. Make certain the instruction label on the lower handle can be read from the operating position.
2. Using a pair of pliers, squeeze one leg of the lower handle against the handle mounting bracket. Insert the hairpin clip into the inner hole on the weld pin. See Figure 1. Repeat on other side.

➤ **NOTE**
There are two (2) holes in the handle mounting brackets. Place the hairpin clip in the inner hole for operation. Outer hole is for storage.

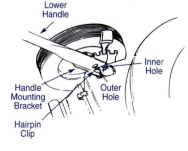

FIGURE 1.

Exhibit 7.8 (concluded)

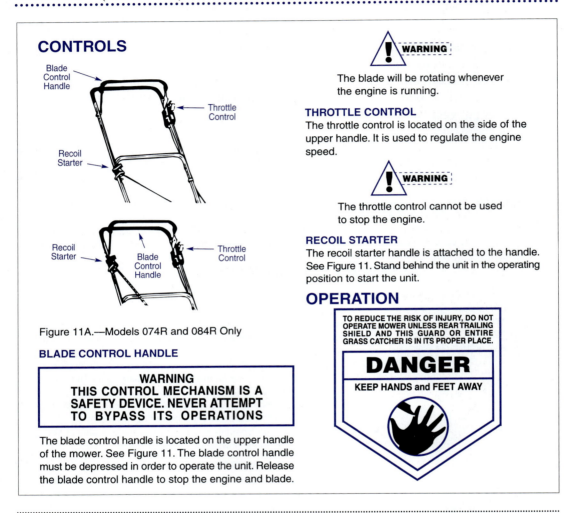

Source: Adapted from *Sears Craftsman Owner's Manual 752074*. Chicago: Sears Roebuck and Company (March 22, 1991). Used with permission.

What is important is that the reader be able to easily differentiate between the different levels of warnings and that serious warnings JUMP OFF THE PAGE! According to research on product liability, warnings must:

- Identify the gravity of the risk.
- Describe the nature of the risk.

- Tell the user how to avoid the risk.
- Be clearly communicated to the person exposed to the risk.[3]

Because some people do not read the instructions that come with potentially hazardous products or because workers may not have had adequate instruction in using dangerous equipment, some manufacturers now place their warnings directly on the product or machine. Your writing strategy should be to *always place your warning before the hazard occurs and as close as possible to where the hazard occurs in the process.*

WRITING STRATEGY: PLANNING INSTRUCTIONS

Many people refuse to read product manuals or instructions. They may glance at the illustrations and only use the text instructions when a problem occurs.

That is why writers who prepare instructions use line drawings and photographs so frequently—to clarify and reinforce key steps in the instructions so that the reader can carry out the task correctly.

Planning the Stages for Instructions

To start thinking visually about your instructions, take a piece of paper and fold it into squares. The following is an 8½" x 11" sheet of paper folded into a planning grid:

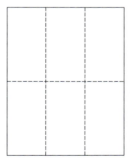

At the top of each "box," place the name of that key stage for your instructions and fill in information about what needs to be done for each stage. A first draft written on how to barbeque steaks is shown in Exhibit 7.9.

---

[3] Gretchen H. Schoff and Patricia A. Robinson. *Writing & Designing Operator Manuals: Including Service Manuals and Manuals for International Markets*. Belmont, California: Lifetime Learning Publications, a division of Wadsworth, Inc. (1984).

220   Part III: Building Communication Skills

Exhibit 7.9   Sample of a First Draft

How To Make Perfect T-Bone Steaks

Preparing Work Area:

Set barbeque grill away from house, shrubs, and trees [add warning]

Need lead time for making fire & cooking

Add to introduction: read all instructions, especially notes and warnings

Gathering Materials:

Barbeque setup [no gas or kerosene]

List cooking equipment [house and outside]

List cleanup equipment

Preparing the Steaks:

Bring steaks to room temperature

Trim fat to avoid flare-ups

Marinate 2 hours before grilling

Lighting the Fire:

Pyramid coals and light [no firestarter]

Spread out coals when they reach the white ash stage [add warning]

Cooking the Steaks:

Place meat on grill

Sear for 3-5 minutes, according to cooking times for rare, medium, well-done meat

Cleaning Up:

Scrape grill w/brush after it is thoroughly cooled

Add troubleshooting guide

Chapter 7: Writing for Co-Workers **221**

As you map out what needs to be done, you may find yourself changing the order, adding more stages or details, and jotting down notes, cautions, or warnings. Your goal is to use this planning grid to anticipate your reader's needs—for information and for graphics.

After you have completed your rough draft of your instructions, look over the draft and ask: Where will my reader have the most difficulty? Where will my reader benefit from seeing an illustration of either tools, supplies, processes, or actions?

Planning the Graphics for Instructions

Most memos and reports use single-spaced text with occasional supporting graphics placed in the text, attached to the memo, or placed in an appendix. For instructions, notice how the single-spaced text is divided into columns and that visuals are used much more frequently.

Most readers in the United States read from left to right. Thus, the first item the reader sees has greatest priority. Which is better, text or pictures? Some readers learn from reading words, others from reading pictures. You can decide which should come first, the pictures or the text, depending on your reader's needs.

Notice how pictures and text reinforce each other in an excerpt taken from Puget Power's brochure written in English, Laotian, Vietnamese, Thai, and Chinese (Exhibit 7.10). This style of instructions is often called "illustruction," because the reader relies as much on the photograph as on the text for meaning.

Deciding on a Basic Layout

Exhibit 7.11 shows four draft page layouts (also called thumbnail sketches or mock-ups). Which one seems most appropriate for a set of instructions?

Use margins and columns to divide the page into boxes for text or illustrations. You will need a margin of at least 1½" on all four sides of your page, and any graphics and headings must stay inside these margins. As you decide where to place your graphics and where to place your text, you're creating a draft or thumbnail sketch of the final format you'll use.

When you are designing layout for a set of instructions, you don't want the reader to be distracted by changes in page layout, unless such changes reinforce an important concept. Thus, once you select the page layout for your instructions, this layout will not change throughout the whole document so that the reader can concentrate on the information and not on the page design.

Part of page design involves looking at "white" space—the space on the page that is *not* covered up by text, headings, or illustrations. You want your text and illustrations to look balanced on the page (top and bottom, left and right). You

Part III: Building Communication Skills

■ Exhibit 7.10 **Excerpt from Brochure**

If you use your kitchen oven or range burners to heat your home.

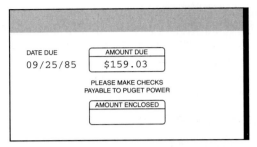

You will be spending more money than necessary to heat your home. You could also cause a house fire by using the oven or burners to heat.

---

want your work to look uncluttered and easy to read. Look at the samples of laboratory safety guidelines in Exhibits 7.12 and 7.13. Which of the two page layouts do you prefer for instructions?

Although writers of instructions are frequently pressed to limit the space they use to because of the costs involved in producing the final manual, especially when that manual is going out to hundreds of thousands of consumers, most well-written instructions use "white" space as a part of page layout for easy reading. The following guidelines are helpful for graphic design of instructions.

- Remember to use one idea for each illustration.
- Don't overwhelm your readers with too much information!
- Avoid using too many **fonts** and *special features;* they compete with each other!

Chapter 7: Writing for Co-Workers   **223**

Looking at Layouts   Exhibit 7.11 ■

Term Paper

←— Double-spaced text

←— Supporting graphic as close to main point as possible

Technical Report

←— Single-spaced text with frequent headings and subheadings

←— Supporting graphic as close to main point as possible

Instructions

Single-spaced text integrated with graphics →

Text divided into stages or steps →

Graphics reinforce steps: →
- What
- How
- Why

Newspaper Article

Headline →

Columns for easy reading →

Single-spaced text →

Graphics added for reader interest or understanding →

Chart

Photo

Part III: Building Communication Skills

■ Exhibit 7.12   **Sample 1**

GUIDELINES FOR POLYMER LABORATORY SAFETY

Follow these rules whenever you are working in the Polymer Lab:

1.   Wear safety goggles during all chemical experiments.
2.   Keep any flammable reagent away from motors, ovens, or any hot area.
3.   Use a well-ventilated hood when working with all hazardous reagents.
4.   Protect reaction vessels by wrapping them in metal screen or wire mesh.
5.   Collect unused hazardous reagents in the proper waste can; do NOT pour unused hazardous materials into the sink.
6.   Find where key safety equipment is kept (fire extinguisher, fire blankets, eye wash, and other safety materials).
7.   DO NOT SMOKE or EAT in the lab.

■ Exhibit 7.13   **Sample 2**

GUIDELINES FOR POLYMER LABORATORY SAFETY

BEFORE WORKING IN THE POLYMER LAB:

1.   Wear safety goggles during <u>all</u> chemical experiments.
2.   Protect reaction vessels by wrapping them in metal screen or wire mesh.
3.   Find where key safety equipment is kept (fire extinguisher, fire blankets, eye wash, and other safety materials).

WHILE AT WORK:

1.   Keep any flammable reagents away from motors, ovens, or any hot area.
2.   Use a well-ventilated hood when working with all hazardous reagents.

BEFORE LEAVING YOUR WORK STATION:

1.   Collect unused hazardous reagents in the proper waste can; do not pour unused hazardous materials into the sink.

> DO NOT SMOKE or EAT in the LAB!

As you read through this chapter, notice *where* and *how* graphics are used in sample instructions. Sometimes these graphics (typically line drawings or photographs) show:

- **What** the machinery or equipment looks like (for example, outside, detailed, or inside—especially useful for a novice).

- **How** the parts fit together (useful for an overall understanding).

- **How** to do a specific task (very useful for new or complex tasks).

SELECTING GRAPHICS FOR YOUR INSTRUCTIONS

Selecting the correct illustration for your instructions involves several key steps:

Plan Where Visuals Will Go:

1. *Decide at what key points your reader most needs visual clarification.* This will help you decide where the illustration belongs.

2. *Sketch what you want the illustration to look like*, using a computer-designed graphic or using a drawing that has already been prepared (remember to give credit if someone else has prepared the drawing). Larger companies may have graphic design people on staff; consult with them!

3. *Decide on the page layout* you want: Instructions? Report style? Manual style?

Prepare the Visual:

1. *Place the illustration so it is "framed" by "white" space.*

2. *Make sure each illustration is properly titled and labeled* so that readers know what they are looking at and why.

3. *Use only one idea for each graphic* to avoid creating too much clutter.

4. *Limit special treatments* (such as bold, underlining, italics, or varying the font size) so the graphic strategies reinforce rather than compete with the information.

Following these guidelines will help your reader to understand your instructions more easily!

Evaluating Instructions   Exercise 7.6

1. Use the above discussion to evaluate how graphics are used in any set of instructions you may find either at home (a lawn mower, a bicycle) or at work. Check to see if all parts of the instructions as listed above appear in your instructions. If not, does this affect the quality of the instructions? How does the layout affect readability? Do the instructions help the new as well as the experienced reader?

2. Bring your comments and your set of instructions to class for discussion with your group. Then, as a group, make a list of criteria for using graphics based on your experiences. Turn in your final suggestions about what criteria results in "good" document design to your instructor.

Part III: Building Communication Skills

USING A PROCESS FOR WRITING INSTRUCTIONS

The following process can be used to draft and test your instructions, revising them until they meet the needs of both novice and expert readers.

1. Draft the Instructions

To meet the needs of your novice readers, approach the task as if you yourself were a beginner. Start by actually doing the task, observing **what** you are doing and **how** you are doing it. If possible, take the time to very slowly repeat more complex steps. Organize your notes into **steps** and draft your instructions. Review your draft to make sure each of these four key areas has been checked:

Content.   Is the information complete? Correct? Are notes, cautions and *warnings* appropriately placed?

Organization.   Is the information put into a sequence that meets the reader's needs? Is the actual instruction section organized logically into stages and steps?

Style.   Is the document clearly written? Concise? Have you used *commands* to emphasize the actions you want your reader to take?

Document Design.   Is this document easy to read and use? Does it use enough "white" space? Do headings help the reader by highlighting key steps or concepts? Are there sufficient illustrations? Do illustration follow guidelines for document design?

2. Have Your Instructions Tested by a Novice Reader

Once you are satisfied that your draft meets these general guidelines, have someone use your draft to carry out the task. This is called running a "usability" test. Observe this novice closely, taking notes when the person hesitates or has trouble finding information. Then make revisions as needed.

3. Have Your Instructions Tested by an Expert Reader

Consider how checklists or a special "quick-start" section may help an expert reader. Revise your draft, if needed, and retest your instructions with an expert reader. Observe closely how the expert uses your instructions to complete the task. After the expert is finished, ask what would be most useful in these instructions for an expert reader.

4. Revise Your Instructions

Revise and retest your instructions until you are satisfied they can be used by a range of readers without any problem. If possible, have your work team review your work.

The larger the audience for your written instructions, the more important it is that you actually carry out usability tests on your instructions with both novice and expert readers.

ANTICIPATING THE CONTEXT IN PREPARING INSTRUCTIONS

Even if you follow all the guidelines for preparing instructions, three additional issues need to be considered as you prepare instructions—defining your readers' needs, considering the location where the instructions will be used, and protecting your customers against hazardous conditions (and your employer against product liability litigation).

Defining Your Readers' Needs

When you prepare instructions, your biggest challenge will be to anticipate the needs of your readers. Most of the time, instructions for a product or a process are used by readers with very different levels of experience.

You can begin to define your readers' needs by gathering different kinds of information about your readers. This could include:

- Demographic data (age, gender, income level, education level, geographic location).
- Psychographic data (information about personal values, behavior, learning style).
- Consumer data (choice of products, prior experience with this type of product or process).

You may want to develop a *user profile* that lists the characteristics of a typical user. For example, the user profile for someone who uses a lawn mower is going to be quite general, whereas a user profile will be necessarily quite narrow for people who use an oscilloscope to measure changes in voltage or current or those who use a seismograph to measure changes in the movement of the ground.

How do you design a document so that it can be used efficiently by both expert and novice? Writers of product manuals grapple with this problem constantly, because many consumers either choose not to read or cannot understand the manuals that have the information they need. Offering customers a toll-free number to help them solve problems can be expensive for your company.

USE A PLANNING FORM TO GET STARTED WITH INSTRUCTIONS

Task to be carried out: _____

_____

_____

Audience: _____

_____

Materials or equipment needed:

_____   _____

_____   _____

_____   _____

_____   _____

Notes, cautions, and warnings needed: _____

_____

_____

_____

Steps (with thumbnail sketches): _____

_____

_____

_____

_____

Steps to set up the task: _____

_____

_____

_____

Steps to carry out the task: _____

_____

_____

_____

Steps to clean up after task: _____

_____

_____

_____

Tests to see steps have been carried out (performance criteria):

_____

_____

_____

List of what to do if something goes wrong (troubleshooting guide):

_____

_____

_____

Chapter 7: Writing for Co-Workers **229**

Considering Where Instructions Will Be Used

Where your instructions will be used can affect how you design your document. For example, what size type would you use for a set of instructions that would be used in an office? For the same instructions to be used on an assembly line where workers move up and down the line? In a work room next to heavy equipment?

Not all work areas have the same level of lighting or good desk space, are quiet, or are even free of wind. These factors may affect the font size, the weight of the paper, protective covering on your instructions, or even color. Your need to understand the conditions in which the instructions will be used is another reason to talk directly to your readers.

Protecting Readers from Hazards

Product liability is another important reason to understand both who your readers are and how much information they already know. When you are preparing instructions for workers or consumers, your employer faces a very real risk for litigation should an accident occur if your instructions did not specify the hazards involved clearly enough.

Companies have a legal responsibility to make products safely; they also have a responsibility to provide instructions for their products that include clear warnings of any hazards.

Lawsuits brought for product liability very often use examples from instruction manuals and safety warnings to show that the manufacturer is liable for the damage or injury that occurred. Additionally, under current Occupational Safety and Health Administration regulations, companies may be fined thousands of dollars for workplace conditions that create safety hazards for workers.

EDITOR'S CHECKLIST FOR WRITING INSTRUCTIONS ■

Use this checklist to help focus your revision before you complete your instructions. Ask someone in your discussion group to act as your peer editor and to answer the following questions about your rough draft.

Encourage your peer editor to either set aside about 20 minutes to work through the list with you, discussing each question, or to write out sentence-length responses to each question.

Check Audience and Purpose:

    ___ 1. Who is the audience for these instructions? Is there more than one audience?

    ___ 2. Why do they need these instructions? Should the instructions be written or spoken?

Part III: Building Communication Skills

___ 3. Will the audience accept or resist these instructions?

___ 4. How and where will these instructions be used?

Check Content and Organization:

Checking the Introduction

___ 1. Does the title clearly describe the task these instructions cover?

___ 2. Does the introduction have an overview of the process?

___ 3. Are the materials and equipment lists complete?

___ 4. Are warnings previewed in the introduction?

Checking the Command Steps

___ 1. Have special conditions, notes, cautions, and warnings been highlighted where they are needed throughout the document?

___ 2. Do command steps begin with active verbs? Are they divided into manageable "chunks" (short and easy-to-understand amounts of text)?

___ 3. Are tasks presented in the exact order they should be completed?

___ 4. Are illustrations placed exactly where they are needed to clarify **what** or **how?**

Checking the Conclusion

___ 1. Is there a troubleshooting guide? Does it clearly identify both the major problems that could occur and how to solve them?

___ 2. Are further resources made available?

___ 3. Did the writer run a usability test? If so, were the results incorporated into the instructions?

Check Style:

___ 1. Does the title clearly describe these instructions?

___ 2. Are the instructions written clearly, concisely, and correctly?

___ 3. Is the tone appropriate for the intended reader?

___ 4. Do headings use parallel form?

Check Document Design:

___ 1. Does the format (narrative, checklist, or combined) fit the audience, purpose, and task?

___ 2. Do all graphics have descriptive titles and are they labeled?

___ 3. Do descriptive headings divide the task into manageable sections?

___ 4. Are there no more than seven steps in any major section?

___ 5. Does the size of the font seem correct for this document?

Allow Time for Editing and Proofreading:

___ 1. Have instructions been edited for correctness in spelling, punctuation, and grammar?

___ 2. Have instructions been proofread for typing errors?

What's Wrong With These Instructions?   Exercise 7.7

Read the following instructions on how to change oil from a car service manual.[4] Draft a list of revisions that could be made to improve these instructions. Share your list with your discussion group.

If your group would find it helpful, bring several car manuals to class so you can compare different writing and document strategies.

If asked to do so, use your draft list (with additions from your group discussion) to write a memo to the person preparing these instructions. After reviewing the memo with your discussion group, turn it in to your instructor with your notes.

PERIODIC MAINTENANCE

Engine Oil Change

To drain engine oil, remove the drain plug in the center of the oil strainer cover. Oil should always be drained when hot. Let oil drain for at least 10 minutes. When oil has drained, remove nuts securing the oil strainer and remove the strainer. See **Figure 10** [not included]. Clean all strainer parts in solvent, then reinstall them with a new gasket. Do not overtighten nuts. Install drain plug. Remove the oil filter cap (see **Figure 11**) [not included] and add 5.3 pints (2.5 liters) of a suitable oil selected from **Table 3** [not included]. Check level on dipstick. The level should be between the marks or slightly above the top mark.

Source: Honda Operator's Manual, 2nd Edition, 1988, Honda Corporation.

Part III: Building Communication Skills

DESIGNING AN EFFECTIVE CHECKLIST

Most checklists are written for expert readers who are already familiar with the terminology, the process, and the context of the checklist.

Some checklists are routine forms for simple tasks that include many steps. Many of us, for example, are familiar with the apartment check-in sheets used when we move into a new apartment. At other times, checklists can be used as "condensed instructions" for very complex tasks (i.e., where workers must follow procedures exactly or where hazardous conditions exist).

Checklists list the most important steps to follow—in the order the steps should be followed. Checklists are:

- *Concise*, considering their expert readers.

- *Logically organized*, so that tasks are presented in the order they should be completed.

- *Correct and complete*, even if tasks are described concisely.

- *Easy to read*, even if this list of tasks is "just a checklist."

Many times, workers must "sign off" after completing a task on a checklist. The checklist may be placed next to the equipment where it serves to document what has happened to this equipment. The checklist may simply itemize the tasks to be done as a readily available reference (see Exhibit 7.14).

At other times, checklists become a permanent part of a file (such as a patient's chart that contains a list of specific treatment or a loan file that contains a credit checklist, summarizing information that has been obtained to support a loan).

Designing a checklist will include understanding both the task and the people who will be using your checklist.

When designing the **content,** reconsider *what* the key steps in the process are, *what order* they occur in, and whether there are any *hazards* the reader must know about. Make sure that you describe actions and objects as specifically as possible. For example:

General description: Turn switch on.

Specific description: Turn SE/A MEMORY switch to ON position.

When designing the **format,** remember that readers will not want to waste their time reading one extra word. If possible, make your checklist no more than one page, using the following guidelines:

- Clear title
- Descriptive headings
- Readable type font and column size
- Balanced use of "white" space

Chapter 7: Writing for Co-Workers **233**

CEES Exposure System Checklist   Exhibit 7.14 ■

CEES CO_2*O_3 Exposure System Checklist

DATE: _____ EXPODAY: _____

COMMENTS: _____

DAILY CHECKS

		AM	PM
	INITIALS:	_____	_____
	TIME:	_____	_____
1.	Status display/scanning present	_____	_____
2.	All chambers show "nominal" CO_2 status	_____	_____
3.	All chambers show "nominal" O_3 status	_____	_____
4.	Computer shows correct time and date	_____	_____
5.	Chamber doors and panels secure	_____	_____
6.	Exhaust fan on	_____	_____
7.	CO_2 dilution tank at 5-10 PSI	_____	_____
8.	CO_2 tank supply adequate	_____	_____
9.	CO_2 flow at .7 LPM for both monitors	_____	_____
10.	O_3 dilution tank 5-10 PSI	_____	_____
11.	O_3 flow rates >1 LPM for both analyzers	_____	_____
12.	Old graphs indicate previous data OK?	_____	_____

WEEKLY CHECKS

1.	Chambers clean of dust and dirt	_____
2.	Sample line filters clean	_____
3.	Printer paper supply adequate	_____
4.	Disk space adequate	_____
5.	Data disk backed up	_____

Source: Environmental Research Laboratory, U.S. Environmental Protection Agency, *Controlled Environment Exposure System (C.E.E.S.) Standard Operating Procedures.* Corvallis, Oregon: Environmental Research Laboratory (1994).

Your checklist should take advantage of clusters of tasks organized under descriptive headings. If there are more than seven steps on your checklist, consider grouping steps into stages with headings!

A final comment about the format of the checklist: Notice that the checklist is meant to be used. The reader may want to check off the steps as they are completed (or may be required to do so), or the reader may want to make notes. You may need to leave space for such checkmarks or comments.

When **testing** your draft checklist, follow the same process discussed in this chapter on how to test instructions. To test your checklist, use people who represent a range of the types of workers who will be using this list.

Exercise 7.8 Evaluating a Checklist

1. Use the editor's checklist for writing checklists to evaluate a checklist you may be using at work or in one of your technical classes. Draft a list of your checklist's strengths and weaknesses. Pay particular attention to how the content in your checklist is organized and presented. How readable is your checklist? Would it be easy to use at work or in a classroom?

2. Bring your draft comments and your checklist to class for discussion with your group. As a group, summarize the characteristics of a "good" checklist from your sample checklists and discussion. Turn in your group definition to your instructor with your checklists and draft comments.

■ EDITOR'S CHECKLIST FOR WRITING CHECKLISTS

Use this checklist to help focus your revision before you complete your instructions. Ask someone in your discussion group to act as your peer editor to answer the following questions about your rough draft.

Encourage your peer editor to either set aside about 20 minutes to work through the list with you, discussing each question, or to write out sentence-length responses to each question.

Check Audience and Purpose:

___ 1. Who is the audience for this checklist? Is there more than one audience?

___ 2. Does this audience need formal, expanded instructions (novice or new skills) or a checklist (audience already familiar with skill or process)?

___ 3. Will the audience accept or resist using this checklist?

___ 4. How and where will this checklist be used?

Check Content:

___ 1. Does the checklist have a clear title?

___ 2. Does the opening sentence tell the reader who, what, where, when, and why?

___ 3. Does the introduction include an overview of the parts of your checklist? What categories and why?

___ 4. If special conditions, equipment, warnings, or cautions are needed, are they placed exactly where they are needed?

Chapter 7: Writing for Co-Workers 235

Check Organization:

____ 1. Are the steps in your checklist organized into "clusters"?

____ 2. Are steps presented in the exact order they should be completed?

Check Style:

____ 1. Can you tell from the title and the introduction who this checklist is intended for?

____ 2. Is your wording as clear as possible?

____ 3. Is each step on the checklist written as clearly as possible? As concisely and correctly as possible?

____ 4. Is the tone appropriate for the intended reader?

____ 5. Do headings use parallel form?

Check Document Design:

____ 1. Have you used a list format? Boxes and headings? Columns?

____ 2. Does your format include a place for checkmarks? For notes?

____ 3. Are there no more than seven steps in any major section?

____ 4. Does the size of the font seem appropriate for this document?

Allow Time for Editing and Proofreading:

____ 1. Has the checklist been edited for correctness in spelling, punctuation, and grammar?

____ 2. Has the checklist been proofread for typing errors?

WHAT'S COMING NEXT?

This chapter has helped you to practice writing informative memos and to plan and revise instructions and checklists. This kind of instructional writing is very common in the workplace and is often needed to document how tasks should be done. Whether you are writing for your co-worker or your customer, people need to know what to do and how to do it.

Your challenge, as your skills and experience increase, is to always remember that instructions are designed to help the audience complete the task more efficiently and effectively. This may mean adapting informational memos, instructions, or checklists for either novice and expert readers.

Part III: Building Communication Skills

In the next chapter, Chapter 8, Writing for Customers, you'll expand your skills in writing letters by focusing on customer needs. You'll practice writing and editing informative and persuasive letters. You'll write to angry customers and to customers who don't want to pay their bills. You'll also see how every letter can help build customer loyalty to your company.

....

■ CONCEPT REVIEW

Can you define the following key concepts from Chapter 7 without referring to the chapter? Write your own definitions in a journal entry or review them out loud.

policy	procedure
product manual	document design
instructions	checklists
narrative style	performance criteria
illustruction	informative memos
negative tone	usability test
reader resistance	notes, cautions, and warnings
functional heading	descriptive heading
clustering	parallel form
planning grid	thumbnail sketches
demographic data	product liability

....

■ WRITING SKILLS REVIEW

To build your skills in editing and proofreading, continue your systematic review of *Appendix C: Writing Skills Review.* Your assignment:

__ Read "Proofreading" in *Appendix C.*

__ Complete Exercises C.37, "Start a Spelling List," and C.38, "Reviewing Commonly Confused Words."

__ Prepare a paragraph summarizing your progress for your instructor or use the form your instructor provides. Be sure to highlight any areas you need to further review.

Chapter 7: Writing for Co-Workers 237

SUMMARY EXERCISE ■

Read through the following checklist. What's wrong with it? Prepare a memo for the writer that lists revisions that should be made before this checklist is distributed to all tank operators. Review your draft memo with your discussion group and make final revisions to your suggestions. If requested, revise the checklist and turn the final checklist in to your instructor.

Liquid Nitrogen Transfer Procedure:

___ 1. Take A&B tank readings.

___ 2. Turn ECV switch to ON position.

___ 3. OPEN local vent valve.

___ 4. CLOSE vent valve to transfer line.

___ 5. Proceed to customer tank.

___ 6. Take C tank reading.

___ 7. Turn ECV to ON position on customer tank.

___ 8. Allow gas to vent until line becomes frosty.

___ 9. Close line urge vent.

___ 10. OPEN top fill.

___ 11. OPEN tank vent 5 turns.

___ 12. Transfer should take about ½ hour.

___ 13. When C tank is at 117" stop transfer.

___ 14. Turn ECV switch at C tank to OFF position.

___ 15. Close TOP FILL

___ 16. Close C tank vent

___ 17. OPEN transfer line vent to wide position

___ 18. Return to plant

___ 19. Get A&B tank readings

___ 20. Turn ECV to OFF position

___ 21. OPEN tank vent to transfer line.

___ 22. Close local tank vent

Note: Contact the superintendent if you have any questions.
Remember safety.

Part III: Building Communication Skills

■ ASSIGNMENTS

Assignment 1: Write an Informative Memo.   Think about the requirements of your most difficult class. Write an informative memo to students who will be taking this class next term. Your memo will give them instructions on how to be successful in the class. Consider what students need to know to be successful at the beginning, the midpoint and the ending of the class. Consider the assignments and types of tests students will complete. Consider class policies and procedures. Consider performance criteria, both stated and unstated. As you develop your draft, review it with your discussion group to follow guidelines for writing an informative memo.

Assignment 2: Design a Checklist.   Design a checklist for using a tool, a machine, or a process that you are familiar with—either at work or around your home. Have your checklist evaluated by both a novice and an expert user and report your findings in a separate memo to your instructor. Turn in all drafts with your final checklist.

Assignment 3: Prepare a Set of Instructions.   Select a process or a tool that you are very familiar with and prepare a two- or three-page set of instructions for an actual audience, following the guidelines in this chapter.

With your final instructions, include a user profile, three draft layouts for your instructions, and all rough drafts. Include a separate memo that reports on the results of your usability test of your draft instructions.

■ PROJECT

Which Is Best? Evaluating Instructions

Working with one or two classmates, gather examples of three product manuals for the same product or service. Analyze them, describing how they are organized, what the writer of the instructions assumed about the user/reader, how graphics are used, and how document design is used. Comment on the overall strengths and weaknesses of each manual. Finally, prepare a set of revised instructions for one small section from one of the sample manuals.

Prepare a briefing (no more than 5 minutes) for the class that tells them which product manuals you selected, why you selected them, and what you discovered about them during your analysis. Turn in your final report (with photocopies of the pages you analyzed) to your instructor.

Chapter 7: Writing for Co-Workers **239**

··· ■

JOURNAL ENTRY: DECODING INSTRUCTIONS

Have you heard these two expressions: "Do it, even if it's wrong!" or "When my boss tells me to jump, I ask how high on the way up!" When you think of directions and instructions you are given on the job, what do these two expressions mean? What kinds of problems or strengths might be involved by following this folk wisdom? Record your reaction, experiences, or analysis in two or three short paragraphs in your journal.

8

Writing for Customers

Informing and Persuading

Chapter PREVIEW

This chapter will help you decide what writing strategies to use when you need to write to your customers. You will work on planning, drafting, and revising strategies that result in timely, courteous, concise, and accurate letters for customers. You'll practice writing *routine* and *persuasive* letters that are frequently used in the workplace.

Journal ENTRIES

Begin by thinking about how to solve problems for your customers and how to build customer loyalty. Write a short one-paragraph response to any of the following questions in your class journal or prepare them for class discussion, as directed by your instructor.

1. What does customer service mean to you? Think of four or five key concepts that should be a part of any customer service policy.

2. Think about your own experiences as a customer. What kinds of written documents do customers get? Are they easy to use and read?

3. What specific skills do people have to have when they work with the general public?

4. When you think of a customer, what picture comes to mind? What special needs do customers have in your field? Why do they stay loyal?

5. What if a co-worker says, "Oh no, here comes trouble. Will you find out what they need this time?" How would you respond?

Chapter READING

As you read this next newspaper article by Jan Norman, consider the hidden costs of poor customer service.

BEFORE GOING TOLL-FREE, CONSIDER WHO MIGHT CALL

Jan Norman

Consumers made 15 billion calls to toll-free telephone numbers last year.

While this so-called 800 service is fairly cheap, companies paying for these calls need to make sure they're getting their money's worth.

The majority of companies with 800 numbers are small businesses, but they don't think about how to get the most out of them, says Phil Seitz, who oversees Custom 800 for Pacific Bell.

Plan Ahead.   Do your homework before activating an 800 number.

"Allow a good month's lead time to think it through. What is your objective (in having a toll-free line)? Who is your audience?" says Mary Weyand, president of TMW Marketing, a Fullerton telemarketing-service agency. Her firm handles calls to toll-free lines for about 25 companies of all sizes.

Many small firms want a toll-free number to get orders from far-flung customers. But according to Purdue University, only 35 percent of calls to 800 numbers are for orders. The rest want information or to complain, comment or suggest.

The 800 number owner needs to develop specific ways to satisfy all those callers and possibly get useful marketing information.

That may require training employees who will answer the phones, Weyand says, and developing scripts to help them get the most out of that call.

TMW, for example, develops different scripts of questions and answers for its operators depending on the things callers are most likely to say.

In just seconds, a trained operator can draw out of the caller a name, address, phone number, age, income and whether he owns a home, explains Anne Griffith, head of business development at TMW.

A toll-free number is just part of an effective marketing plan, she says. "You're constantly adding to information you have about a customer," she says. "Are they really interested? What is the time line when they might buy? Are they ready for (an in-person) sales call? How did they feel about the sales representative who visited them?"

Analyze the Information.   Too many small-business owners fail to analyze information gleaned from callers to toll-free lines, Griffith says.

The data can tell you whether a specific ad is effective, whether your audience is whom you targeted, the demographics of those responding to ads, whether a caller is a real sales lead or just a lookie-loo, and more.

The data can alter company plans for growth, says Diana Hill of J.G. Banks Inc., a Los Alamitos seminar firm. "I thought I knew my (customer) demographics,

Chapter 8: Writing for Customers **243**

but (800-number data) gave me a broader idea of my customer. And we added a workshop on living trusts."

Building a database of caller information is vital, Hill adds.

An 800 number can test your advertising, adds PacBell's Seitz. "You'll discover exactly how effective your advertising is . . . and which medium pulls the most."

The data can also help plan company growth, he says. If your Santa Ana firm is getting a lot of customer interest from Ventura, maybe it's time to open a sales office there.

Who Answers the Phone. If a small business receives one call a month to its toll-free number, it needs no outside help answering the phones, but 30 calls a day at all hours require some help, says Weyand of TMW.

"The problem for small businesses is who is going to answer the phone. Inbound toll-free calls ebb and flow. When the phones don't ring for a while, small businesses tend to assign (phone operators) to do other chores and then there's no one to answer the calls."

Mary Turner of Turner Video in Newport Beach has turned her toll-free phone answering over to TMW because, "I do mail order all over the country and internationally and I never want to miss a call. If you expand, it's hard to answer the phones yourself."

Purdue University's research found that toll-free number callers expect the phone to be answered within four rings (not two, as industry wisdom suggests). They don't want to be put on hold more than 30 seconds or to spend more than five minutes on the entire phone call. And, they want help and answers from the first person they talk with rather than being transferred two or three times.

If the small business cannot satisfy those caller desires in-house, they probably need help from a phone-answering service that can.

Fees vary by the amount of work provided and volume of calls, but a service's fee averages around 85 cents to 95 cents a minute, Jody Martinez of TMW says.

The increase in business is worth the price, says Hill of J.G. Banks Inc. "My response rate has tripled since hiring TMW."

Her firm tried to handle its own customer inquiries, but it used an answering machine instead of live operators and had just three phone lines, which were inadequate for the volume of calls that inevitably came immediately after its radio ads aired, Hill says.

However, she believes it is important for operators at such service agencies not to pretend to be a department within her company.

"If you try to be something you're not, it doesn't work," she says.

And using an outside agency requires preparation, Turner adds. "Only I know my products well enough to explain them to people, so we had to write scripts for operators to follow. They also do order fulfillment, so we had to set up a computer program to invoice. I have to apprise them when my ads run so they will know how to respond."

244 Part III: Building Communication Skills

Don't Keep It Secret. It seems too simple to state, but customers must know about a toll-free number before they will respond.

"It's amazing how many companies have an 800 number, but don't put it in the Yellow Pages ad, in advertising on their business cards and Rolodex cards," PacBell's Seitz says. "It's like it's a secret."

"You have to advertise it."

Many companies considering a toll-free number worry about getting too many calls, PacBell's Seitz says, not because they're worried about handling those calls and orders successfully, but because they're worried about cost.

Most telecommunications companies charge for 800 service by the hour, he says. The most expensive rate is $11 an hour; the average toll-free call costs a company 50 cents to 75 cents.

Compare that price with the more than $300 that an in-person sales visit costs.

Source: Jan Norman, "Before You Sign Up for a Toll-Free Number, Consider Who Will Call and Why—and Who Will Pick Up," *Orange County* (California) *Register,* May 22, 1995, p. D3. Reprinted with permission of the *Orange County Register,* copyright 1995.

Analyzing THE READING

To build your reading skills, use the following questions to analyze the article and to develop your reaction to it. Write a one-page entry in your class journal or prepare your answers for class discussion.

Reading for Content

1. What is the main idea of this article? Can you find a sentence that introduces the main idea? Does this article have a conclusion? Why or why not?

2. How and why does the writer use factual support in her article? What effect does this have on you as the reader? As the possible user of telemarketing services?

3. What purpose do the headings serve? Do the headings match the discussion in every instance? Does each section begin with an introduction and end with a conclusion? Why or why not?

Reading for Reaction

1. Should small businesses use a telemarketing service? What key factors would help a small business owner to make this decision?

2. What are the advantages of using a toll-free number? Are there any disadvantages? Consider both the business owner and the customer.

3. What communication skills would someone need to work well with customers?

WRITING FOR CUSTOMERS

Your customers want quality products or services at a reasonable price and on time. How you serve your customers will, to a great extent, determine whether your customers will return to you or whether they will switch to another company providing better service.

Many companies value customer service so highly that they publish policies and organize publicity around the theme of customer service and customer satisfaction. Maytag's television commercial showing a service technician waiting for a call that never comes is one well-known example. A few companies are working to develop written performance standards for every aspect of their company to gain international certification through the ISO 9000 program. As part of its efforts, Barwon Water wrote this about its customers:

> Our goal is to embrace Total Quality Management in all our products and services. By integrating the principles of continuous improvement in all our systems and processes and by working in partnership with our customers and suppliers we will provide services that fulfill customer expectations.
>
> We acknowledge that the customer is the focus of all our activities. Increased customer satisfaction will be the result of quality leadership, setting our strategic direction, implementing policies to ensure that we are meeting the targets we have set for ourselves and by the elimination of all wasteful work practices.[1]

Defining Customer Service   Exercise 8.1

Think about a job you would like to have (or have had) in your field. List the kinds of communications that can happen between your customers and your employer. Describe in one paragraph some strengths and weaknesses you have observed when working with customers. What have you learned from these experiences?

Bring your list and draft paragraph to your discussion group for review. Your group needs to draft a policy statement of three or four sentences that would help new workers understand the importance of customer service. Turn in your policy with your draft list and draft paragraphs to your instructor.

All of your interactions with your customers—spoken or written—can follow the same guidelines for "good" writing, which we've been working on throughout this book. They should give customers:

Thoughtful Content

- Anticipates and meets the customer's needs.
- Provides just enough supporting information.
- Solves the customer's problem reasonably.

---

[1]Water Barwon, "Quality Policy" (May 4, 1995). Internet citation: http://www.barwonwater.vic.gov.au/WWW/BarwonWater/BWquality.html. E-mail: webmaster@barwonwater.vic.gov.au

246 Part III: Building Communication Skills

Logical Organization

- Presents information in order of importance to the customer.

- Uses an introduction, discussion, and conclusion.

- Puts the main idea first for routine writing (direct order).

- Puts the main idea last for persuasive writing (indirect order).

Clear and Courteous Style

- Is concise and easy-to-understand.

- Avoids overused wording.

- Emphasizes the "you" of the audience instead of the "I" of the writer.

- Avoids negative wording or excessive apologies.

Easy-to-Understand Format

- Follows conventional format.

- Uses short paragraphs and lists to present information.

You can anticipate many different kinds of writing for your customers, from routine, fill-in-the-blank kinds of forms to more persuasive kinds of letters. You will need quick turn-around time for writing to your customers. This chapter will help you to practice each of these writing guidelines.

WRITING STRATEGY: WRITING LETTERS TO CUSTOMERS

When you are writing to your company's customers, you are a representative of your company. You are the primary contact between your company and the customer—regardless of where you work in the company. For this reason, you indirectly become a salesperson, reselling the client on the quality of your company's products or services.

Every part of your letter can affect your company's reputation. What would you think if you received a letter from your major supplier filled with content and typing errors? Would you (as a customer) worry about other errors that may affect your billing or your deadlines? Exhibit 8.1 shows a promotional letter recently sent by a small copying business to all its clients. What is your reaction to this company's management or services?

When we're working with customers, sometimes a phone call or a face-to-face session can solve the problem faster than a letter. Because the costs of preparing and producing a letter are more than $20.00, always ask: What is the most effi-

Chapter 8: Writing for Customers **247**

Sample Promotional Letter   Exhibit 8.1 ■

INSTAPRINT SERVICES

San Antonio's Independent Photocopying Dealer
400 North Travis Avenue ■ San Antonio, TX 78205
(210) 555-4832

July 12, 1996

Dear Instaprint Customer,

This is the month that we all feel patriotic about our great country. The fourth of July marking our independence and the running of the Olympic torch through our grand city. Like our country did over 200 years ago Intstaprint has claimed its independence from authorized dealers. We do business the American way! Have a great July and support our Olympic team with all your enthusiasm.

Sincerely,

Charles Fulton
President

cf:sk

---

cient way for me to get this information to my customer? Which is better, a phone call, a face-to-face interview, a form letter, or a personalized letter? Why? Try to choose a balance between meeting your customer's needs and using your employer's resources wisely.

Deciding on how to best send information to customers or what content, style, or format should be included in a letter depends on your understanding of the customer. Exercise 8.2 will help you define who your customers are.

Who Are My Customers?   Exercise 8.2

Write a profile of the typical customer in your field. After you have described this person's interests, needs and uses of the primary products or services in your field, list two or three different types of customers you may need to work with. Consider and list a few of their *demographics* (external, measurable factors like age, income level, gender, marital status, or educational level) and their *psychographics* (internal and difficult-to-measure factors like personal values, beliefs, and preferences).

Compare your list and description with those from members of your discussion group. How could a company use such a customer profile? Summarize your group discussion for your instructor, turning in all customer profiles.

248   Part III: Building Communication Skills

UNDERSTANDING THE PARTS OF A LETTER FORMAT

Over the last several hundred years, writers preparing business letters have developed a very specific letter format and conventions for writing letters. This next section gives you guidelines for each part of a letter.

Let the Reader Know Where the Letter Came From.   The *company letterhead* or *return address* usually comes first and shows who the letter is from.

Date Your Letters.   The *date* is always used to document when the letter was prepared. Undated letters can result in confusion. When was that order placed? When a letter is longer than one page, some letter writers suggest including the date (and page number) at the top of each page.

Use a Formal, Inside Address.   The *inside address* shows the person's name, title, and department (if known), the company, the street address, and the city, state, and ZIP code, exactly as it appears on the envelope. This repetition of what appears on the envelope confirms to whom the letter was sent.

Use a Subject Line.   The *subject line* previews what's in this letter in five to seven key words that clearly describe three elements: the subject, the purpose of the letter, and the urgency of the information (if needed).

Do you want your reader to read the entire letter before he or she understands what this letter is about? Using a subject line generally improves the efficiency with which your letter is processed. The reader can use the subject line to decide when to read the letter, to make sure the letter goes to the right person, or to make sure the letter is filed properly. Two examples follow:

RE: VACATION REQUEST 8/5/95 THROUGH 8/25/95

SUBJECT: REQUEST APPROVAL FOR NEW WORKSTATION

You should be able to clearly determine the **subject** and the **purpose**—just from reading the subject line. The following are two examples:

SUBJECT: Our Moving Costs to Lansing

SUBJECT: Recommend Acme Transport for Lansing Move

Both of these subject lines could be effective. The first subject line tells the reader that the memo will be about moving costs to Lansing. Although we don't know if these costs are our costs or the charges from the various moving companies, the introduction to the memo can clarify these points. The second subject line recommends that Acme Transport move us from our home office to Lansing. Readers may not agree with your recommendation, but the subject and purpose are very clear.

Some letter writers believe subject lines are only used in memos. However, using a subject line in all letters helps the reader quickly understand what the letter is about. Some writers place the subject line immediately after the inside address; other writers place the subject line after the salutation.

Choose a Greeting Style That Works.   The *salutation* often seems too friendly to student writers. How can you say "Dear Fred," for example, when you don't really know Fred?

Without a friendly salutation, however, business letters can have a cold tone. If you don't feel you know the person well enough for "Dear Fred," address the person by his or her last name.

Dear Fred,	Personal and friendly.
Dear Mr. Ronzini,	Formal, yet friendly.
Dear Ms. Ronzini:	Acceptable for all female readers unless previous correspondence shows your reader prefers a different salutation.
Dear M. Ronzini:	Some writers use this approach when they don't know whether the person is male or female.

You'll also need to decide if you want to use a comma after the salutation (considered informal) or the colon (considered formal).

Organize the Body of the Letter.   The *body of the letter* usually contains several paragraphs for the introduction, discussion sections, and a conclusion. In a business letter, you can expect to see shorter paragraphs (usually no longer than five lines of text) and more frequent headings and lists. Notice how readable the paragraphs appear in the sample letters shown in this chapter. Short paragraphs invite the audience to read your message.

Use a Formal Closing.   The style of the *closing* or *signature box* of your letter is often decided by your company. Here are some commonly used styles:

- Sincerely,

  COMPANY NAME

  Your Name
  Your Title

  This closing emphasizes that the person signing this letter is acting as an agent of the company.

250 Part III: Building Communication Skills

- Cordially yours,   Most people use "sincerely" as their closing, but other wording still sets a personal and courteous closing to your letter.

  Your Name
  Your Title

- Respectfully yours,   This closing may be too humble in tone for many letter writers, but this style is used in request letters.

  Your Name

Letter writers typically allow about three to four blanks lines where they write their signature. You don't want your signature to appear crowded, and most readers respond to a signature that is easy to read—not a scrawl.

Deciding on a Conventional Letter Format

Not following the conventional parts of a letter immediately affects your credibility in the eyes of your reader. You may notice that some letter writers drop the salutation and the closing just before the signature. Most readers feel that these changes give the letter a cold tone.

However, writing styles and letter formats do change over time. In the 18th and 19th centuries, people signed letters: "Your humble and obedient servant." Writers in other countries also follow formal conventions dictated by their cultures. For example, Japanese letter writers typically begin and end formal business letters by referring to the changing weather and the passing seasons of the year.

Notice what formats and styles writers in your industry use and follow these guidelines. Your employer may dictate the exact format and style for you to use.

You can see all these parts of a letter at work in the following routine letters (Exhibits 8–2 and 8–3). Notice how both letters use "white" space to balance the text on the page. Also notice how the writers have worked to make the letters easy to read.

Block Style.   The letter from Worldwide Resort (Exhibit 8.2) uses a **block style,** the oldest format still in use. For efficiency, typists were told *not* to indent five spaces for each paragraph. Each line starts flush with the left margin. The result is a somewhat formal and impersonal style.

Modified Block Style.   The letter from MicroSupplies (Exhibit 8.3) uses a **modified block style** to respond to a request for product information.

Notice that the modified block letter format has indented paragraphs. The date line is lined up exactly with the signature block, and each paragraph is indented

Block Style Letter Exhibit 8.2

July 1, 1997

The Office Supply Company
425 West Third Avenue
Columbus, OH 43212

Ladies and Gentlemen:

Please send me the following items, which are described in your spring 199- catalog.

```
4 doz.    #32867    memo pads          $7.25 doz.    $ 29.00
10        #16703    desk calendars      8.95 ea.       89.50
15        #75079    desk calculators    6.50 ea.       97.50

                                                     $216.00
                       Shipping and Handling           13.00
                                                     $229.00
```

Charge the order to the company account. Please process this order immediately so that we can anticipate delivery by July 30.

Sincerely,

Kenneth Lopez
Training Department

1500 Magnolia Blvd., Williamsburg, Virginia, 23185 ● 714. 555.1212

252　Part III: Building Communication Skills

■ Exhibit 8.3　**Modified Block Style Letter**

MicroSupplies
2327 Lancaster Street
Pittsburgh, PA 30417
412/555-4739

August 17, 1997

Mr. Evert L. Eagleston
1133 Newhart Blvd.
Palo Alto, CA 94303

Subject: Prices for Surge Suppressors

Dear Mr. Eagleston,

　　Thank you for writing about pricing for our surge suppressors. The 1996 prices range from $250 to $489, including delivery.

　　Your request for prices for surge suppressors delivered in 1998 will be available in November of 1997. Prices for 1998 are impossible to estimate and guarantee at this time, although I can say that our prices have been rising at 4.5% per year for the last four years.

　　Your subscription to MicroSupplies catalog has been renewed for the next three years at no cost to you. Our catalog will enable you to follow our prices and products over the next several years.

　　Thank you again for writing MicroSupplies. Please contact us if we can further help you and your new business.

Sincerely,

Mark Dragott
Regional Sales Manager

---

five spaces. This widely used letter format sets a professional yet friendly tone, and it is often used in letters to customers or friends or with resumes.

These sample letters (and the other sample letters in this chapter) also show the writer's professional style, which is clear, courteous, and concise.

Writing with a "You" Focus

Another strength of the letter from MicroSupplies is how the writer has emphasized the needs of the reader in presenting the information. Notice how frequently "you" appears, and notice how the writer gives the reader a subscription to help solve the reader's problem.

Writing with a "you" focus (also called an audience focus) can be very difficult because most of us tend to think in terms of ourselves. We usually begin writing from our point of view, but, as we revise, we need to consider our reader's needs first. The following suggestions can help you to write with an audience focus:

- Use "you" more than "I" or "we."
- Think about the audience's problems, needs, and point of view.
- Provide specifics that help the reader.
- Consider the reader's feelings.
- Try to state negative information positively.
- Provide reasons when you must say no.

Exhibit 8.4 presents some examples that contrast writer focus and audience or "you" focus.

Contrasting Focuses Exhibit 8.4 ■

Writer Focus	Audience or "You" Focus
You will not qualify for the student membership rate of $45 a year unless you are a full-time student.	If you are a full-time student, you will qualify for our student membership rate of $25 a year.
Your order will be sent within the next few weeks.	You will receive your order by May 5, 1997.
We are happy to extend a credit line to you of $5,000.	Your new credit line of $5,000 has been approved.
Please do not hesitate to contact the undersigned if I may be of any possible further assistance to you in this matter.	Please call me at 503/555-4000 if you have any questions at all.

Revising for a "You" Focus Exercise 8.3

Practice your skills in revising for a "you" focus by rewriting the body of this letter sent requesting an appointment.

> Dear George,
>
> I respect your immediate workload, but I would again request to meet with you on October 18. I would appreciate a 20-minute visit, and I could make that at your convenience. I wouldn't mind having breakfast or lunch with you. I would like to present our London transmitter line and, if time permits, I would like to discuss our new 590 series Cheshire recorders. I will check with you early in the week by phone.

Bring your revised draft letter to your discussion group for review. As a group, identify the strengths and areas needing further revision for each draft letter. Then, revise again if requested by your instructor, and turn in the completed letter to your instructor with all drafts and comments.

WRITING TO INFORM OR CONFIRM

In many ways, *informative letters* are the easiest of all to write. Your customers may want to buy something, or they may simply want information. Your job is to give the customer what is needed as promptly, correctly, and courteously as possible.

Using the Direct Pattern to Say Yes.   Routine requests for information about your employer's products, services, operations, or staff come in daily. Customers may want an exchange or adjustment, or they may want to place an order. Customers also frequently request exchanges, returns, replacements, or refunds from your company when there has been a product defect of some sort.

If the amount of money involved is small, many companies simply honor the customer's request to build good will. However, each return or refund costs the company in lost sales and in lost customer good will. Using a **direct** pattern can help you respond to such routine requests quickly:

- **Introduce** the subject and purpose in the first paragraph.

- **Provide** details or explain any action taken in the discussion section.

- **Resell** the customer on your company's products or services.

- **Conclude** courteously and tactfully.

Many companies use *forms* and *form letters* to ensure that all the information needed is provided quickly and easily. Exhibit 8.5 is an example of a form letter sent by a bank to confirm account balances. Notice what elements the form letter includes to answer the customer's request.

"Filling-in-the-blank" can be a time-saving approach for form letters. Once you have drafted a form letter, double-check your writing purpose, what the customer might need and why. Ask if this form can be designed to prevent phone calls. That's the secret agenda behind form letters. They save time for your customers, your employer, and you.

Customers often receive forms from companies to confirm or change orders or shipment dates, to announce new products or upgrades in service, or to handle billing. Sometimes these forms are generated by computer; sometimes your co-workers have filled in the blanks and have dropped a card in the mail. Other times a "blanket" letter (one letter for many readers) is written, and sometimes workers have used word processing to personalize the letter. A "blanket" thank you to a company's dealers is shown in Exhibit 8.6.

All these form letters have two purposes: to serve the customer as efficiently as possible and to build customer loyalty. Thus, all the information is presented in a way that is easy to read and that is relevant to the customer's needs.

Chapter 8: Writing for Customers 255

Example of Routine Form Letter   Exhibit 8.5 ■

```
Dear Customer:

Good business practice and standard banking procedures require
that we regularly verify the balance of various accounts. We
accomplish this by selecting accounts at random and by obtaining
direct confirmation of the account balance from the customer.

Please compare the following information with your records.
Then SIGN and RETURN this letter in the enclosed envelope.

Thank you,

Anne L. Bergstrom
Auditing Department

Branch of Account:        San Lorenzo
Account Number:           473-94-431223-X
Account Type:             Time Certificate
Issued:                   7/22/92
Matures:                  7/22/98
Account Balance:          $28,753.27

This information is:    ___ Correct    ___ Incorrect

Please explain any incorrect information on the reverse of this
letter.

Signature: _____   Date: _____
```

Although form letters can meet customer's needs efficiently, sometimes customers feel they are being processed because the form is poorly designed or has an impersonal tone.

Responding to Difficult Requests

If the company cannot easily respond to a request, you may be asked to say no to a routine request. In such a situation, letter writers frequently personalize their routine responses, clarifying the circumstances and following an indirect pattern before saying no.

Using the Indirect Pattern to Say No.   Notice how the letter in Exhibit 8.7 uses an **indirect** pattern of organization by starting with a buffer that softens the "no," that **explains** why the company cannot say yes to the request, that **suggests**

Part III: Building Communication Skills

■ Exhibit 8.6 "Blanket" Thank-You Letter

Dear friends,

Every so often, an event occurs or a milestone is reached that causes one to stop and reflect. Recently, our bank reached such a milestone. On October 17, 1996, our total dealer loans exceeded $9 billion.

While we are, quite understandably, proud of this achievement, the overriding sentiment is one of gratitude. We are very well aware that without the long-term support and loyalty of the dealer community, we would not today be celebrating this milestone.

To you and to all of our dealer friends, we would like to say thank you. It is our intent that in the coming months and years, we will continue to provide you with the same commitment and loyalty that you have afforded us.

Again, a very sincere and heartfelt thanks.

Sincerely,

■ Exhibit 8.7 Indirect Form Letter

Thanks for your questions about our company. Unfortunately, we receive many similar requests and we aren't able to give individual attention to each query. We have enclosed information to give you a sense of what we and our industry are about. The material demonstrates our commitment to performance, innovation, and teamwork. Hopefully it will also answer some of your questions (and maybe others you haven't thought to ask).

We greatly appreciate your interest and regret we don't have the resources to better serve you.

Best wishes for a successful project.

Employment Center

Revising a Form Letter   Exercise 8.4

Sometimes, form letters are written for a very large audience. Analyze the following letter sent directly to students in the fourth grade. What are its strengths and weaknesses? How would you revise it? Consider your primary audience (the kids) and your secondary audience (the parents and the teachers). Rewrite the letter if instructed to do so by your instructor. Otherwise, bring your ideas to class for discussion.

❦ My Own Reading Club
2817 East McArthur Street
Jackson City, MO 65101

Dear Reader,

    Thank you for starting the year with MY OWN READING CLUB books.

    I'm extraordinarily sorry to say that the publisher of the book you ordered, *Abraham Lincoln's Frontier Days,* is unable to send any more copies. In its place we are sending you two other popular books, *All About Dinosaurs* and *Exploring Our Space Frontiers.* We think that you will be very happy to get two books to show you how sorry we are.

    If you are not completely satisfied with this substitution, please return this letter to the address shown herein and request a credit or refund.

    Thank you for your understanding in this matter. We look forward to serving you again soon to your complete satisfaction.

                                          Sincerely,

                                          John Chadwick
                                          MY OWN READING CLUB

---

options that could help the reader, and, finally, that attempts to **build good will** by providing information that will help the reader respond positively to the company.

    Notice in particular the courteous and helpful tone. Such letters can also document the kinds of requests and problems your customers face with your products or services. If a pattern of problems emerges for a particular product or kind of customer, your company can then decide what should be done.

258   Part III: Building Communication Skills

EDITOR'S CHECKLIST FOR WRITING AN INFORMATIVE LETTER

Use this checklist to help focus your revision before you complete your persuasive sales letter. Ask someone in your discussion group to act as your peer editor and to answer the following questions about your rough draft.

Encourage your peer editor to either set aside about 20 minutes to work through the list with you, discussing each question, or to write out sentence-length responses to each question.

Check Audience and Purpose:

___ 1. Do I know *who* the audience is and *why* they need this information?

___ 2. Do I know *why* my company is sending this information? Is my writing purpose informative and persuasive?

Check Content and Organization:

___ 1. Does my introduction clearly state the purpose of this letter?

___ 2. Have I presented information in order of importance to the reader?

___ 3. Does the discussion section give the reader the information he or she needs, without being wordy?

___ 4. Do I explain any action that my company has taken or that the reader needs to take?

___ 5. Have I included information about other products or services that would be helpful to this reader?

___ 6. Have I provided options if I cannot resolve the reader's problem?

Check Style and Document Design:

___ 1. Is my tone positive and courteous?

___ 2. Have I emphasized "you" over "I"?

___ 3. Are the word choices clear, avoiding jargon?

___ 4. Have I used short paragraphs and lists to ensure easy reading?

___ 5. Does my letter follow a conventional format?

Allow Time for Editing and Proofreading:

___ 1. Has my letter been edited for correctness in spelling, punctuation, and grammar?

___ 2. Has my letter been proofread for typing errors?

Chapter 8: Writing for Customers   **259**

WRITING TO PERSUADE

You can decide whether to use routine forms and letters or more persuasive letters by thinking about the needs and attitudes of your customers. Also, consider their importance to your company.

If your customer needs some information, has no reason to complain about your products or service, has not requested special service, and a relatively small amount of money is involved, then a routine form letter sent promptly will be your best choice.

If your customer needs complex information, is upset or concerned about your products or services, needs some special service, or is a major client, you will need to prepare a persuasive letter. If your customers have not paid their bills or you must refuse to grant them credit, you'll need to write a persuasive letter.

This kind of writing is no longer routine. It involves more than just saying yes or no and explaining why. Now, your letter writing skills must persuade the reader to act—and to remain loyal to your company. When you use an **indirect** pattern:

- **Start** with a buffer that builds goodwill with the customer or states some area for common agreement in the first paragraph.
- **Explain** the problem with enough detail so the customer understands the factors involved and then is prepared for a compromise solution or a "no."
- **Say no** indirectly, if possible.
- **Offer** the customer options by either suggesting a compromise or by proposing other solutions when you must refuse the customer's request.
- **Resell** the customer on your company's products or services—tactfully!
- **Conclude** positively and courteously.
- **Restate** any action you need your reader to take.

Are these guidelines at work in Exhibit 8.8, a form notice of a late payment? Why would a company need to use such a strong negative tone?

Compare the notice from First Midwest Bank with the form letter from J. Wentworth & Company (Exhibit 8.9). Which would you rather receive? Which do you think would be more effective with customers? In what ways does the situation influence the tone of the message?

Notice that as J. Wentworth & Company attempts to motivate the customer to pay the bill, every effort is made to continue a cordial relationship with the customer.

Because both of these letters are attempting to collect money from the customer, they are a special category of letter (the collection letter), which is discussed in a later section of this chapter.

■ Exhibit 8.8 Letter with Negative Tone

FIRST MIDWEST BANK
Simi Valley Branch, California

April 7, 1997

ACCOUNT NUMBER 05327439752

TO: M. Huford
       473 North Plains Road
       Simi Valley, CA 93097

Your VISA payment is now past due.

Any portion of your minimum payment not received within 15 days of your payment due date will be subject to a late charge, the lesser of $5.00 or 5% of the amount past due.

If the 15-day period expires on a Saturday, Sunday or a legal holiday, payment is required by the end of the next business day.

This notification is provided to you as required by CAS 708.480(3).

Sincerely,

FIRST MIDWEST BANK

Exercise 8.5 Drafting a Routine "No" Letter

Your supervisor has asked you to write "no" letters to several summer job applicants. Use the guidelines listed above to draft a form letter that lets the reader know someone else was selected as this year's summer intern. Revise your draft for a courteous tone.

Bring your draft to your discussion group. Review the rough drafts, highlighting strengths and weaknesses of each, listing revision suggestions at the bottom of the letter. Turn in the draft form letters and comments to your instructor.

Chapter 8: Writing for Customers **261**

Letter with Positive Tone   Exhibit 8.9 ■

J. WENTWORTH & COMPANY
3475 Rio Vista Drive
Albuquerque, NM 87201
(505) 555-4632

March 4, 1997

In reply, refer to: 827-353-575-028

L. MacDonald
2714 N.W. Buena Sierra
Albuquerque, NM 87220

A FRIENDLY REMINDER! Realizing that most often a missed payment is an oversight, we
have learned that our customers appreciate it when we bring these matters to their attention.

As of December 8, 1996, the amount due on your account was $47.00. If you have not already
paid this amount, please either mail us your check or money order for the amount due today or
pay this bill at your local J. Wentworth & Company store.

If your payment for the amount due and this letter crossed in the mail, please accept our apology
for this reminder and THANK YOU for bringing your account current.

Please be assured that you are a valued customer at J. Wentworth & Company. If you would like
to discuss your account or if I can be of any assistance, please contact me at the address or
telephone number shown.

R. Hovelle
Account Representative

WRITING STRATEGY: RESPONDING TO CUSTOMER COMPLAINTS

Sometimes our customers can become very angry at us—for reasons that may
not be connected to our company. It may be easier for customers to become
angry at a distant "company" than to take out their anger on a real cause.

If you follow routine policies for resolving customer complaints, stay calm
and behave professionally, your chances for solving customer complaints—even
from very angry customers—are very high. These guidelines hold true whether
the angry customer has sent you a letter or whether the customer is standing in
front of you.

However, sometimes customers cannot be reasonably satisfied. Your courteous treatment of the customer, even if you cannot approve the customer's request, can still defuse the situation.

Exercise 8.6 **Responding to Angry Customers**

1. Think of a time when you received unsatisfactory service or bought defective merchandise. Analyze customer needs for a particular product or service. Imagine you have purchased this product or service and it has not met your expectations. Draft an "angry" letter to the company, requesting a full refund.

2. Draft a company response to your "angry" letter. Bring both drafts to your discussion group for review. Discuss how your understanding of the problem may have changed as you wrote the different letters.

3. In your discussion group, select one letter as a "good" model, listing why this complaint letter and/or response letter is so effective. Consider content, organization, style, and document design. Turn all of the letters with your group discussion notes in to your instructor.

The customer may be right, and your company may be at fault. Try to emphasize the solution and the future. Mistakes happen to everyone. Customers will appreciate your positive tone, especially if you avoid over-apologizing. Besides reminding the customer of all that went wrong, sometimes too many apologies may lead the customer to make more demands or to take legal action.

WRITING STRATEGY: WRITING TO COLLECT MONEY

Most customers pay their bills on time—or reasonably on time. Sometimes customers will need a reminder of exactly how much money they owe and when it is due.

Approximately 5%-15% of your company's customers, however, will delay payment well past a reasonable time. This means a double loss for your company; not only is the product or service not paid for, but it could have been sold to someone else who would have paid for it.

Managing accounts payable (bills that customers need to pay your company) is a responsible job. In small companies, many times this job is given to a new person because it can be tedious. Yet the more customers pay "late," the more important it is to have a consistent and fair policy for late payments.

If you understand the process that can be used to collect money from "late" customers, you can responsibly manage accounts payable. Because companies rarely get results with the first collection notice or letter, it is important to *consistently follow through with each late customer.* The process outlined in Exhibit 8.10 is commonly used.

Revising for Positive Tone Exercise 8.7

Read over the following letter drafted by one of your co-workers. Revise the letter, focusing its style (check clarity, conciseness, courteousness, and completeness).

Bring your draft revision to class for your discussion group to review. As a group, identify the strengths and weaknesses of each revision. Underline or highlight particularly effective revisions and turn in the drafts from your group to your instructor.

Super Carpet
1800 Ridge Road
Homewood, IL 60430
708/555-1212

Dear Sam Suriega,

   Thank you for calling us yesterday. We appreciate hearing from our customers, and we humbly apologize for not installing carpet in your closet, but when our carpet planner Steve visited your home to measure the rooms for the new carpet installation, your wife told him not to bother with the closet. I am sure that the pleasure you experience as you use your new Olympian blue carpet installed throughout the house at a cost of $6,784 will offset that nasty feeling that hits you every time you open your closet. This unfortunate oversight was really your fault. Why don't you talk to your wife about what should have been measured and don't blame us! When you're ready to go ahead on the kitchen remodeling job, please call us to set up an appointment. We're sorry this wasn't perfect, but you really can't blame us. We did use carpet remnants to patch together a partial covering—so we're the good guys!

                                Sincerely,

                                Your friends at SuperCarpet
                                Steve, Tom, and Paula

   In the early stages of the collection process, the writer should work to build rapport with the customer by emphasizing cooperation, fair play, and even sympathy for the customer's situation. Flexibility can result in payment. However, in the later stages of the collection process, the writer is less effective in motivating the customer by appeals to fair play alone. Usually in the later stages, the writer must be more demanding and appeal to the customer's self-interest, embarrassment, and even fear.

Part III: Building Communication Skills

■ Exhibit 8.10 **Scheduling a Collection Process**

Time	Action
Immediate	Give customer detailed bill for goods or services. Offer discount for early payment. Show charges if payment is late.
30 days	Send monthly billing. Restate amount due for goods and services. Show charges if payment is late.
60 days	Send monthly billing. Add "overdue" stamp or other note. Show late charges.
90 days	Send first reminder letter with monthly billing. Show accumulated late charges. Invite customer to contact you. Contact customer by phone if possible.
120 days to190 days	Send second and third collection letters. Contact customer directly (phone or visit). Turn over to collection agency if payment is not made.

Look again at the collection letters from First Midwest Bank and J. Wentworth & Company. Can you tell now which stage in the collection process these letters are from? Does this explain the difference in tone?

Notice how the tone of these letters changes from courteously inquisitive (J. Wentworth & Company) to a demand for action (First Midwest Bank), followed by threats if payment is not made. Notice also how the "last chance" letter very clearly restates the payment needed and the deadline date, so the reader knows exactly what is required to avoid legal action or being turned over to a collection agency.

Your goal in preparing a series of collection letters is to persuade your customer to pay the bill. Because collection agencies typically take 50% of any billings they collect for you, most companies use collection agencies as a very last step.

From your customer's point of view, though, being threatened with a collection agency is very serious because the customer's inability to pay will be permanently documented on his or her credit history. A negative credit rating could mean future problems in borrowing money—or in getting future accounts opened. Because information about how customers pay their bills could be damaging, every effort must be made to keep these records private.

You can save your employer money by writing collection letters systematically, that is, by using a process that is fair and consistently managed.

Chapter 8: Writing for Customers   **265**

EDITOR'S CHECKLIST FOR WRITING A LETTER THAT SAYS "NO" ■

Use this checklist to help focus your revision when you are planning to write a letter that says "no" to your audience. Ask someone from your discussion group to act as your peer editor and to answer the following questions about your rough draft.

Encourage your peer editor to either set aside about 20 minutes to work through the list with you, discussing each question, or to write out sentence-length responses to each question.

Check Audience and Purpose:

___ 1. Do I know the audience? Do I know why the audience is neutral or may resist my proposal or presentation?

___ 2. Can I tell why this letter is being written? Is the writing purpose informative and persuasive?

Check Content and Organization:

___ 1. Does my opening create a buffer (a neutral ground, an idea or starting point that my reader and I can agree upon)?

___ 2. Have I described the problem objectively so that a reasonable person would understand what's involved?

___ 3. Have I offered the reader alternate solutions or other resources to solve the problem?

___ 4. Have I said "no" indirectly, if at all?

___ 5. Can the reader easily understand what the next step is—and by what date?

___ 6. Does my letter end positively?

Check Style and Document Design:

___ 1. Is my tone clear, concise, and courteous?

___ 2. Have I avoided using "you" with negative information? For example, avoid saying, "Your shipment was late again." Say: "The shipment expected June 15 was delayed by three weeks."

___ 3. Are the word choices clear, avoiding jargon?

___ 4. Have I used short paragraphs and lists to ensure easy reading?

___ 5. Does my letter follow a conventional format?

Allow Time for Editing and Proofreading:

___ 1. Has my letter been edited for correctness in spelling, punctuation, and grammar?

___ 2. Has my letter been proofread for typing errors?

Exercise 8.8 Developing a Collection Policy and Procedure

If a customer owes money to several businesses, who would be paid first? Why? Do some companies or some industries expect "late" payments? How would this affect a "late" payments policy? What procedure should a company follow to persuade "late" customers to pay their bills?

Have you had any experience in working with customers who have not paid on time? What kinds of action did your company take? Were there any successes? Problems?

Prepare your answers to these questions and discuss them with your group. Then, as a group, prepare a memo outlining the collection policy and procedure you would want a company to follow. Turn in your draft policy, procedure, and notes to your instructor.

WORKING WITH CUSTOMERS OVER THE PHONE

When you are talking with customers over the phone, you cannot see their faces for clues about whether they understand you. The following suggestions can help you to ensure good two-way communication.

Before the Phone Call:
1. **Gather** the information the customer needs, so you can present it concisely and correctly.
2. **Clarify** possible solutions with your supervisor before making a commitment to your customer.

While on the Phone:
1. **Introduce** yourself by name and department.
2. **Listen** carefully to what the customer wants.
3. **Ask** questions to gather or clarify the information you need to solve the problem (name, telephone number, exact problem, model number or date of service, and any deadlines).
4. **Speak** clearly, avoiding interruptions.
5. **Use** a helpful, courteous tone of voice.
6. **Give** a clear time line for when you expect to resolve the problem or give your customer the information.

After the Phone Call:
1. **Follow through** on your promises to the customer.
2. **Confirm** in writing any needed information, thanking your customer for working with you.

Following these suggestions will help you provide consistently good service to your customers. You can also analyze any pattern of customer complaints to find out ways to improve your company's policies or procedures.

Chapter 8: Writing for Customers   **267**

WRITING STRATEGY: WRITING TO SELL PRODUCTS

One of the most important factors to the success of your employer is the ability of the sales staff to sell the company's products and services. Personal contacts can build relationships between the sales staff and the customer, but sometimes only a sales letter can reach your potential customers.

The same persuasive indirect plan is useful in writing the sales letter—with a few changes for this special situation. The most effective sales letters generally follow these principles:

- **Grab** the reader's attention in the very first paragraph!

- **Introduce** the product and create interest in it by describing the product's features very specifically and by introducing a central selling theme.

- **Build** the reader's desire for the product by emphasizing how the product will meet the needs of the reader and by using vivid language to involve the reader.

- **Motivate** the reader to take action by giving the reader a reason to act, by making such action easy, and by ending on a positive note.

These sales principles have been a part of American sales since the late 19th century. Notice how they work in a letter (Exhibit 8.11), written just before World War II to sell a device to make hernia ruptures less painful.

The American public is used to persuasive sales letters. Therefore, companies often try a wide variety of strategies to get the consumer to open the envelop (for example, making the envelope look as if it contains a check or an important document, inserting a free sample, or using a large font to promise a gift).

Understanding What Makes People Buy Products

Depending on the price of the product, more or less effort must be made to convince the consumer to try the product. You might expect, for example, that different strategies are used to sell a candy bar versus an expensive sports car.

Many times sales letters are organized around Abraham Maslow's hierarchy of needs. Maslow, a famous psychologist, identified a ladder of needs and said that people cannot satisfy their needs for self-actualization until they are safe and secure.

Sales letters have also been influenced by ideas from Aristotle, a Greek philosopher in the 4th century BC, who proposed that the most persuasive writing was based on three different kinds of appeals for readers:

- *Pathos* = appeals based on emotion (This candy bar will make you feel good).

- *Ethos* = appeals based on character (This candy bar is manufactured by a company with the highest qualities and the most prestigious reputation).

- *Logos* = appeals based on logic (This candy bar is nutritious).

Part III: Building Communication Skills

■ Exhibit 8.11 Letter Using Several Sales Principles

Chas. Cluthe and Sons

Proprietors of The Cluthe Truss for Rupture
Bloomfield, New Jersey

Feb. 28, 1939

Mrs Thos. A. Jones,
Corvallis, Oregon

Dear Madam;

Here is our book "CLUTHE'S ADVICE TO THE RUPTURED."

You will find it contains much helpful information and sound advice.

As you read this book, please bear in mind that it is the result of our years of successful day-after-day experience with every form and condition of reducible rupture.

It tells you how to escape the curse of wearing leg straps and cutting elastic belts or coiled body springs. It shows you how to guard against buying contraptions which do not hold and so do more harm than good.

It puts you on guard against deceptive "free" trials and the snare of worthless liniments; it stresses the importance of proper holding—a condition absolutely necessary for any possible improvement. The book tells how thousands, by following the advice given in this very book, have found perfect holding with ease and comfort for the first time since ruptured, and how many have reported complete recovery, thus making an operation unnecessary.

It tells truthfully and plainly how most rupture appliances are made in large quantities—one like another—and sold with little regard for the particular needs of the wearer. And then the book shows how each Cluthe Truss is completely made up especially for each case to meet each person's individual requirements.

It tells how a properly fitted Cluthe Truss means proper holding and safety under most difficult conditions of strain; no pressure on hips; no leg straps. HERE is a guaranteed holder that is sanitary, waterproof, perspiration-proof, light, yet durable and made for real service.

To enjoy the advantages of a made-to-order Cluthe Truss, simply fill out the enclosed blank carefully, returning it to us with the amount quoted on back of the blank, subject to our liberal 60-day trial plan.

Yours very truly,

Chas Cluthe & Sons

Notice how these ideas are contained in the letter (Exhibit 8.12) written by Kreta Cooper of Jitney Delivery as she sells package delivery to small businesses in Louisville, Kentucky.

Jitney Delivery   Exhibit 8.12 ■

Jitney Delivery

Sept. 15, 1997

Ms. Nancy McDonald
Second Glance
110 3rd Street
Louisville, KY 55330

Dear Ms. McDonald:

2 can get you 10 . . . FAST! For only $2, let us help you speed your packages (10-pound maximum) to your clients anywhere in Louisville's greater metropolitan area. Jitney Delivery is your new, local package delivery service with unmatchable speed, personal attention, and price.

When time matters. . .

Who else gives you the delivery times you want? Guaranteed. Jitney's technologically advanced delivery units are equipped with the most modern mobile telephone equipment. Trained drivers of our three jitneys keep in touch by mobile telephone so that delivery routes can be adjusted to meet your needs throughout the day.

When customer satisfaction matters. . .

What can you expect from us? Your satisfied customers are important to us. Our courteous drivers deliver your packages person to person. With set delivery times, packages don't have to sit on doorsteps unattended. Customers don't have to wait until the next day for their packages because they missed an unscheduled delivery. Your clients will receive their packages when they want them. **Guaranteed.**

When price matters. . .

Why deliver that package yourself? For only $2—much less than what it would cost you in time and gasoline—let us deliver your packages (10-pound maximum) to your clients in the greater Louisville metropolitan area. We can also deliver packages up to 50 pounds (rates vary per weight).

<u>Find out for yourself how we can help you!</u> Take a moment, right now, to order our service by phoning (513) 555-1122. One of our courteous staff members will take your order. Or fill out the attached postage-paid card and mail it in for additional information.

Sincerely,

Kreta Cooper
President
Jitney Delivery

Exercise 8.9   Analyzing a Sales Letter for Its Appeals

Examine the advertisement below. Use the principles for writing a persuasive sales letter that are summarized in this section and the ideas from Maslow and Aristotle to identify the strategies this writer used to convince female students at Oregon State University that they need to buy a Harley American moped.

List the strategies the writer used and decide if this sales letter is effective. Bring your analysis to your discussion group and be prepared to discuss the strengths and weaknesses of this sales letter.

Is This Your Idea of Glamour?

The Female Cyclist? Arriving on campus, sweaty and disheveled? Crusty Levi's, soggy sneakers?

There are probably 10,000 bicycles stowed around OSU, and for some students that's the answer for the 10 to 15 miles the average student must travel each day.

But have you just preferred to walk? Ending a rainy February day sodden, gloomy, tired? Shoes ready for Goodwill. More ready to flip on the TV than to take on the next academic challenge?

At American Motorcycle Classics, the HARLEY AMERICAN MOPED means utilitarian transportation:

    155 MPG   SIMPLE   RUGGED   RELIABLE   AFFORDABLE

155 MPG—that's less than 2 cents a mile, cheaper than shoe leather! On special this month at only $399, with normal financing, your payments will be only 58 cents a day.

You'll be proud of the quality and practicality of your HARLEY AMERICAN MOPED: The frame is strong single-unit construction, built like a motorcycle. The four-stroke engine is long wearing. At 30 MPH, you can move with city traffic and have power to spare. With two luggage racks you're ready for trips to laundry and grocery as well as class. A headlight that comes on automatically and handy turn signals are designed for safety. Front and rear suspension give comfort and control . . . the specifications are detailed in the accompanying brochure. The Four-Stroke Difference.

On weekdays you can arrive on campus in style, cool and composed, with the polish of a professional woman. On weekends, your HARLEY AMERICAN MOPED is as much fun to ride as it is practical on weekdays.

    SPIRITED   SLEEK   LIGHT-WEIGHT   READY TO GO

For winter clearance prices and to be ready to explore country roads on spring days, come by to test drive the HARLEY AMERICAN MOPED during February.

                                            THE NEW HARLEYS . . . RIDE ONE.

                                            Peter Fonda
                                            Sales Manager

Chapter 8: Writing for Customers   **271**

EDITOR'S CHECKLIST FOR WRITING A PERSUASIVE SALES LETTER

Use this checklist to help focus your revision before you complete your persuasive sales letter. Ask someone in your discussion group to act as your peer editor and to answer the following questions about your rough draft.

Encourage your peer editor to either set aside about 20 minutes to work through the list with you, discussing each question, or to write out sentence-length responses to each question.

Check Audience and Purpose:

___ 1. Do I know the audience? Do I know why the audience is neutral or may resist my proposal or presentation?

___ 2. Can I tell why this letter is being written? Is the writing purpose clearly persuasive?

Check Content and Organization:

___ 1. Does my opening "grab" my reader's attention?

___ 2. Have I created interest in my product or service by describing its features around a main idea?

___ 3. Have I used vivid description to motivate my reader to care about my product or service?

___ 4. Have I emphasized the benefits of my product or service for my reader?

___ 5. Have I made it easy for the reader to say "yes"?

___ 6. Does my letter end positively?

Check Style and Document Design:

___ 1. Is my tone positive, courteous, and professional yet enthusiastic?

___ 2. Have I emphasized "you" over "I"?

___ 3. Are the word choices clear, avoiding jargon?

___ 4. Have I used short paragraphs and lists to ensure easy reading?

___ 5. Does my letter follow a conventional format?

Allow Time for Editing and Proofreading:

___ 1. Has my letter been edited for correctness in spelling, punctuation, and grammar?

___ 2. Has my letter been proofread for typing errors?

272   Part III: Building Communication Skills

WHAT'S COMING NEXT?

This chapter has helped you think about ways to respond to different kinds of letters that solve the needs of your customers. Whether you are writing to customers, to co-workers, to people outside your employer, or to government agencies, you must balance the needs of your audience with your employer's needs—and your needs.

How important is this audience or this request to your company? Is this a routine letter that requires informative writing or a special project involving analytical and persuasive writing? How much time should you allow for this task?

Once you master the writing skills, your main challenge will be time management. How can you set priorities efficiently and effectively? Time management experts suggest the following tips:

- Set aside the first 15 minutes each day to anticipate and plan your daily work schedule.

- Use a daily and monthly planner, preferably one that does not require that you recopy tasks each day.

- Prioritize your daily "to do" list.

- Highlight important deadlines, especially when working with a team.

- Review second- or third-priority work once a week. File it, delegate it, or trash it!

You can successfully manage a large workload by tightening up how you organize and complete individual tasks. After all, every large construction project—a new bridge or a building—starts with a drawing.

Sometimes expected work or deadlines create pressure. You can reduce stress by tackling tasks one at a time. But only you can manage your workload by saying that most difficult word: No. You can make it easier for your audience to accept your "no" by identifying another resource that could help solve the problem. This approach will help you meet your own needs as well as the needs of your co-workers, your staff, and your supervisors.

The next chapter, Chapter 9, Writing for Managers and Teams, focuses on an area that requires very skillful communication skills; however, often this area is not talked about. How do we best communicate in a production environment? How do we work well on writing projects in teams? How do we present formal proposals to our supervisors? These writing and speaking strategies will be reviewed in Chapter 9.

Chapter 8: Writing for Customers   **273**

CONCEPT REVIEW ■

Can you define the following key concepts from Chapter 8 without referring to the chapter? Write your own definitions in a journal entry or review them out loud.

buffer	"you" focus
customer base	product mix
customer loyalty	consumer
demographics	goodwill
market-driven	service sector
guide letter	form letter
collection letter	dated action
confidential	proprietary

SUMMARY EXERCISE ■

What's Wrong With This Letter?

Read over the following draft of a response to a request for a refund prepared by one of your co-workers. Make a list of the changes in content, organization, style, and format that could improve this letter.

Share your list with your discussion group. If your instructor wants you to revise the letter, please do so, turning in the final copy with all notes to your instructor.

Robotics International
Miss Sandy Smithers
Claytonville, OH

Dear Sandy Smithers,

I am incredibly sorry your order for our Model 280XTRA automatic lawn sprinkler which you said you needed right away was lost in the mail. I am sure that anyone could have lost this in the mail, but it just so happened it was your turn to experience the fickle hand of fate (ha ha). More seriously, our legal department informed us that to get a refund of the original price of $280, you must without fail send us a certified copy of your check within 10 days of the date of this letter. Do not fail to send this documentation in support of your refund. I will hold your file in abeyance until receipt of the documentation and I must deduct a service charge of $35, so don't expect a full refund. Please do not hesitate to contact the undersigned if you have any further questions. Enclosed for your perusal is our new spring catalog, with breezy spring lawn decorations, gardening tools, and watering systems. You may place your order with a certified check. I am very sorry for any inconvenience this may have caused you, and I am sure you will continue to be a valued customer!

Very Sincerely,

CUSTOMER SERVICE
Ronnie de Selva

274   Part III: Building Communication Skills

■ ASSIGNMENTS

Assignment 1: Write a Persuasive Sales Letter.
Working alone or with another person, identify a product or service you are familiar with. First, write a short definition (about one paragraph; you can use a list) that defines a very specific audience for this product or service. Your audience must be a specific targeted group such as, for example, Peter Fonda's female student bicyclists.

List the product features, develop a central selling theme, and identify the benefits of your product or service for your targeted audience. Then, using vivid language, draft a sales letter, following the principles listed in this chapter. Use the editor's checklist as you work with your discussion to review your rough draft.

When you have finished revising, editing, and proofreading your letter, turn in the final version, including all drafts and preliminary notes.

Assignment 2: Analyze Customer Service.
Working alone or with another person, make a list of the key areas for which you think customers should expect good service. Visit two companies that provide similar products or services (for example, restaurants, hardware stores, department stores, gas stations, or shoe stores). Observe how customers are treated, how the layout of the store serves customers' needs, and observe the quality of the product or services. Summarize what you observed and what you conclude about customer service in a memo to the class.

Assignment 3: Interview a Supervisor.
Considering class discussion, your own experiences as a customer, and the field you are planning to enter, list the questions you have about the importance of customers.

List three "what" questions (e.g., "What do customers need most?"), three "how" questions (for example, "How do you serve customers?"), and three "why" questions (for example, "Why are customers important to your business?").

Set up an interview with a manager of a store or office related to your field. If you are in nursing, for example, you could interview someone who works on the admitting desk of a local hospital or a nursing home administrator. Ask the questions you have listed, and encourage the person to share his or her impressions with you. Write up your findings and conclude your written report with advice for the class. Present your findings to the class in a three-minute informal report.

Assignment 4: How Do I Build Customer Loyalty?
Think about and list the experiences you have had when working with customers. Be specific about the problems with customers and actions you took to solve such problems. Were there specific strengths and weaknesses in customer service at any employer you worked for? What caused these strengths and weaknesses? If you could change any process or policy regarding customers, what would it be and why?

Write a short report about your conclusions for your class and your instructor. End your report with a list of the key employee skills you believe are most important for building good customer service. You may also present your conclusions in a 4-6 minute report to the class, including time for questions.

PROJECT: WE CAN DO A BETTER JOB! ■

Your assignment: Working with a partner, visit a store that provides consumer goods of any kind and analyze the quality of customer service. Write a memo to the manager of this store that suggests ways that customer service could be improved. Include the following items in your memo:

- Begin with a tactful description of the problem.

- Specify the benefits and weaknesses of any changes you recommend. Emphasize what is best for the audience you select.

- List performance criteria that a new system would have to meet. Consider employee courtesy, responsiveness to customer requests, employee availability, and any other factors you think are important in providing goods or services to the general public.

Suggestions for getting started: Work with another person to draft an outline. Draft questions that will be answered in each section of the outline, and then draft the answers to each question. After you are satisfied with your rough draft, share it with your discussion group for feedback. Revise, turning the final copy in to your instructor.

JOURNAL ENTRY ■

Do you agree with this popular saying: "The customer is always right!"? Write a journal entry that explores your reaction to this concept. Support your answer with two or three examples from either personal or work experience. Plan to discuss your journal entry with your discussion group or the class as suggested by your instructor.

9

Writing for Teams and Managers

Meeting and Presenting

Chapter PREVIEW

When we are at work, nearly everything is influenced by verbal communication. Whether through formal meetings or a hallway chat, most people find out what is happening by talking to their co-workers and supervisors.

If you have written a proposal to purchase a major piece of equipment or want to redesign how work is done in your work area, your ideas may be presented on paper. However, you will need to verbally convince the decision makers who will say yes to your proposal.

This chapter will help you develop your meeting skills by understanding what happens in a meeting, how to use an agenda, how to moderate discussion in a meeting, and, perhaps most important, what to do after the meeting is over.

This chapter will also help you practice several presentation skills. How do you prepare for a meeting? How do you decide which visuals or handouts to use for a meeting?

Journal ENTRIES

Begin by thinking about your work in small groups. How was information presented to you and how did you present information? Write a one-paragraph response to any of the following questions in your class journal or prepare the answer for class discussion, as directed by your instructor.

1. What does the phrase "just another meeting" mean to you? Describe a meeting in which everything seemed to go wrong. What were the key factors?

2. Describe a meeting in which everything seemed to go right! What were the key factors?

3. What is the difference between a formal meeting and an informal meeting? Why do people need formal meetings? What is the purpose of an informal meeting?

4. Some meetings are called so supervisors can make decisions (termed *top-down* planning). Some meetings are called so that co-workers can solve problems (termed *bottom-up* decision making). Give an example of each type from your own experience. Could the same goals be met without a meeting?

Chapter READING

When a company must change how it operates or is faced with drastic change in markets, technology, or staffing, workers often find out what's happening or how to prepare for change at meetings.

Part III: Building Communication Skills

In the following article by David M. Upton, meetings are used as a key vehicle for planning and carrying out change. As you read it, notice how the responsibility for change is turned over to work teams—and how such change is coordinated through meetings at Mead Corporation's Escabana paper mill.

COMBINING COMPUTERS AND PEOPLE TO BUILD FLEXIBILITY

David M. Upton

Managers and engineers have long preached that computer systems should complement, rather than replace, the skills of operators. In the end, however, most have embarked on a path to computer integration that caused them to place machine over man and resulted in less flexible rather than more flexible factories.

What accounts for the disparity between idea and execution? The problem rests with how—or whether—managers define the type of flexibility to pursue and then choose the appropriate computer systems, work practices, training programs, incentives, and measures. To create a highly flexible computer-and-people-integrated manufacturing system, managers of each individual factory have to come up with their own unique formula. The managers of Mead Corporation's Escanaba Mill in Escanaba, Michigan, which makes coated fine paper and employs 1,300 people, did just that.

In the early 1990s, the Escabana complex, like the paper industry in general, was struggling: The market was in a deep slump, there was a glut of capacity, and price cutting was rampant. To make matters worse, the mill was facing intensifying competition from a growing number of competing mills with bigger and newer machines. Finally, a spurt of imports from Europe raised concerns that foreign producers, which had never posed a threat, might be planning a major assault on the North American market.

Like most of their peers in the paper industry, Escanaba managers had long believed that the key to competitiveness was to achieve the lowest possible costs through long production runs and few product changeovers. But suddenly many in the industry were forced to rethink those assumptions and look for new ways to distinguish themselves from competitors.

The answer for Escanaba, its managers decided, was to be more responsive to customers than their competitors were. They would accomplish that goal by being highly mobile—able to change production schedules quickly so that they could fill orders much faster than competitors could. At that point, the industry's customers might have to wait as long as two weeks to receive an order, even in slack times. Escanaba's managers set out to slash that time to one or two days.

The new strategy of emphasizing responsiveness demanded a new way of working in the mill: It required faster product changeovers and nimble decision making. When managers analyzed how the mill had been operating, they came to a sober conclusion: The plant was slow in switching from one product to another and in changing schedules to accommodate new customers' requirements. It took too long to execute a production-schedule change, and grade changes often generated

Chapter 9: Writing for Teams and Managers **279**

product-quality problems or mistakes. They also realized that the plant was not inherently incapable of quick changeovers; changeovers hadn't improved, because improving them had not been a high priority. Managers had been judged primarily on their success in maximizing the mill's capacity utilization and product quality, and they simply focused on long production runs to achieve their performance targets. The long runs meant that operators did not have to learn how to improve changeovers. They also produced a culture that placed very little value on responsiveness to customers. "We make it; you [the sales force] sell it—that's the way the manufacturing people used to see things in this mill," recalled Henry Swanson, the mill's manager of process control and information.

The mill's managers also realized that computerization in itself was no panacea. Some of the machines in the mill were relatively highly computer integrated. "Even though we had a lot of computers on the plant, sometimes we'd make a couple of hours' worth of production before someone realized that something was wrong," Swanson said. "It was just too easy to trust that the computers had got it right." In addition, the opaque computerized system prevented workers from learning how they might improve operations.

So one of the first steps that managers took to transform Escanaba was to rip out the old millwide computer systems. They replaced them with a new system called QUPID (for Quality and Information for Decisions). Unlike the turnkey systems that previously coordinated manufacturing processes, QUPID was custom designed to support operators in each operation; the operators controlled the manufacturing process and would be free to make changes, depending on what they saw happening on the production line. In other words, the system was designed from the outset to help workers make better decisions rather than to cut them out of the decision-making process.

To that end, the mill's managers insisted that operators be intimately involved in the system's design and development. "If the previous system taught us anything, it taught us that we didn't want black-box computer integration that only the vendor really understood," said Glendon Brown, the mill's vice president of production technology. "We needed an architecture that we were part of and that was much more open and easy to change."

Rather than depending on a single supplier, the mill bought the system's building blocks from several different sources. There were two fundamental criteria for choosing those sources: Each had to be a leader in its area of expertise and be willing to customize the system for Escanaba.

Significantly, Escanaba designed the system's interfaces in-house. Each function designed its own interface to ensure that its people got the information they felt they needed to do their jobs and in the format that was easiest for them to understand. "If we were going to succeed in the longer term, we needed the ability to make changes 24 hours a day, seven days a week—times when most managers aren't around," Brown said. "We needed to provide useful information to our operators at the lowest level in the organization, so that *they* could make decisions."

The mill also overhauled its training programs. But it took several attempts before managers hit on the right approach. First, Mead made a common mistake: It used technical people to explain to the operators how the new computer systems worked. "It was a disaster," Swanson said. "The technical guys knew too much about the systems and told people more than they needed to know about technical issues and not enough about the business issues and why we needed to work differently."

The mill's managers dropped that program. In its place, they created one designed to help workers understand what they had to do to satisfy customers. The program explained why quick changeovers were critical to the mill's long-term success. And it emphasized that the systems were only a tool to help them perform their jobs better. Operators who had long assumed that keeping the machines running was all that mattered began to look at their work differently.

That program has since evolved. Originally, a professional trainer conducted classes in a seminar-like setting. However, managers came to believe that learning and training should be part of all their employees' jobs. With that goal in mind, managers trained a team of the mill's supervisors and operators to teach others on the floor and in classrooms.

The result of the training program is a plant that now excels at learning. Armed with an intimate knowledge of different jobs and their challenges, the new trainers have played an instrumental role in helping operators become increasingly adept at carrying out quick changeovers and responding to the demands of new customers. Operators are no longer mere machine tenders.

Managers also wisely realized that the mill's measurement and incentive systems had to bolster its new strategy. Accordingly, measures and incentives aimed at maximizing capacity utilization and output and minimizing costs have given way to measures and incentives aimed at maximizing responsiveness and customer satisfaction.

Each year, the mill surveys customers to identify what it is doing well and how it needs to improve. Using the survey, a team drawn from managers and employees throughout the mill then generates specific operating goals. For example, the team might challenge Escanaba to reduce changeover losses by 25% within six months. (This kind of focused goal is much more effective than simply declaring, like managers at all too many factories still do, "We should strive to improve flexibility.") The compensation of manages and superintendents is based on the mill's success in achieving those goals.

The results have been astonishing. The mill's responsiveness and customer satisfaction have increased dramatically. In 1993, it sold the most paper in its history. Escanaba is now the most productive mill in Mead's fine-paper group and has dramatically increased its market share. By emphasizing what workers need to do and providing them with the information they need to do it, Mead has proved that a factory can increase flexibility and, at the same time, boost productivity and lower costs. It's just a question of figuring out precisely what kind of flexibility one wants to achieve and giving people the support they need to achieve it.

Source: David M. Upton, "What Really Makes Factories Flexible: Combining Computers and People to Build Flexibility." *Harvard Business Review* (July/August 1995), pp. 80–81. Reprinted by permission.

Chapter 9: Writing for Teams and Managers **281**

Analyzing THE READING

To build your skills, use the following questions to analyze the content of the article or develop your reaction to it. Write a one-page journal entry in your class journal or prepare your answers for class discussion.

Reading for Content

1. What does the phrase "highly flexible computer-and-people-integrated manufacturing system" mean?

2. Why did the managers at the Escabana paper mill want to change their operating strategies? What changes did they want to make? What were the key problems preventing the changes they wanted?

3. Upton says that the culture at the Escabana mill "placed very little value on responsiveness to customers." What does this mean? What is a "corporate culture"?

Reading for Reaction

1. Upton says managers at the Escabana mill decided they "needed to provide useful information to . . . operators at the lowest level in the organization, so that they could make decisions." This kind of teamwork to diagnose problems and come up with solutions is increasingly common in the workplace. What skills are necessary for such teamwork?

2. Swanson, a key manager at Escabana, said, "The technical guys knew too much about systems and told people more than they needed to know about technical issues and not enough about the business issues and why we needed to work differently." What changes would you suggest that the technicians make in their meeting and presentation strategies?

3. Upton points out that at Escabana "managers came to believe that learning and training should be part of all their employees' jobs." The team of supervisors and operators was then set up to train their own co-workers. What are the advantages of such an approach? What problems would such a work group face? What specific steps did the company take to help support the changes?

4. Why would reducing the "turnaround time" or improving customer service be important to any company?

5. The article ends by emphasizing how much the manufacturing workplace is changing. What are the specific ways that the writer suggests that the manufacturing workplace is changing? How can workers prepare for these kinds of change?

Part III: Building Communication Skills

TRANSFORMATION IN THE WORKPLACE

Increased technology, improved ways of communicating with each other, and the gradual shift from manufacturing to service kinds of jobs have changed the American workplace. Everywhere, workers who can work on teams and work with computers are needed for newer, more technical jobs.

These changes in how the average worker works are happening everywhere—from nursing staff in a hospital to the police officer working out of a patrol car, from a process operator in an electronics company to a inventory parts clerk working in a warehouse.

We'll see a transformation of the workplace as companies switch to new technologies (especially the use of computers). Workers will have access to more information and greater control over processes; they also will have the ability to manipulate the information about these processes for more sophisticated analysis.

To keep a job and to move up in responsibility, you will need:

- Good communication skills to share and control the information.

- Abilities to work well independently and in teams.

- Strong statistical skills to analyze and use the information.

- Great flexibility to adapt to continued changes.

This chapter will focus on two key concepts: (1) understanding communication styles and the "corporate culture" where you work, and (2) using the meeting skills you'll need for successful teamwork.

TEAMWORK AND CONTINUED CHANGE

As companies cut operating costs by shifting to computerized processes and downsizing wherever possible, we can anticipate that the greatest changes will happen in *how* the work is actually done. Every aspect of work is changing—from inventory management to cost accounting. These changes will require higher skills and greater flexibility of all workers.

Changes in Manufacturing

From a Manufacturing Line to Flexible Work Teams.   Henry Ford radically changed production in 1908 when he invented mass production by standardizing and "deskilling" the assembly of automobiles. Instead of each worker starting from scratch and completing each car (the skilled craftsman

Chapter 9: Writing for Teams and Managers

approach), Ford had each well-paid worker completing only a few standardized tasks on the car as it passed by on a moving assembly line. The results were products with reasonable quality and low prices. The "line" could be sped up. Standardization of parts and the deskilling of each job led to interchangeable workers and high production.

Today, manufacturing companies of all sizes are adapting their assembly lines to meet the needs of customers through *flexible production*—producing high-quality but different products at reasonable prices that meet individual customer's needs. Today, workers on assembly lines are just as likely to work in small, roving teams that first check the production quantity or quality of increasingly complex machines and then reprogram the machines for the next run. Such work teams are made up of very skilled maintenance people, technicians, and process operators.

Changes in Inventory

Inventory Management Requires Good Communications. Another example of increased innovation is how companies manage inventory. Keeping enough inventory stock on hand to meet customer demand without investing too much money in too much inventory is a problem faced by all companies with tangible products.

Many companies use some kind of "just-in-time" (JIT) inventory management system, in which the right amount of inventory is delivered to the company at just the right moment. Through a JIT inventory delivery system, the company can save money in two key ways: smaller warehouse facilities are needed and fewer products need to be stockpiled.

In order to work, JIT management depends on good communication at all levels of the company. Imagine operations not knowing when marketing made a major sale of a special kind of product. How could operations deliver what marketing sold? How much money will the company lose if manufacturing must shut down because it doesn't have the right raw materials to process into finished products?

Computer technology is being used to further refine JIT inventory systems. One trend is the tighter connection between companies and their *vendors* (companies that provide the raw materials, goods, and services needed by the employer to produce a product). Often, they share a computer network system to track materials and ensure their delivery "just in time." Such systems are called *vendor-managed inventory systems.*[1]

---

[1]Mitch Betts, "Manage My Inventory or Else!" *Computerworld* 28, no. 5 (January 31, 1994), p. 93.

Part III: Building Communication Skills

A few companies have linked suppliers, distributors, and customers electronically so that purchase orders can be written, authenticated by the purchaser, and delivered by the supplier—all by means of the computer.[2] This changes how the customer and the company communicate.

Computers also have been used to tighten the control managers have over how much inventory is ordered. For example, clerks at K-Mart's 2,300 stores nationwide use hand-held laser scanners to track inventory by "reading" bar-coded labels on products right on the shelf. Every night, the inventory information is sent electronically to company headquarters in Troy, Michigan, where teams of managers decide what products are reordered and reshipped. This information also can be used to track how profitable each store is and to analyze customers' preferences in different parts of the country.[3]

This ability to track customers' preferences *daily* gives regional and national managers greater control over inventory costs and new tools to measure the profitability and productivity of individual stores.

WORKING WITH A TEAM

You may have worked as part of a team before you graduated from high school or during the last several years at work. What does working with a team in a highly technical workplace require?

Think how a nurse in an operating room or a paramedic in an ambulance is part of a team of highly skilled workers. Think of a welder assigned to work on fabricating an experimental airplane. This welder is part of a team responsible for one part of a very large job. Each team planning and building that airplane is connected, most likely by computers. As more and different kinds of computers are used to plan, measure, control, and test the work being done, we will need more skills to work in this environment. Not only are the tools we work with changing, but also the ways in which we work are changing. We must be ready to:

- Follow orders yet work independently.

- Use a variety of computers and technical processes.

- Plan schedules and meet deadlines.

- Solve problems as they occur.

- Work flexibly with different teams of workers (from technicians to supervisors).

---

[2]James Daly, "What Happens When 'Close Enough' Isn't Close Enough Anymore? *Computerworld,* 27, no. 1 (December 28, 1992), p. 38.
[3]"Remote Control: Retail Technology." *The Economist,* 327, no. 7813 (May 29, 1993), p. 90.

You will also need the writing skills we have been working on throughout this text. You may need to write *descriptively,* by efficiently producing a variety of forms, memos, and instructions or manuals that will support your team's work by summarizing and documenting routine work, such as troubleshooting, inventory management, and equipment use.

You may need to write *persuasively*, by preparing proposals or short reports that recommend changes in equipment or policies and procedures that control how the work is done.

You can improve the productivity of your team by understanding how different individuals respond to new situations, or, in other words, by understanding their learning styles.

Understanding Learning Styles

You will work with a great mix of people who have very diverse communication styles. Not only will your co-workers represent different kinds of jobs and different levels of authority, but they will also have different personal styles that affect how they learn new tasks and how they communicate.

Some of your co-workers and your supervisors will be easy to work with because their learning and communicating styles are compatible with yours. Others may seem reserved or hard to talk to. For such people, you need to put forth some effort so you can understand exactly what their needs are and how to reach them.

Knowing your own communicating and learning styles and the styles others prefer can help you to develop strategies to improve your communications and processes. Some larger companies schedule workshops on improving communications or understanding learning styles for their employees in an effort to improve group productivity.

What's Your Learning Style?   How do you find out about your learning style and what other learning styles are possible? Formal tests such as a Learning Skills Inventory or the Myer-Briggs Personality Test can be used to provide you with much information about how you learn and what communication style you prefer.

Although it's always risky to oversimplify, you can gain insight into yourself and others by answering the following four questions honestly:

Situation: You need to learn a new and complex skill for your job. Read the following questions and pick your first choice:

1. **What?** I want someone to present the information to me in some detail so I can understand exactly what the new job involves and how the new process works before I begin.

Part III: Building Communication Skills

2. **How?** I want to be given a manual and a sample, and then I want to be left alone so I can work this out by myself in a "hands-on" setting.

3. **Why?** I appreciate the opportunity to find out how to use this new skill, but I want to ask questions that probe beneath the surface. My favorite question is "why." I want to know why we need this change and how it will fit in with everything else we're doing.

4. **What If?** I want to get more information about this new job, but I already see there may be other ways of doing this that are simpler or might work better. I'd like to try them out. What if we switch our training to every other Friday after work? What if we change how we start using this new skill?

If you can see yourself in one of these basic categories, recognize that as the situation changes, your basic reaction may also change—but your preference will still be the same. You may prefer to work out new problems with a hands-on process, or as soon as you understand the problem, you may want to jump to possible new solutions.

Understanding your own preferences (which generally are set by age seven) could be helpful in understanding how you—and your co-workers—learn new material, make decisions, or solve problems.

What's fascinating about these learning styles is that they tend to be relatively evenly distributed across the population. This means that any team you work with is likely to be made up of folks from these four categories. You can see that people who ask "what if" too quickly for the team will probably make people who are concerned with "what" and "why" a little nervous.

Learning Styles of Your Co-Workers.   As you work with different teams, notice how people react to new situations and how they prefer to work.

You can adjust your own style to others or present information in ways that will help your co-workers be successful, by anticipating who will have *what, how, why,* and *what if* reactions. Please note these basic reactions that people have to something new can be complicated by the person's command of English, length of time on the job, ability to think quickly under pressure, or any number of other factors.

Asking these kinds of questions routinely of yourself and your work team can help you develop your own flexibility in responding to others and to different perspectives as they relate to the problem you are trying to solve.

UNDERSTANDING CORPORATE CULTURE

Every workplace has its own personality, its own way of doing things. The ideas, common values and goals, and shared experiences of all the workers at a particular employer create a community of like-minded people. The result is a *corporate*

Applying Learning Styles to the Workplace Exercise 9.1

Describe a situation when you or someone you know had to learn a new skill. What made this experience successful or unsuccessful? What learning style was the most helpful? Did the person focus on one type of question (*what, how, why,* or *what if*) at some stage in learning the new skill?

After you have written a one-paragraph summary, consider what you learned from this experience. List two or three insights you gained about how people learn a new skill.

Then, work within your discussion group. After appointing a recorder, review the results of your thinking about learning styles with your discussion team. Be sure to include some discussion about how people on teams use different learning styles. Turn in your one-paragraph summary and the discussion notes to your instructor.

---

culture, which is an unstated common belief in why people behave as they do at this particular employer.

Understanding the corporate culture where you work can help you to clarify your communication environment. You can observe some *physical clues* to define an employer's corporate culture by noticing how employees dress, how offices are furnished, or who has been given computers.

You can define an employer's corporate culture by observing other more subtle clues, called *process clues,* that track how people interact. For example, how closely do employees follow work hours? What kinds of social interactions are there between employees? Can anyone talk to anyone else at any time? Is your supervisor's door always open? How flexible is the employer in responding to employees' personal problems?

All of these decisions are influenced by the common values of the company's employees, and they are influenced by the company's history and its leaders. Daily processes and small decisions reinforce a company's sense of "community." Corporate culture is what makes the difference between "us" and "them."

This corporate culture will affect everyone you work with—including your supervisor. No matter what the corporate culture is or how it changes, you will always need to consider many different audiences, purposes, and levels of information. You will always need to pick out the essential information for your audience by asking:

- Who is the decision maker?
- Why does this person need this information?
- How will this information be used?
- What needs to happen next?
- What are the deadlines?
- How important is this task or information to my department's or company's goals?

Exercise 9.2 Defining a Corporate Culture

Think of a company you have worked for or a group you have worked with (a summer camp or a church group, for example). Your writing goal is to define the corporate culture for this company or group. Start by listing brief answers to the following questions:

- Did this company or group have strict working rules and supervisors who "bossed" the workers? Was this company or group organized around teams with few supervisors, where the workers set the work rules?
- What was the attitude of your co-workers? How did they dress? Were they on time? What penalties were there if they were late? Did people talk socially while on the job? How serious were employees about working when the supervisor was absent?
- What mottos or key words did your co-workers repeat? For example, at a major electronics manufacturer, key managers might say over and over in different settings, "We are committed to continuously improving our performance." Retail managers might say, "The customer is always right!" Why are these certain words repeated?
- How would you describe the goals of management? What were their attitudes, behaviors, or communication with the workers? Were they considered as distant or part of the group? How did this perception contribute to the corporate culture of this company or group?

Share your answers with your discussion group, noticing the differences between small groups and different types of companies. How would your group define *corporate culture* now, after completing this exercise? Draft a definition of corporate culture, and turn the draft in with your answers to the questions as a team project for your instructor.

DEVELOPING MEETING SKILLS

At the managerial level, careers can be made or broken in a meeting. Relationships between workers can be healed or stressed during a meeting. Used positively, a meeting can focus the work of your team and build positive relationships between co-workers. This next section will combine writing and speaking exercises to develop your meeting skills.

Why Are Meetings Necessary?

Many people dread meetings because they seem like a waste of time; however, meetings can be key ways for technical people to exchange information about production, deadlines, priorities, and problems.

Most companies follow some sort of "meeting protocol," which has been built over long periods of time. You can find out what value your employer puts on effective communication by noticing some of the factors that influence meetings (see Exhibit 9.1).

No meeting is ever perfect, but if each person takes responsibility for thinking about the factors that affect meetings and the quality of the work being done, the result can be focused, quality work in a short time span.

Chapter 9: Writing for Teams and Managers   **289**

Factors That Influence Meetings   Exhibit 9.1 ■

Physical Factors Can Affect Meetings

How large is the meeting?

The larger the meeting, the more difficult it is for members to communicate. A group of 5-7 people is an ideal size for a work team; it is large enough to encourage critical discussion, uneven in number to prevent people from "taking sides," and small enough for everyone to participate.

Does the meeting start and stop on time?

When meeting beginnings and endings are "fuzzy," people may feel their time is not being respected.

Does the meeting room support quality discussion?

Too small or too large of a meeting space can affect the quality of discussion. Even the shape of a table can make a meeting more formal (and it can make it more difficult for people to share their ideas).

How long is the meeting?

Once people work for more than one hour, their attention span begins to droop, unless they have been well prepared ahead of time.

Planning Factors Can Affect Meetings

Is the meeting set at the right time?

Scheduling a meeting when other priorities must be met can ensure that people will not be able to attend or will be distracted.

Is better information gained from informal conversations?

The main purpose of a meeting is to share information as a base for action or decision making. If people feel the best information can be gained from other sources, the purpose of the meeting is undercut before it begins.

Is too much time spent on routine issues?

Because time is so limited, every person has a responsibility to bring important issues to the larger group. If meetings are perceived as "rubber stamps" or vehicles for one-way communication, commitment to the meeting is affected before it begins.

Personal Relationships Can Affect Meetings

Is the tone of the meeting professional?

Personalities sometimes clash. A good meeting can rebuild bridges between people by focusing on the common goals people need to achieve.

■ Exhibit 9.1   (concluded)

Is discussion encouraged?	For people to solve problems in groups, flexibility must be emphasized. Ideas must be encouraged, and everyone's opinion must be respected.
Does politics dictate the agenda?	Most of the real work of the meeting occurs before the meeting actually happens. Politics (how those with real power affect what gets done) is inevitable; however, good communications require a dialogue, not a one-way conversation.

Using meetings effectively is such an important skill that companies will often train their employees so that everyone will follow the same general "rules." Of course, it's simpler for everyone when people at all levels agree about how to share information and make decisions, but the larger your company is, the more likely there will be disagreement about the best way to work.

We often cannot come to agreement unless we share a common commitment to the company's overall goals. We all need these clear goals to guide our day-to-day work. That effort to build a working team is the hidden agenda behind every meeting you attend.

Exercise 9.3   Analyzing a Meeting

Write a one- or two-paragraph summary that describes any meeting you have attended recently. Introduce the group by identifying its purpose. Describe the members and where and when it meets. Include other details that would help the reader understand this group. Then describe the meeting you observed and the cues the group gave about how work would be completed.

Use the list of factors affecting meetings (Exhibit 9.1) to help you describe the group's communication style, how readily it shared information, and how conflict was handled (if applicable). Review the material in Chapter 1: Writing in the Workplace to identify the roles group members play and what phase the group is in (forming, norming, storming, performing, etc.).

Write another one- or two-paragraph summary that explains your reactions to what you observed. Based on what you observed, how successful and efficient do you think this group will be? Conclude your summary by analyzing what will happen next in this small group or by giving your personal reaction to this group.

Bring your draft summaries to your discussion group for review. Turn your final summaries in with supporting notes to your instructor.

What Happens in a Meeting?

Meetings occur when people from different work areas or levels in the company need to share information or make decisions together. Some work groups meet daily; other meetings may be scheduled weekly, every other week, monthly, quarterly, annually—or when an emergency occurs.

Chapter 9: Writing for Teams and Managers **291**

When a meeting is successful, everyone arrives on time and is prepared for thoughtful discussion. People contribute their ideas without fear of criticism, and when the meeting is over, everyone has a clear idea of what decisions have been made and what actions must be taken. In reality, this kind of meeting rarely happens, but you can help create the "ideal" meeting by being aware of the practical steps you can take to make meetings effective and efficient.

Formal Meetings.   In *formal* meetings, a strict agenda may be used, and people may be asked to make presentations that meet a specific time limit. These formal presentations may be supported by handouts, computer-generated graphics, slides, flip charts, or overhead projectors. Each one of these presentation technologies requires skill, practice, and experience to use well. For example, Robert's rules of order may be followed to guide the presenting, seconding, and voting on motions, and to moderate discussion. Formal meetings may be very large (from 20 to 50 people), and they are more likely to involve people from outside the company.

Informal Meetings.   *Informal* meetings are likely to be smaller and involve people from inside the company. You may or may not see an agenda, and when people present their ideas, presentation technologies tend also to be informal (such as flip charts instead of computer-generated slides). People making presentations at a smaller meeting can expect more discussion, and they need skills for anticipating and responding to questions (sometimes even hostile questions).

MANAGING THE MEETING

Two of the most difficult communication skills are to present information at a meeting and to facilitate the work of a meeting.

A key part of any meeting occurs when individuals summarize verbally the key information that is needed by this particular work group. If you are asked to make such a summary, you will need to prepare for a thoughtful discussion before the meeting begins.

If you are the chair of a committee or you are leading this work team, you have an added responsibility to facilitate discussion in a way other participants do not. After the meeting is over, you may need to summarize the highlights of the meeting either verbally or through written minutes of the meeting.

In this section, we will consider:

- Preparing for a meeting.

- Selecting graphics to support a presentation.

Part III: Building Communication Skills

- Preparing an agenda.
- Facilitating discussion.
- Writing minutes.

Preparing for a Meeting

If you are a participant, the best preparation for a meeting is to understand the goals of this particular meeting and to prepare for the discussion by reading the agenda and reviewing any supporting material *before* the meeting begins.

Consider the participants in the meeting, the areas of the company they represent, and their needs for information. Sometimes it's helpful to think about the results participants want from the meeting or the project—to clarify the materials you will need.

If you have questions that you need answered, telephone or meet with people ahead of the meeting to try to gather the right material.

If you need to make a presentation, consider how formal the meeting is, anticipate your audience's needs, and prepare the right supporting material. For example, if your audience needs statistical information, use handouts. Don't expect your audience to write everything down!

Deciding on How to Make a Presentation. Most people feel a little uncomfortable when they think about making even an informal presentation. Most of the time, your workplace presentations will be informal, but the "discomfort meter" may go up when you must present to a larger group or you need to persuade a small meeting of your supervisors to approve your project or proposal.

The first step in preparing your presentation is to choose how you will deliver it. The following are four commonly used modes of delivery:

- **Memorize.** Some people feel most comfortable when they memorize every word. Informal meetings, however, can shift direction quickly, and your co-workers may interrupt you at any time. Under the stress of the moment, people have been known to forget where they are. This is not funny if you are the presenter!

- **Read.** Audiences, especially small work groups, may resent being read to. There is little opportunity for interactivity when the presenter reads the paper. This may be appropriate for an academic conference; however, even in this case, discussion is highly valued.

- **Speak from notes or an outline.** This is the preferred delivery method for both informal and formal presentations. Even in large groups, you want to connect with your audience, to be responsive to their needs. Using notes or an outline means you will remember everything you want to say, but you also can quickly shift direction if needed.

Chapter 9: Writing for Teams and Managers   **293**

- **Speak spontaneously.** Impromptu speaking or delivering your presentation without any advance preparation can be risky, especially for a longer presentation. However, you may be called on suddenly to comment. Take a moment to think about the audience, the purpose of the meeting, and your goals before responding.

Once you decide *how* you will deliver your presentation, your emphasis shifts to preparing that presentation. These next sections will help you think about how to prepare an outline and how to select visual aids.

Remember, practice does help you prepare for informal and formal presentations. This may mean practicing out loud and timing your comments to stay within time limits.

Reviewing Materials with a List.

The best way to prepare for a meeting is to set aside some quiet time to review the materials that will be discussed and to make a list of the points you want to cover or the questions you need to ask.

As you make your list, consider the goals of this project and how they relate to the company's goals. This will help you to prioritize your presentation in terms of other work the group is doing, and it will help you anticipate your audience's questions. The following questions will help you to make an effective list.

- What key information does this audience need? Why?

- How do I know I am giving the correct information to this group?

- How much factual support will this audience need?

- Do I need a decision or a recommendation from this meeting?

- Can I expect support or resistance? From whom? When? Why?

- How important is this information to this group? Why?

- How important is this information or project to the company? Why?

- If a decision is needed, how much time is needed to make the decision?

Sometimes our jobs don't allow such quiet time. You will need to find some time either on the job or outside of work hours to think about the "why" before you are under pressure to present or lead a discussion. Make a short list, using the following ideas to clarify your key points.

Develop the Content of What You Want to Say by
- Focusing your analysis on the issue.

- Selecting and organizing the specific support you need.

- Linking your presentation to the overall goal of the team and company.

Support Your Presentation by

- Helping you prepare for the meeting.
- Reminding you of what you want to cover at the meeting.
- Using the meeting time most efficiently.

The real value of such a list is that it is short, contains key words, and helps you remember the complexity of the issue being discussed. You will need to write the list in two steps. First, make sure the issues are complete. Second, revise the list so the ideas are presented as concisely as possible and in some logical order. The example on the following page shows both first and second drafts of such a working list.

Exercise 9.4 Analyzing the Context for a Meeting

Reread Dan's second draft list on the proposed tuition increase and answer each of the following questions to help you anticipate how the audience will respond to Dan's presentation.

- What is the most important information for Dan's audience? Why?
- How much factual support will Dan's audience need?
- How does Dan know he is giving the correct information to this group?
- Does Dan's discussion support the action the board needs to take?
- Can Dan expect any support or resistance? From whom? Where? Why?
- How important is this information to the board? Why?
- Because Dan is requesting action, has he given the group enough time to make the decision?

Can you make any suggestions for Dan to help him improve either the content or order of his presentation? Bring in your suggestions for review by your discussion group.

Audiences.
Additionally, the information you summarize must be useful to different kinds of audiences. Most of the time, you can prepare for very diverse audiences in the workplace by thinking of three different levels:

- **Novices** are beginners, workers who may be new to the company or the project and who need definitions and background before understanding your information or recommendations. Sometimes we forget that novices can appear any time or at any level, depending on their involvement in a particular project.

- **Technicians** are typically experts, workers who can skip background or definitions, but who understand the technical details and implications of your information or recommendations. Technicians can be valuable resources.

- **Decision makers** are usually managers, and they actually make the decisions. They need to see the connection to productivity or overall profitability, what is commonly called the *bottom line*.

Chapter 9: Writing for Teams and Managers   **295**

Situation: Dan is preparing a 10-minute presentation for a community college board on a proposed increase in student tuition

First Draft: Proposed Increase in Student Tuition

- A tuition increase of $2 per credit made next term (or $4 per credit fall term) will solve the problem.
- Our last increase three years ago was $2 per credit.
- OSU, Greenville University, and Lewis & Clark increased tuition every year for two years (average range $7–$12 per credit).
- Our operating costs are exceeding income by $600,000 each year.
- Our tuition is lower than any of surrounding colleges by $24 per credit.
- We don't have a separate computer or technology fee as does OSU & Greenville ($50 per term per student). (Lewis & Clark doesn't either.)
- Students were OK about the last increase but may be concerned now.
- We have a history of lower tuition than other schools in our area.

Dan's first draft has much useful information. He must put these ideas and facts in an order that will be most persuasive for his audience. Notice how he used headings in his second draft to shape his discussion. The headings will help him find information quickly when he is under pressure at the meeting.

Second Draft: Proposed Increase in Student Tuition

Problem:	• Our operating costs are exceeding income by $600,000 (60% set aside for computers for students).
	• Students OK about last increase but may resist now.
Tuition:	• Our tuition is historically lower than other area schools by $24 per credit.
	• Our last increase was three years ago—$2 per credit.
	• OSU, Greenville University, and Lewis & Clark have increased tuition during the last two years (range $7–$12 per credit).
Computer Fee:	• We don't have a separate computer fee.
	• OSU & Greenville charge computer fees of $50 per term per student.
	• Lewis & Clark does not have computer fee.
Discussion:	• Increase tuition? A tuition increase of $2 per credit made next term (or $4 per credit fall term) will solve the problem.
	• Set a technology fee? Technology fee of $30 per student per term adds more computers and upgrades software on an ongoing basis.
Action:	• Decision needed by next board meeting.

Anticipate the questions your audience will have by considering who will attend the meeting and why.

Exercise 9.5 Using a List to Prepare for a Meeting

You have had many experiences as a college student. Prepare a list to support a five-minute presentation on any of the following topics (or another topic of interest) for your discussion group or the larger class, as your instructor directs. Use the questions in this section to analyze your audience and to prepare (and revise) a list that will support your presentation.

- Tips to surviving registration.
- How to avoid writing a term paper over the weekend.
- Which computer skills do students need most?
- How do you decide what career to follow?
- Should computer skills be required for graduation?

Review your list with your discussion group. Turn in both rough and final drafts. Only make the presentation if your instructor assigns it!

Presenting Technical Information.   Many times technical information is needed for a group to make a decision. However, the group may need the information but may not understand it without a translator. This is your job as the presenter—to analyze the audience and determine exactly what kind of information they need and then to translate the information so your audience can best understand and use it.

Plan Your Technical Presentation by:
1. Identifying the range of people who will attend the meeting.
2. Remembering the kinds of information people need to be successful at their jobs.
3. Anticipating the kinds of questions people will have.
4. Observing how others present this kind of technical information.
5. Noting what kinds of questions people ask and the reactions the audience gives.

Prepare for the Meeting by:
1. Identifying and defining any key terms.
2. Linking the goal of this group to the purpose for this information.
3. Using visual aids to present technical information.
4. Highlighting rather than reading the information.
5. Checking details about any problems, actions taken, or overall impact on the group, so you are prepared for questions.

Chapter 9: Writing for Teams and Managers   **297**

6.   Having a colleague review your visual aids for clarity and usefulness.

7.   Planning some time for questions.

Notice that one of the suggestions here, using visual aids, can be very useful for any complex presentation. The next section summarizes the types of visual aids commonly used for workplace presentations.

Using Visual Aids in a Meeting.   Using visual aids in a meeting can be as simple as stepping up to a blackboard or preparing a handout ahead of time. You may find your employer prefers formal presentations that are supported by slides.

Visual aids can make your presentation easier to understand and remember. The more complex or detailed the information, the more likely a visual aid will be helpful. The following list includes the most commonly used visual aids.

- *Handouts* give your audience an immediate summary of your presentation. Agendas, lists, charts, or outlines are often handed out at the beginning of a presentation. These can be used for notes by both small and large audiences.

- *Boards* (chalkboards or whiteboards) can be prepared ahead of time or used by the presenter during the discussion to capture key concepts or questions, especially for smaller groups. Whiteboards can be very useful in encouraging group discussion.

- *Flip charts* (tablets of large paper attached at the top and mounted on a table or floor easel) can also be prepared ahead of time or used during the discussion to record key concepts. Flip charts can be difficult for larger audiences to see.

- *Overhead transparencies* are commonly used in business meetings. You can photocopy anything onto a transparency (a sheet of transparent film) and then project it onto a screen. Be very careful that the print size is large enough to be read by your audience.

- *Slides* and *multimedia presentations* (frequently computer-managed "shows" that combine slides with color, video clips, and sound) are very useful for larger, formal audiences and training programs. These more complex media require good preparation by the presenter and may be cumbersome and expensive for smaller, problem-solving groups. However, new software that simplifies how such "slides" are prepared and presented may result in higher standards for informal presentations.

Some of these visual aids encourage interaction; others reinforce a more passive role for the audience. Always work to match the visual aid you choose with how much involvement you want from your audience.

The key idea to remember is that visual aids *support* or *supplement* your presentation; they don't replace you! The following are some suggestions for preparing visual aids:

- Match the type of visual aid to your audience's needs and size.

- Use consistent titles for all the visual aids in your presentation.

- Double-check that visual aids are easy to read—in a large print and an uncluttered format.
- Use one idea for each visual aid.
- Consider using humor to break up the intensity of your presentation.
- Use visuals to *highlight,* not replace, the substance of your presentation.

Exercise 9.6   What's Wrong with This Visual Aid?

"Read" this visual aid and bring your comments to class on ways to improve this graphic developed for an annual meeting of shareholders. If your teacher instructs you to do so, prepare a revision of this visual.

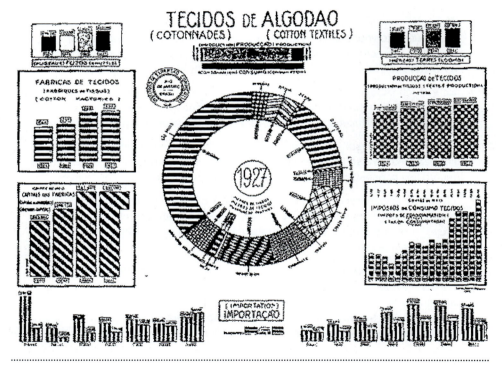

Preparing an Agenda.   Agendas always explain when and where the meeting will occur. They usually include the topics of the meeting, the time allotted for discussion, and the order in which each topic will be addressed. Agendas also may explain who is attending this meeting, and they should be sent to participants *before* the meeting. Meeting reminders and agendas (with supporting material) are often sent by e-mail.

Agendas can be powerful tools because they can be used to prepare people for the meeting and to schedule the discussion of controversial issues at the "best" time. For example, should salary cuts be discussed at the beginning or end of a meeting? This would depend on many factors—If rumors have already started to circulate throughout the company, then scheduling this discussion earlier makes sense. However, if management wants to make an announcement and to squash discussion, the item might not show up on the agenda at all.

Agendas and supporting materials should always be sent before the meeting. Exhibit 9.2 shows a formal agenda sent outside the company confirming a meeting to hold a demonstration and review of specifications for a new client.

Remember that the purposes of an agenda are to set objectives for the meeting, to focus the group's attention on the topics of the meeting, and to schedule the time allotted to each topic. An agenda can also remind key participants of what they need to do to prepare for the meeting.

Agenda Memo   Exhibit 9.2 ■

IDG MANUFACTURING   **INTEROFFICE MEMO**

January 23, 1997

TO: City of Beaverton visitors: Robert Knoll, Susan Daley, Marty Weinstad
FROM: Your IDG Representatives: Marsha Levitt, Cornell Heitz, Carl Wright

---

Meeting Notification

Meeting Date: Tuesday, February 23

Meeting Time: Start 1:00 PM   End: 3:30 PM

Meeting Location: IDG's headquarters
1476 Industrial Park, Missoula

Bring to Meeting: Please bring your final performance specifications for IDG representatives Marsha Levitt and Carl Wright.

Agenda

Meeting Purpose: We'll give you a "hands-on" tour of the two ambulances (Model X-17 and Model X-37) you are interested in so that you can evaluate the design, engineering, and performance of each model and refine your specifications.

Item	Who	Time
1. Review of your specifications and questions	Cornell Heitz	30 minutes
2. Comparative test drive of both ambulances	Carl Wright, Marsha Levitt	90 minutes
3. Debriefing and cost comparisons	Marsha Levitt	30 minutes

Using an Agenda. Unless there are compelling reasons to change the order of the items on the agenda, most people advise following the agenda, step by step. A typical meeting follows this pattern:

1. Review the agenda overall, adding any new or old business.
2. Review the first topic.
3. Discuss and reach agreement.
4. Decide on action for the first topic.
5. Move on to the next topic.

Routine issues can be covered quickly. Very technical or controversial issues (for example, explaining the company's overall performance or a change in computers) require careful preparation before the meeting.

Responding to Hostile Questions. Sometimes hostile questions can disrupt a meeting. Each person at the meeting needs to have the sense that he or she can contribute to the discussion. Questions that appear to be hostile to you personally or to the project often point to issues that must be aired and resolved.

The best way to respond to hostile questions is before the meeting. If possible, meet with people you have already identified to clarify their opposition and perhaps resolve it before the meeting. People may have unanswered questions that have not yet been aired. Cultivate your supporters as well. Encourage people in a one-on-one situation to think about the issues.

If hostile questions emerge in a meeting, to forestall an all-out attack, try to recognize the question and postpone the discussion. If possible, redirect the audience away from personalities and toward the overall mission or goal of the particular task. If people are reminded why some action or decision is needed, they are less likely to be sidetracked by emotion. However, never underestimate how thoughtful discussion can move the group to a new understanding or position.

Exercise 9.7 Planning for a Meeting and Preparing an Agenda

You've been asked to moderate a meeting on staff training. The following people want to share information.

- George Bookmann wants to present extensive background on the previous training the company has provided.
- Anna Robson needs to review the budget allocated to staff training and is very concerned about limiting expenditures.
- Alberto Garcia-Sanchez has information from an employee survey that suggests employees are very unhappy with the current review and promotion process. (Alberto shared the results of the survey with you because your office is next to his and you've become friends.)

Chapter 9: Writing for Teams and Managers **301**

- Your company's president, Rita Wilson, would like to talk to the committee about the importance of training and then leave. Rita is very well respected.

The purpose of this work team of 12 representatives from every area of the company is to recommend specific training for the company. This is the first meeting and will last one hour. You hope to have recommendations for the president within three meetings.

List the steps you would take before the meeting and the advice you would give each person about the allotted time he or she will have at this first meeting. Draft an agenda. Bring your list and draft agenda to your discussion group for review. Make notes and polish the agenda to final format if asked to do so by your instructor. Give the completed project (with all notes) to your instructor.

Leading a Meeting. Leading a meeting takes a different kind of preparation and skill. As a leader, you are responsible for setting the agenda for the meeting and for facilitating the discussion so that the goals for the meeting are accomplished.

If you are working for a company that values collaborative decision making as part of its corporate culture, then you will need to include others in your goal setting for the meeting and for the project. All key decisions will need to be reviewed and approved by the group. Although such involvement is time consuming, it often leads to more support for the group's decisions.

Meetings often are better than the "sum of the parts." The leader of a meeting has the overall goal of keeping the group's work focused. Some committees meet for years without achieving anything. If you are asked to chair a committee or lead a meeting, make sure you know *what* the immediate tasks are, *what* the long-term goal of the meeting is, and *what* the deadlines are for the work done at the meeting.

The following are additional suggestions for facilitating discussion at a meeting:

- *Use an agenda to guide the meeting.* If possible, distribute the agenda (and any supplemental materials) ahead of the meeting.

- *Follow start and end times exactly.*

- *Appoint a recorder to take minutes.* This job may rotate among members of the group.

- *Encourage people to participate equally.* The larger the group, the more difficult it will be for some to participate.

- *Keep the group focused on the task.* You may need to say, "Are we moving away from today's goal?"

- *End the meeting with a summary of what's been accomplished.* People need to know exactly what's been done and what needs to be done next.

- *Set the next meeting* only if you have a concrete task, time, and agenda.

Writing Minutes. Once the meeting is actually in session, if you are not leading the meeting, you may be asked to take notes or prepare minutes. Meeting minutes most often list who attended, what issues were discussed, what decisions were made, and what follow-up actions may be needed. Readers need minutes that accurately summarize what did happen and what should happen next.

If you are the recorder, notice who is speaking. Jot down key words to capture who is speaking, what was said, and what was decided. You can always call someone after the meeting to verify numbers, dates, or important statements—but only if you know who said what.

Minutes should be easy to read and concise. In Exhibit 9.3, notice how the format reinforces the action needed in this example of meeting minutes. Headings are particularly useful in pointing the reader to needed information.

These meeting minutes are not very long, but they do give readers a clear picture of what happened at the two-hour meeting and what should happen next. Employers often have guidelines or an established format that you can follow as you prepare meeting minutes.

Exercise 9.8   Writing Minutes

Follow guidelines given in this chapter to prepare formal minutes for a meeting or class you need to attend. Start by listing the main ideas, then organize the minutes into several short paragraphs that capture the essence of the meeting and the decisions made or follow-up actions that are needed.

Experiment with different formats (lists, caption headings) to highlight the main ideas and any action needed. Make any revisions after considering three different audiences who will read your minutes: someone who is knowledgeable, someone who is new to the group, and someone who will make a decision based on the minutes.

Bring your draft minutes to your discussion group. Analyze which minutes would most effectively document the achievements of the meeting or class and why. Turn in your notes with draft minutes to your instructor.

WRITING STRATEGY: USING SLANG

Much of your communications every day will be verbal and face-to-face. The same principles that you follow to write effectively will be useful in planning your verbal communications, for example, following an introduction + discussion + conclusion pattern.

However, when we talk to each other in the workplace, often we slip into using slang. Using informal wording (like, "Way to go!" instead of "Outstanding job!") can build a team and strengthen personal relationships.

In your writing, even in less formal memos, you'll need to drop the "speech markers" like "well" or "you know" or "right."

Regardless of the slang you may hear on the job, underneath the casual wording are company goals. As you build your team, you become an important player in meeting those objectives.

Chapter 9: Writing for Teams and Managers **303**

Minutes Exhibit 9.3 ∎

LANGUAGE ARTS BOARD MEETING
Chicago Phoenix Inn
March 2, 1996

MEMBERS PRESENT:

Janice Stevens	Lance Rowan	Nat Cooley
Jamie Jones	Rick Howard	George Hardfield
Ann Reynolds	Beth Gordon	Steve Hunnicutt
Alex Campbell	Amy Bervin	Kay Chin
Debbie Teitlebaum	Beverly Baines	Alan Higgins

The meeting followed the spring conference: Multicultural Connections, beginning at 3:15 PM, and was chaired by Kay Chin.

NATIONAL LANGUAGE ARTS ASSN: Special guest Beverly Baines, President of the National Language Arts Association, spoke to us about the NCTE/IRA Standards for English to be released March 12. She urged us to read and talk about them in our schools and communities.

ANNOUNCEMENT: *School Talk,* new elementary publication. Call Rick Howard to subscribe.

MINUTES of the December quarterly meeting were approved.

TREASURER'S REPORT: Beth Gordon (handout). Approved.
Checking:	$ 6,576.54
Mutual Asset	$12,462.25
Mutual Cash	$ 8,063.24

PROGRAM COMMITTEE: Janice Stevens reported that the fall conference would be at McClelland High School again. Possible speaker: Anne Ruggiero. Theme ideas are needed.

RESEARCH COMMITTEE: Ann Reynolds reported only one proposal had been received and it was not acceptable. Suggested keeping budgeted $200 to add to next year. Discussed process to invite previous award winners to give conference presentation.

SCHOLARSHIP: The scholarship committee is currently reviewing applications from 15 high school students for annual award.

STATE DEPARTMENT OF EDUCATION REPORT: Amy Bervin reported on meeting to revise common curriculum goals last month. She handed out new copies of State Standards paper being circulated to all districts with April 8 deadline for comment. FAX comments to Amy.

1997 REGIONAL CONFERENCE IN CHICAGO. Rick Howard reported the costs of rooms downtown at Hilton and Marriott hotels (available March 1–4) at $270 and $260 per night. All other major hotels are already booked. The convention center is already booked completely. Discussion involved need for sufficient conference meeting rooms, budget costs. Overcrowding was a major concern at the last regional conference. Motion by Nat Cooley to have Rick return to Hilton/Marriott and alert them that the room guarantee is too high and to try for better terms that will add conference meeting rooms. Motion seconded and passed.

NEXT MEETING: May 4, Illinois State University after Writing Festival.

Adjourned 5:30 PM.

Respectfully submitted,

Ann Reynolds, Secretary

REFINING YOUR COMMUNICATION SKILLS

Common sense can help you solve routine problems of misplaced or missing information. If you and your co-workers work on the same shift, routine slips in communications can be dealt with face-to-face, the preferred way to improve communication. Even if you work in different buildings or work zones, or seven-day operations and 24-hour shifts, you can develop a sense of the company overall and build good working relationships.

Some people cannot communicate clearly by speaking. They may be terrified of speaking in front of even a small group. How does needed technical information travel within the work group? How does someone with good technical skills but less fluent English contribute to the work group? How does a work team function if half of its members are not really committed to getting the job done?

Some people have never worked in a group. Suddenly, they've been asked to perform in a group, troubleshooting with senior managers or technicians as if they were equals. Entry-level workers have to be able to speak up and participate in the group's decision making. Experienced people have knowledge to contribute, but newer people often see things others do not.

The most important skill you can have when starting any job is to not be afraid to ask questions—especially when you don't understand what you are supposed to do.

Even after you become more experienced, you can avoid the most common pitfalls of poor communication by

- Listening carefully to understand the task and the issues.
- Communicating as clearly as possible.
- Avoiding too much technical jargon or "company speak."
- Giving feedback when you understand what you have been told.
- Asking questions when you are not sure about the task.

Finally, consider your interpersonal skills. Friendly, professional people are easier to work with. People who are antisocial are easy to work around and ignore.

Exercise 9.9   Analyzing Communication Barriers

List some barriers to communication you might encounter in the workplace. In addition to what has been briefly summarized here, consider listing generational, emotional, cultural, and language barriers.

Meet in your discussion group and appoint a recorder. Discuss which are the most important barriers and offer ways to improve communication in three of the problem areas. Turn in your discussion notes to your instructor.

Chapter 9: Writing for Teams and Managers **305**

When Something Goes Wrong.   Minor crises occur every day. Sometimes, you will notice a crisis because company goals must shift to meet the needs of the customers. Sometimes, the sales staff sells orders faster than production can fill them, although this is a problem few supervisors would complain about! Ideally, customer's demands and production's capacity are neatly matched. Other problems may surface because equipment breaks down (software, hardware, actual production equipment) or supplies are not received (JIT inventory is *not* received in time).

Teams of workers need to be able to respond quickly when something goes wrong. Being successful on the job means being able to read and interpret technical information. In a manufacturing setting, workers must constantly run diagnostics, checking to see if production matches performance expectations. If not, teams—made up of production operators, working with maintenance technicians and engineers—must figure out why.

This is not so different from what happens on the shop floor of a local garage where equipment and cars are becoming increasingly automated. Supervisors constantly need to read and interpret data coming in from people working in the field—whether they are truck drivers, paramedics, or police officers. Your ability to work well in a team setting can help to solve problems.

EDITOR'S CHECKLIST: PLANNING AND FOLLOWING UP MEETINGS ■

Before the Meeting:

___ 1. Schedule the meeting at a time that matches your purpose and the needs of the attendees.

___ 2. Check meeting room size, setup, and number of chairs.

___ 3. Confirm key attendees ready for meeting.

___ 4. Order visual aides: overhead projector, VCR, computer hookups (and supplies).

___ 5. Order refreshments, if appropriate.

___ 6. Send agenda with attachments to attendees *before* the meeting.

___ 7. Prepare extra copies of all materials for the meeting.

During the Meeting:

___ 1. Appoint a recorder.

___ 2. Ask if there are changes to the agenda and then follow the agenda.

___ 3. Notice who's talking and facilitate the discussion for even participation.

Part III: Building Communication Skills

    ___ 4. Use the agenda to moderate the pace of the meeting so the meeting can end on time.

    ___ 5. Summarize discussion and ask the group to consider action.

    ___ 6. Take careful notes throughout to make sure you know what needs to happen next, who's responsible, and what the deadline is.

    ___ 7. Schedule the next meeting—only if needed.

After the Meeting:

    ___ 1. Thank participants and staff who helped set up or run the meeting.

    ___ 2. Prepare the minutes as quickly as possible.

        a. Emphasize *summary* (what was discussed and decided) and *action* (what needs to be done now).

        b. Revise for clarity and conciseness.

        c. Double-check key information (names, dates, deadlines, amounts).

    ___ 3. Use mail or e-mail to distribute the minutes to everyone involved.

WHAT'S COMING NEXT?

Many people are concerned that computers will replace jobs. Certainly, computers have changed how people do work. High-risk jobs (generally jobs in production or manufacturing, jobs that are dangerous, and jobs that have high injury rates) and repetitive jobs (jobs that are boring or tiring) have been the first to be automated. In fact, nearly any task that does not require interpretation can be automated.

Although a few people wish that computers would go away, they will continue to transform how we do our work. Not only will the tools and processes change, but also the span of responsibility that individual workers assume will change.

Computers are useful in providing information, but the ability to interpret what's wrong can still be best provided by human workers. Computers can give us tighter control over what work is actually completed, but we have to be able to put all the pieces together.

Meeting skills thus become crucial for success. Companies need people who can work with others and who can motivate high productivity through teams. Companies need people who are curious, who want a challenge, who can shift into new work situations, and who can be responsible for a bigger picture (i.e., pulling together the tools, processes, and tasks needed to meet the company's goals).

In the next chapter, we will investigate how to create a career plan, considering some of these changing aspects of the workplace.

Chapter 9: Writing for Teams and Managers 307

CONCEPT REVIEW ■

Can you define the following key concepts from Chapter 9 without referring to the chapter? Write your own definitions in a journal entry or review them out loud.

agenda	minutes
conflict resolution	group process
performance expectations	performance criteria
chart or grid	flip chart
formal presentation	informal presentation
overhead	multimedia

SUMMARY EXERCISE ■

What's Wrong With This Meeting?

Read the following description of Shawn Roberts' meeting, Then, working alone or with a partner, write a memo to Shawn that summarizes suggestions to improve his meeting skills. Focus on ways to improve what Shawn did before the meeting, what he should have done during the meeting, and what he needs to do now that the meeting is over.

Shawn's Big Meeting

Shawn has called a meeting of the department heads in every area of Mass Marketing, Inc., a national telemarketing company that provides equipment and support services. This is the major presentation he's been worried about for weeks. Today's goals are to gain approval of the a new product line—an automated telephone caller—and to begin work on finalizing the schedule for production. Senior management is enthused about the product and has directed Shawn to spend all his time on it.

Shawn arrives late for the meeting; he opens the meeting by talking about the problems he had with the photocopy machine. He's ready to hand these 20-page packets out right now, but people don't seem ready to review them. They start bickering about the price of the automated telephone caller, the timing of the entry into the market (before or after New Year's Day), whether customers will accept the new product because it requires an upgrade in their telephone system, and whether Mass Marketing should expand its products to automated telephone callers in the first place.

Part III: Building Communication Skills

After 20 minutes, Shawn decides he better change his "to do" list for the meeting because he doesn't want to waste the group's time. He realizes they need the "big picture" before hearing about Victor's marketing study and Marsha's review of production. So he starts his 30-minute slide show, a part of today's presentation that he's invested nearly 60 hours in. However, the right computer support material is not available, so he must hold up his working copies (printouts of his slides, complete with editing comments), and the people at the far end of the table can't really see what he's talking about. After 15 minutes of Shawn's discussion, Marsha leaves the meeting.

At the end of the presentation, nobody has any questions. It's now 5:30 PM; the meeting has lasted longer than the original hour scheduled, and people are ready to leave. Shawn says, "I guess if there are no questions, that means everyone's in favor of our new automated telephone caller. I'll call another meeting to review the important results from marketing and to set deadlines for manufacturing. I'll let you know when the meeting will be. It's probably going to be another late afternoon meeting." As soon as Shawn says this, people start to leave. The big meeting is over, and Shawn is not sure what to report to management.

■ ASSIGNMENTS

Assignment 1: Class Presentation.   Working alone or with one other person, plan a short class presentation on how to solve problems that could prevent an effective meeting. Research this topic to find one or two articles to support your presentation; you may also use your own experience.

Prepare an outline of key concepts your presentation will cover and a list of questions you believe the class will have in response to your presentation. Allow 3–4 minutes to present content and 2–3 minutes for questions. Prepare one visual aid to go with your presentation. After your presentation, turn in your outline and list of questions to your instructor, along with one paragraph assessing the strengths and weaknesses of your presentation.

Assignment 2: Research Conflict Resolution.   Use your library's CD ROM resources or the Internet's World Wide Web (for example, the Alta Vista search engine) to find two or three articles about conflict resolution in the workplace. Look for articles that thoughtfully define the problem and propose solutions. Consider such questions as: How extensive is conflict in the workplace? What types of conflict are there? Over what issues? How do people resolve personal conflict in the workplace? How should a worker respond if conflict affects his or her work team? You may want to expand the topic to include violence in the workplace.

Summarize each article and draw conclusions from the information you have gathered. Present the results of your research in a memo to the class. Include a correct citation for each article and a photocopy of the article with your memo.

Assignment 3: Interview a Worker about Communication Skills.

Working alone or in a small group, list 8–10 questions you have about how workers communicate—formally and informally. Consider how workers give instructions, communicate in meetings, conduct evaluations, and communicate with teams. Consider how communication is affected when deadline pressure is high.

Revise your questions to gather information about how workers read, write, speak, and listen. Refine your questions in your discussion group. Be sure to include questions that ask *what, how, why,* and *what if.* Your goal in gathering this information is to understand in a practical way exactly how workers communicate on the job.

Use your final questions to interview (either in person or over the phone) someone who is working full-time at a local employer. You can contact the company's personnel office or interview a friend or neighbor. Each person in your group should interview someone different.

Summarize your interview results in a memo for your class that introduces who you interviewed (give the person's title and employer), that summarizes what you learned, and that concludes with your reaction to what you discovered. If you are working with a small group, review these interview results to discover common themes. Prepare a brief oral presentation for the class that summarizes the work of your group (3-4 minutes) and invites questions (1-2 minutes). Turn in your individual summaries to the instructor with interview questions and one paragraph assessing the strengths and weaknesses of your presentation.

Assignment 4: Preparing a Briefing.

Your experiences in this class have prepared you to give very helpful advice to the students who will take this class next term. Your assignment (working alone or with a partner) is to prepare a short oral and written report for these students that will help them be successful in this class.

Consider barriers to learning that students experience as well as the types of reading, writing, listening, and speaking experiences you have had in this class. If possible, consider your workplace experiences to identify the most difficult and most valuable parts of this class.

Working before class, prepare a rough draft list of advice. Then, working with your partner or your discussion group, develop several categories of advice, checking for completeness (Does this list fully reflect the content and intent of this class? The needs of the students?). Double-check to make sure that your list is complete, specific, and useful for your intended audience.

Rewrite your advice into a formal memo and turn in all rough drafts with the final project to your instructor. Prepare a brief oral presentation to the class that summarizes the work of your group (3–4 minutes) and invites questions (1–2 minutes). Include one paragraph with your formal memo that assesses the strengths and weaknesses of your presentation.

310   Part III: Building Communication Skills

■ PROJECT: WHO'S ON FIRST?

Part of analyzing any object or process can involve thinking visually. First read the following article by Lynne Curry-Swann, which gives creative suggestions for handling the responsibilities of a self-directed work team. Then, working alone or in a small group, design three charts that could be used to:

a. *Track a team's progress* in planning, researching, drafting and revising a project that needs to meet deadlines.

b. *Evaluate an individual team member's performance*. Use the previous work categories (planning, researching, drafting, and revising) and define performance criteria for each work category. You'll need to set the performance output you want, the deadline you want this by, and the quality expected.

c. *Describe the elements of a specific task*. As you define the categories for this chart, consider what the task must accomplish, how the task is carried out, and what format the task must be completed in. Task descriptions like these are being used by companies all over the world to gain ISO 9000 certification (an international certificate granted to companies that calls for extensive documenting of tasks). These task descriptions are usually developed and/or reviewed by groups.

TRY A CHART

Lynne Curry-Swann

Question: We're trying a self-directed work team and are having coordination problems. We become confused when the team feels they're responsible for decision making and others in the organization feel they should have been consulted and allowed to influence the final decisions.

As you can imagine, this is creating a lot of hostility, both among team members and others. We're also having a difficult time on the team when some of the individuals feel they can make decisions independently and others feel they should be involved. We're also having difficulty to let outsiders know what's going on. Got any ideas?

Answer: Try responsibility charting. Create a graph for each of your major team tasks and projects. On the vertical axis, list the tasks and decisions. Along the horizontal axis, list the names of team members and others in the organization.

Put an "r" next to all those who are responsible for the activity or decision. Put a "c" for all those who must be consulted during decision making or the activity itself. Put in "i" for all those who should informed during the activity or following the decision making.

While responsibility charting won't fix real differences of opinion over who should be involved, it will clarify the process and fix most of the inadvertent coordination accidents you're having.

Source: *Anchorage Daily News,* Monday, August 14, 1995, p. D4.

Your goal is to come up with grids or charts that help you analyze and compare data objectively.

Once your group is satisfied with the charts, conduct a debriefing session to identify the hardest part of completing this assignment and to highlight which visuals your group members think best meet the assignment's purpose. Prepare a memo for your instructor that introduces and analyzes each chart and includes a summary of your debriefing.

JOURNAL ENTRY

This journal entry is about meeting matters. How would you deal with someone who constantly interrupts everyone at a meeting you're attending? Would your strategies be different if you were the chair of the meeting? If this were a large formal meeting? If this were a small work team? If you didn't like the person? If the person was new to the company?

Start by listing your ideas, and then, in two or three paragraphs, describe and analyze the best strategies for solving this problem.

WRITING ABOUT YOUR CAREER

Part IV invites you to develop a career plan and shows you how to prepare a resume and an application letter that can be used to find a job or to move up within an organization.

Whether you are preparing to find a first job, planning to change jobs to gain a promotion, or making a major career change, writing is a strategic skill in opening up job opportunities.

10

Writing to Get the Job You Want

Chapter PREVIEW

Finding a job is only one part of the overall career plan you will begin in this chapter and continue developing in the next chapter.

Very few of us today stay with one company or even in one profession. The following steps can help you find the best fit between you and the "right" job.

1. Gather information about yourself by preparing a career plan, and gather and refine information about the positions, careers, and fields of interest to you.

2. Prepare your resume and application letter.

3. Prepare for an interview, set up interviews, and apply for jobs.

4. Interview with companies to practice your interview skills and to find out about the employers.

5. Select the company you want to work for.

6. Negotiate the terms of your employment (place, hours, and salary).

7. Refine your career plan to identify the education, training, or experiences you need for the next stage of your career.

All of these steps require clear planning, strategic thinking—and using your oral and written skills persuasively. This chapter will help you by reviewing ways to complete guidelines, develop a resume, develop an application letter, and prepare for an interview.

Journal ENTRIES

Begin by thinking about the steps you have gone through (or have seen others go through) to get the jobs you have wanted. Write a short one-paragraph response to any of these questions in your class journal or prepare them for class discussion.

1. Where is the greatest demand for people in your field today?
2. What skills do you think will be most important during the next five years for success in your field?
3. How do people really get jobs? Think about what you already know about the job-hunting process. Has this system worked for you or people you know? Why or why not?
4. How is job hunting different today than it may have been 5 or 10 years ago?
5. What would you put on a job application if you were a recovering alcoholic, had a prison record, or were seriously ill? Does the potential employer have a "right" to this information?

Part IV: Writing About Your Career

Chapter READING

Sometimes when we finally get that perfect job, we find it's not quite what we wanted. Part of developing your future could involve having the courage to make a job change. That means going through the whole process of preparing a resume, gathering information about potential employers, interviewing, and then making a decision about what company you want to work for.

As you read John Blyth's article, which describes what happens to an employer when employees change jobs often, think about how many times you may change your job during the course of your career.

THE MOBILITY TRAP

John R. Blyth

My comptroller just quit. The third one in five years. Earth-shattering news? No. It won't rate a single line in the business section of the local newspaper. But the implications are profound for my company—a manufacturer of computerized fuel-management systems that I cofounded in 1981. Once again I'm forced to begin a monthlong search for a replacement. With three weeks' notice (that's better than the one or two I normally receive) I'll have to spread her work among other staffers while I look for a successor. I'll go for months without suitable financial reporting. Innumerable small items will fall through the cracks in her absence. I hope there won't be any big ones.

My company, with a staff of 40, will spend thousands of dollars in direct costs in our search for a suitable applicant, and thousands more in indirect costs to train a new comptroller. A small high-tech firm is hit especially hard: hundreds of hours will be used in the headhunting and subsequent training that could be spent in work-related productivity. The person who left was with us for just 18 months before she took off for yet another "opportunity."

Were we unhappy with her performance while she was here? Not at all. She was young, intelligent, competent, trustworthy, and productive. As comptroller, she was in a pivotal position, with responsibility for all financial aspects of the company, including accounts receivable, payables and payroll. She instituted better financial reporting and controls, brought fresh ideas to management, and participated 100 percent in our organization. She was well liked by virtually all employees and respected by her comanagers.

Was she unhappy with us? Quite the contrary. She was paid a very competitive salary. She received a substantial raise six months after we hired her. Her benefits were above average for a small firm. I gave her a free hand to operate her department as she saw fit. We acknowledged her accomplishments frequently. Her departure had nothing to do with issues involving sexual harassment or discrimination.

Her chances for growth and advancement were excellent. A recent buyout of one of the owners, although it caused a temporary pinch in cash flow, provided the right moment to purchase company stock at a low price. My firm's tremendous future growth potential would have enabled her to participate in company expansion and to receive a percentage of company profits. She was due for a large incremental raise in a few weeks.

What happened? She found what she felt was an even better opportunity. She admitted that her decision to make the move was an agonizing one. While my firm is firmly established in the high-tech computer field, her new employer, although larger than mine, is decidedly low tech, providing farm processing services. The new position offered a small increase in pay and a better job title. Maybe there were other factors in her decision that she didn't share.

So what's my beef? The seemingly prevalent attitude among many young professionals that each job is just a stepping stone to that *one great job*. But these stones are so small that they will ultimately become stumbling blocks to those job hoppers who leap too quickly. I'm certain more than half the resumes I'll receive in our search for a comptroller will contain multiple employment histories with an average length of service measured in months that barely reach double digits. I have to assume that most applicants don't have a clue how much each new employee costs a small high-tech business. Perhaps they do and don't care. I haven't reached these conclusions based on one employee. She just happened to be the last straw. Our personnel files are bulging with short-term job seekers' resumes. Regrettably, we've hired some of these short-termers in the past. Now I'm adding length of prior service to my criteria for applicants.

I think a minimum amount of time an employee should remain with a company before moving on is four years. Even at this rate, an individual looking for relatively constant advancement could jump jobs more than 10 times in a 40-year career. Four years provides an opportunity for the employer to recover most of the costs of the employee search. It is difficult, within the framework of most organizations, to promote during the 18 months frequent job hoppers seem to set as their basic timetable for moving on. With a reasonable time spent working at one place, employees might discover that their talents are recognized by employers and rewarded by career promotions and pay increases. Those who commit themselves to a job for a longer period may find it to be financially beneficial, giving personal satisfaction and greater rewards than a series of short hops.

I'm not recommending a lifelong career at any one company. Working for only one organization is an occupation of the past. My parents' generation was probably the last to see that kind of commitment. And many large corporations have shot themselves in the foot with headline-grabbing massive layoffs that certainly do not encourage dedication among employees. Companies do not hire full-time employees with the intent to terminate quickly. Employers recognize the cost of hiring and training good people and, in fact, are greatly concerned about how to retain competent people. I am suggesting that a token term agreement is necessary. The employee would begin to reap the benefits of pay raises and job advancements, and the employer would profit from the fruits of a seasoned worker's time.

318 Part IV: Writing About Your Career

> I don't begrudge anyone the chance to grow. If that means leaving one company for clear-cut advantages that beckon elsewhere, I'd be the first one to encourage moving on. One or two short-term jobs end up on almost everyone's resume for many reasons. But when the search becomes habitual and the resume is crowded with unfulfilled opportunities, perhaps the employee should step back and consider the cost, not only to the businesses through which they have blithely waltzed, but to their own careers.

Source: John R. Blyth, *Newsweek,* April 15, 1996, p. 20. Reprinted by permission.

Analyzing THE READING

Use the following questions to analyze the content of this article and your reaction to it. Write a one-page journal entry in your class journal or prepare your answers for class discussion.

Reading for Content

1. What does Blyth want his readers to remember about job changes? Why?
2. Do you feel that Blyth used enough supporting examples to persuade his audience?
3. What are some of the effects on a small or medium company when a key person leaves the company?

Reading for Reaction

1. How does Blyth define *short-term* employees and *long-term* employees? When you think of future jobs you will hold, how would you describe yourself? As a short-term or long-term employee? Why?
2. Do some employees have "better" job security than others? In what kinds of jobs and why? Is the position you hope to hold one day one of these "protected" jobs?
3. Is it ethical for someone to leave an employer without giving notice? What is adequate notice? What reasons would justify leaving without giving notice from the employee's point of view? From the employer's point of view?
4. What was your reaction to this article? Do you agree with Blyth's ideas of company loyalty? Why or why not? Can you think of any company policies or actions that may cause you to agree or disagree with Blyth that employees should stay at least four years at one company?

WRITING TO GET THE JOB YOU WANT

Whether you are looking for your first job or have many years of experience and are seeking a different job, looking for work can be a very stressful experience. We're never really sure why we don't get a job we really want. Was it our resume? Our lack of experience? Too much experience?

Many times, our first jobs or job changes seem to happen accidentally. We just happen to be in the right place at the right time. Or we might be in the wrong industry at the wrong time—facing plant closings or industrywide contractions in job demand so that we are forced to make a job change.

If you are getting ready to look for a first job, you might ask yourself how you know this is a good direction. If you are investing in a college education or retraining, what kind of job will be waiting for you when you graduate?

The first step in getting the job you want is having a career plan that helps you to understand where you are going and why. This next exercise will help you explore jobs you would like and what's happening in your field. Your career plan will be developed further in Chapter 11.

Clarifying Your Career Plan   Exercise 10.1

Step 1. Freewrite for five minutes in your journal, answering the following questions:
- What is your ideal job? Where? If you can't name a specific job, identify a field in which you would like to work (such as nursing, forestry, or electronics).
- Why is this your "ideal" job or field to work in?
- What are your workplace strengths? Weaknesses?
- What do you take pleasure in doing at work?
- Do you have any special skills in working with technology? With people?

Step 2. Visit your local library or career center. Use the U.S. Department of Labor's *Occupational Outlook* to look up a description of your specific job and industry. Photocopy the page relating to your field, highlighting information that explains future demand in your field and potential earnings. Ask your librarian or career counselor if there is information about these jobs in your state. If so, add this information to your notes.

Step 3. Share your information with your discussion group. What unanswered questions do you have about your future career? How might these questions be asked?

Step 4. Summarize your group's findings in an informal memo to your instructor and turn in all rough drafts and resource materials with your work.

COMPLETING AN APPLICATION

How to get the job you want means knowing what you want and setting up a process to get it. For example, if we wanted to take a vacation in Montana, most of us would go through some sort of planning process before we left. We would pull together all the equipment and supplies we needed, and we might develop some sort of a schedule to organize where we would go, what we would see, and what we would do.

The same sort of planning and gathering together of resources occurs when you are starting a job search.

You may need to develop some kind of career plan, talk to employment counselors, or take an aptitude test to determine your career strengths. You may also need to investigate the kinds of jobs and employers that are available.

Once you have narrowed down the field you want to work in, you'll need to prepare an information sheet to help you make telephone inquiries and fill out applications. You may also need to prepare a resume and application letter to apply to jobs advertised in the newspaper. Finally, and perhaps most importantly, you'll need to prepare for the employment interview.

Filling Out an Application.   Many entry-level jobs do not involve sending in an application letter with a resume. Employers just want you to fill out an application, which is a preprinted form that lists your education, experience, and references in a certain order. Sometimes the application is more detailed than a resume, asking for your salary history or Social Security number, for example.

Because applications can take hours to complete, many people want to just fill in some of the blanks, attach their resume and then write: "See attached for more information." Most employers who have an application process want you to complete the application to make the job of comparing work history and qualifications an easier task for them. This means that your application may not get a fair reading if you just write "see attached."

The application standardizes the information that the employer wants and puts it in a certain order. Exhibits 10.1 and 10.2 show two examples.

Exercise 10.2   Reading Job Applications

Read and analyze the two job applications shown in this chapter and bring your answers to the following questions to class for discussion.

- How are these two job application forms alike? How are they different?
- Read the form completed by Julia Bochy. What do you notice about how she has completed this application form?
- Which of the two forms would be easier to complete? Why?
- Which of the two forms would be easier for the potential employer to use? Why?
- What can I tell about the employer from this application form?

Look over the guidelines for completing job applications and consider what additional suggestions you could give to someone who's looking for an entry-level job.

Chapter 10: Writing to Get the Job You Want   **321**

Practice Application Form   Exhibit 10.1 ∎

FOR OFFICE USE ONLY	
Possible Work Locations	Possible Positions

APPLICATION FOR EMPLOYMENT

(PLEASE PRINT PLAINLY)

FOR OFFICE USE ONLY	
Work Location _____	Rate _____
Position _____	Date _____

To Applicant: We deeply appreciate your interest in our organization and assure you that we are sincerely interested in your qualifications. A clear understanding of your background and work history will aid us in placing you in the position that best meets your qualifications and may assist us in possible future upgrading.

PERSONAL

Date 3-15-94

Name   BOCHY (Last)   JULIA (First)   MARIE (Middle)   Social Security No. XXX-XX-XXXX

Present address   123 (No.) ELM (Street) ST.LOUIS (City) MO (State) 63122 (Zip)   Telephone No. (314) 555-1200

Are you legally eligible for employment in the U.S.A.? YES   State age if under 18 or over 70 N/A

What method of transportation will you use to get to work? CAR

Position(s) applied for SECRETARY   Rate of pay expected $ NEGOTIABLE per week

Would you work Full-Time ✓   Part-Time _____   Specify days and hours if part-time N/A

Were you previously employed by us? NO   If yes, when? N/A

If your application is considered favorably, on what date will you be available for work? UPON TWO WEEKS' NOTICE 19 __

Are there any other experiences, skills, or qualifications which you feel would especially fit you for work with our organization? KNOWLEDGE OF WORDPERFECT 5.1/6.0, LOTUS, dBASE III PLUS, SHORTHAND @ 100 WPM, TYPING @ 60 WPM, SPANISH, TWO YEARS' OFFICE EXPERIENCE, PERFECT ATTENDANCE.

RECORD OF EDUCATION

School	Name and Address of School	Course of Study	Check Last Year Completed				Did You Graduate?	List Diploma or Degree
Elementary	TYRONE 30309 LITTLESTONE KANSAS CITY, MO 64123	N/A	5	6	7	8	☒ Yes ☐ No	N/A
High	HARPER CREEK HIGH 13960 BEACONSFIELD ST. LOUIS, MO 63122	COLLEGE PREP AND BUSINESS	1	2	3	4	☒ Yes ☐ No	H.S. DIPLOMA
College	MILLER COMMUNITY COLLEGE 13011 12 MILE ROAD ST. LOUIS, MO 63122	BUSINESS	1	2	3	4	☒ Yes ☐ No	ASSOC. DEGREE OFFICE ADMIN.
Other (Specify)	SEKICH SCHOOL OF BUSINESS 26001 HOOVER ROAD ST. LOUIS, MO 63122	WORD PROCESSING	1	2	3	4	☒ Yes ☐ No	DIPLOMA WORD PROCESSING

322 Part IV: Writing About Your Career

■ Exhibit 10.1 (concluded)

List below all present and past employment, beginning with your most recent

Name and Address of Company and Type of Business	From Mo.	Yr.	To Mo.	Yr.	Describe the work you do	Weekly Starting Salary	Weekly Last Salary	Reason for Leaving	Name of Supervisor
ACME MANUFACTURING 1011 HALL ROAD ST. LOUIS, MO 63122 AUTOMOTIVE SUPPLIER Telephone (314) 231-5100	6	93	PRESENT		SALES SECY	$280	$320	N/A	GAYLE SUDDICK

Name and Address of Company and Type of Business	From Mo.	Yr.	To Mo.	Yr.	Describe the work you do	Weekly Starting Salary	Weekly Last Salary	Reason for Leaving	Name of Supervisor
RINKY PONTIAC 33000 GRATIOT ST. LOUIS, MO 63122 CAR DEALERSHIP Telephone (314) 270-0110	1	92	5	93	RECEPTIONIST	$200	$250	BETTER OPPORTUNITY	ALL RAQUEPAU

Name and Address of Company and Type of Business	From Mo.	Yr.	To Mo.	Yr.	Describe the work you do	Weekly Starting Salary	Weekly Last Salary	Reason for Leaving	Name of Supervisor
ROSE KIDD ELEM. 16050 GLADSTONE KANSAS CITY, MO 64123 SCHOOL Telephone (816) 575-3400	1	91	12	91	PTO VOLUNTEER	N/A	N/A	RELOCATED	N/A

Name and Address of Company and Type of Business	From Mo.	Yr.	To Mo.	Yr.	Describe the work you do	Weekly Starting Salary	Weekly Last Salary	Reason for Leaving	Name of Supervisor
SALLY WEBB 11201 ROUNDTREE KANSAS CITY, MO 64123 N/A Telephone (816) 313-4113	9	90	12	91	BABYSITTER	$70	$90	RELOCATED	SALLY WEBB

May we contact the employers listed above? __NO__ If not, indicate by No. which one(s) you do not wish us to contact __NO 1__

PERSONAL REFERENCES (Not Former Employers or Relatives)

Name and Occupation	Address	Phone Number
JANET PENROSE, HOMEMAKER	13098 WINONA ST. LOUIS, MO 63122	(314) 555-1310
CALVIN STEIN, FOOD BROKER	450 MAPLE ROAD ST. LOUIS, MO 63122	(314) 555-8391
KATHY BRANCH, PRINCIPAL	600 OAKHILL KANSAS CITY, MO 64123	(816) 555-2420

MILITARY SERVICE RECORD

Were you in U.S. Armed Forces? Yes _____ No __✓__ If yes, what branch? __N/A__

Dates of duty: From __N/A__ To __N/A__ Rank at discharge __N/A__
           Month Day Year   Month Day Year

List duties in the service including special training __N/A__

Have you taken any training under the G.I. Bill of Rights? __N/A__ If yes, what training did you take? __N/A__

Chapter 10: Writing to Get the Job You Want **323**

Job Application   Exhibit 10.2 ■

Job Application

PERSONAL INFORMATION

LAST NAME	FIRST NAME	MIDDLE	DATE
ADDRESS	CITY	STATE	ZIP
PHONE	SOCIAL SECURITY NUMBER		

Are you legally authorized to work in the United States? ____ Yes ____ No

Were you ever convicted of a felony?____ Yes ____ No When? Nature of felony

Do you have physical condition which may limit your ability to perform the job applied for? Is so, please give details.

EDUCATION

NAME OF SCHOOL	LOCATION	COURSE OF STUDY	LAST YEAR COMPLETED	DID YOU GRADUATE?	LIST DIPLOMA OR DEGREE
HIGH					
TRADE/BUSINESS					
COLLEGE					
GRADUATE SCHOOL					

WORKING CONDITIONS

POSITION APPLIED FOR:	DATE AVAILABLE FOR WORK:

ARE YOU APPLYING FOR
____ FULL TIME   ____ TEMPORARY   ____ PART TIME   ____ ANY OF THE ABOVE

ARE YOU WILLING TO RELOCATE?   ____ YES   ____ NO

WOULD YOU CONSIDER WORKING:

ANY SHIFT	___YES___NO
WEEKENDS & HOLIDAYS	___YES___NO
ROTATING SHIFTS	___YES___NO
ON CALL	___YES___NO

MILITARY EXPERIENCE

Did you serve in the Armed Services? ____ YES ____ NO   Which branch? _____

Briefly describe any skills acquired through the military service relevant to the position for which you have applied.

324   Part IV: Writing About Your Career

■ Exhibit 10.2   (concluded)

WORK HISTORY

List all previous employment beginning with the most recent.

EMPLOYER'S NAME _____

ADDRESS _____

JOB TITLE _____

EMPLOYED FROM _____ TO _____   START SALARY _____   END SALARY _____

IMMEDIATE SUPERVISOR _____   TITLE _____   PHONE _____

REASON FOR LEAVING _____

EMPLOYER'S NAME _____

ADDRESS _____

JOB TITLE _____

EMPLOYED FROM _____ TO _____   START SALARY _____   END SALARY _____

IMMEDIATE SUPERVISOR _____   TITLE _____   PHONE _____

REASON FOR LEAVING _____

EMPLOYER'S NAME _____

ADDRESS _____

JOB TITLE _____

EMPLOYED FROM _____ TO _____   START SALARY _____   END SALARY _____

IMMEDIATE SUPERVISOR _____   TITLE _____   PHONE _____

REASON FOR LEAVING _____

CAN WE RUN A DETAILED REFERENCE CHECK INCLUDING BUT NOT LIMITED TO A CHECK WITH YOUR EMPLOYERS? ____ YES ____ NO

LIST ANY EMPLOYERS YOU DO NOT WISH CONTACTED AND EXPLAIN WHY _____

_____

_____

I certify that the facts contained in this application are true and complete to the best of my knowledge and understand that, if I am employed, falsified statements on this application shall be grounds for dismissal.

Date _____   Signature _____

GUIDELINES FOR COMPLETING AN APPLICATION

Many employers use a job application to screen job applicants, selecting only the most qualified applicants for an interview. Job applications put the information the employer must review into a standard form, one that is easy to analyze and compare.

Because your application will be used to make an employment decision, always consider who is reading your application and what kinds of information this reader needs in order to consider you for the job you want. These next suggestions may be useful to you.

Read the Instructions before You Fill Out the Application. Many applications, for example, ask you to type or use an ink pen to make sure the application is easy to photocopy and to read. Following instructions accurately is also an important skill that cannot be easily measured. Expect the reader of your application to look at *how* you completed the application as much as *what* you said on the application.

Fill In All the Blanks. If the question does not apply to you, write in "not applicable (N/A)." Be as neat as possible as you complete the application. This will suggest you are methodical and detail-oriented. If you do not want to answer a particular question because the information may be negative, write "I will explain in the interview" right on the application.

Provide Information as Clearly and Completely as Possible. Fill in names, dates, employer names, addresses, telephone numbers, and any other facts you think might be needed. Because your job application may be compared to other job applications, work to present the information as clearly and specifically as possible—given the limited space. You want to show what experiences you've had and what skills could be useful for this particular job.

Double-Check for Accuracy. Make sure that dates don't accidentally overlap or that you have not left out any employment information.

Double-Check for Completeness. Sometimes you may need to attach photocopies of your driver's license or other certificates you hold.

Sign the Application before Turning It In. Remember that when you sign the application, you are certifying its truthfulness.

Sometimes companies are concerned about hidden productivity or insurance costs, so there may be key questions on the application that are "warning signals" to the employer. For example:

- How often has this applicant changed jobs?

- Are there any gaps in employment that cannot be readily explained?

> - Are there any serious health problems?
> - Does this person have a prison record or driving problem?
>
> If some of these questions apply to you, fill out the blanks on the application form honestly. You might also consider writing on the application: "I will explain in the interview."

Exercise 10.3 Completing an Application

Pick up an application from a local fast-food company or an application from a company you would like to work for. Bring it to class for discussion. If asked by your instructor, complete the application and turn it in.

PREPARING A RESUME

A resume is usually part of an "employment package" that includes your resume, an application letter, and sometimes copies of recommendation letters. Because you never know if a potential employer will read your resume or your application letter first, you'll need to make sure both present you as persuasively as possible.

Even though you will spend hours crafting a perfect resume and application letter, how much time will the employer spend on reading your package? Researchers say you've got six seconds on a first reading. Your resume may be one of several hundred resumes the employer is skim reading. You want the employer to skim read your resume and say yes!

The second or third time the employer reads your resume is where your hard work will pay off. Remember, the purpose of the resume and application letter is to highlight your skills, experience, and education as clearly as possible so that the potential employer will consider you for the job.

Some students are setting up their resumes on the World Wide Web feature of the Internet, and companies are starting to publicize openings through a searchable database. Try to find out if these resources are available to you as you begin drafting and revising your own resume.

Picking a Resume Format

A good resume has useful content, is logically organized, is clearly written, and is correct. Above all, a good resume is easy to read. This means you need to use document design (format and layout) to make the resume both easy to read and persuasive.

DRAFTING YOUR RESUME

Use the following form to draft your resume just by answering the questions in each section. This "skills" resume is designed to highlight the skills you may have gained through past jobs, college courses, and volunteer work.

As you draft your resume, keep these three key writing strategies in mind:

1. Put sections in an order that highlights your strengths.

2. Use active verbs, fragments, and specific details!

3. Consider document layout and overall readability as you plan a balanced resume.

TYPE YOUR NAME HERE
Street Address
Town, State ZIP
503/222-2222

OBJECTIVE: What position do you want to be considered for?
Entry-level? Summer intern?
At what company?
Using what skills?
Leading to what career path?

EDUCATION: List schools attended, degrees earned, and dates of graduation, most recent first. Example: A.S. degree in Electronics, Linn-Benton Community College, Albany, Oregon (anticipate graduation, June 1998).

Consider highlighting key classes taken.
Awards, scholarships, other recognition?
Certificates and/or licenses?
GPA? (OK if over 3.0) GPA in major?

Worked 50% while attending school.
Name of High School, Town, State, Date of Graduation.

EXPERIENCE: Job Title. Name of Employer, Town, State (dates you worked 10/91–9/92). Describe job responsibilities as specifically as possible, starting with action verbs (facilitated, managed, hired, sold, cut, developed). Number of people supervised? Promotions? Do not include salary!

List jobs in chronological order (most recent first).

Part IV: Writing About Your Career

TYPE YOUR NAME HERE   PAGE 2

OTHER
EXPERIENCES: Volunteer activities? Emphasize leadership?

SKILLS: Foreign languages? Computer skills, software, programming languages?
                    Special equipment? Trucks? Special tools? Consider work skills, people
                    skills, academic strengths? Publications?

INTERESTS: Hobbies? (You can combine this section with skills. It is generally good to
                    include a mix of interests.)

REFERENCES: At least three personal, professional, or academic references. For each:
                    Name, title, employer, address, phone.

                    OR

                    Excellent references available upon request.

Preparing a resume involves these key steps:

- Gathering the information.

- Picking a type of resume.

- Picking a resume layout.

- Deciding on graphic features.

- Writing an objective.

- Filling out the parts of the resume.

- Arranging information in a persuasive order.

- Checking final format and layout features.

- Editing for readability, clarity, and correctness.

Follow these steps to prepare your resume.

Gathering the Information.   Pull together information about schools you
have attended and jobs you have held. Include any workshops or training you
have completed. Use the form in "Drafting Your Resume" to help you plan your
resume or update it.

Picking a Type of Resume.   Most employers prefer information arranged
chronologically. When you put the "most recent information" first in the

experience section, for example, you are organizing the information about the jobs you have held by the dates you worked.

With a standardized format, an employer can quickly tell how long you worked at a particular employer, what kind of skills and experience you offer, or if there are any gaps in your education or work experience.

This easy-to-read format is the first choice for most potential employers, but it may not always highlight your skills and experiences in the most persuasive way. Notice how Nataley Bailey's work experience section is organized chronologically, making it very easy for the employer to see how many years of experience she has had at different jobs (see Exhibit 10.3).

Many students prefer a skills section, like the one that Nataley uses, because this format highlights the skills they have gained through their program or on the job. Notice the skills section is placed very close to the top of the resume so that the reader can quickly get a sense of the applicant's strengths.

People who are changing careers also benefit from using a skills section to highlight their knowledge and skills, because it helps the reader to understand how skills gained in a previous field can be used in a new field. Notice how Robert Harding's resume (Exhibit 10.4) supports his career change from logging to building materials. Robert puts his work experience immediately after his objective. He believes that the work experience is his most persuasive qualification.

The key strategy is to arrange the order of your resume sections by what most qualifies you for this particular position.

Selecting a Resume Layout.　　After you've decided what type of resume you'd like, consider what *format* you will use to present the information. Three popular types are described in Exhibit 10.5.

Deciding on Graphic Features.　　Word-processing programs give us many choices today to add graphic design elements. You can use the <u>underline</u>, **bold**, and *italics* features of your word-processing program to add interest to your resume. By changing font styles (such as the font style called Times New Roman, the font style called Ariel, or the font style called Courier New, among many other choices), you can change the personality of your type on the paper. By changing the font size, you can vary impact (font sizes can be size 8, size 12, size 14, size 18, among many other choices).

Many word-processing programs also now include several different standardized forms that you can select and fill out. The result can be a polished-looking resume that takes advantage of format changes (font size, specialized prints). The resumes of Barbara Rosenfeld (Exhibit 10.6), Veronica Schaup (Exhibit 10.7), and Thomas Martinelli (Exhibit 10.8) were all prepared using this feature of a word-processing program.

Part IV: Writing About Your Career

■ Exhibit 10.3   Sample Resume

NATALEY BAILEY
273 Washington Street SW
Pleasantville, NY 15432
(518) 555-2706

OBJECTIVE: To work as a full-cycle accountant

EDUCATION: Associate of Science degree, Accounting Technology
Green River Community College, Pleasantville, New York
Anticipate graduating: March 1998   GPA: 4.0

SKILLS:

Accounting   Full-cycle bookkeeping, worksheets, trial balances, financial reports, general ledger, subsidiary ledgers, accounts receivable, accounts payable, payroll, cost accounting, government accounting

Computer   Computerized accounting, DOS, Lotus 1-2-3, Dbase, Microsoft Works, Word Perfect 6.1, Computerized typesetting

Other   Typing (95 wpm), 10-key calculator (150), shorthand.

WORK EXPERIENCE:

1990 to present   Bookkeeper/Typesetter, Franklin Press, Pleasantville, NY. General ledger, accounts payable, payroll, including quarterly and year-end reports, general job shop typesetting, corespondence.

1/90-3/90   Instructional Assistant, Computerized Accounting Class, Green River Community College, Pleasantville, NY. Assisted students with variety of projects while attending school full-time.

1982-1990   School bus driver, Marylhurst Academy, Pleasantville, NY. Supervised buses, planned routes, and drove routes.

OTHER: Member, National Accounting Association (1985-present) Interests include the Internet, reading, and hiking

REFERENCES: Excellent references available on request.

Chapter 10: Writing to Get the Job You Want   **331**

Sample Resume   Exhibit 10.4 ∎

ROBERT HARDING
6247 NW Alder Street
Tuscaloosa, AL 35643
(205) 573-2426

Objective: Part-time employment in retail sales of building materials

Related Work **Operator Technician** (1989-1994)
Experience: Georgia-Pacific Corporation, Albany, Oregon
- Produced plywood and particleboard resins
- Monitored production of continuous process formaldehyde plants
- Coordinated sales and manufacture of interior and exterior doors, windows, and trim packages

Manager/Sales, Carpenter (1983-1989)
Richard Aames Construction, Albany, Oregon
- Handled direct sales to both residential and commerical contractors
- Rough framed single-family and multiple-family construction
- Read blueprints and installed all types of exterior siding, windows, and doors
- Performed interior finish millwork, including installing doors, cabinets, and specialty wood trim

Education Maryland University (June 1994-present) Major: Business
Anticipate graduating, June 1997. Grade point average: 3.6 out of 4.0

A.S. in Business/Accounting, Lincoln Hills Community College, Tuscaloosa, Alabama. Grade point average: 3.8 out of 4.0

Additional training:
- Owens-Corning Fiberglass seminar on roof installation
- Moore Paints workshop on all aspects of sales and service of paint products

References: Excellent references available on request.

Part IV: Writing About Your Career

■ Exhibit 10.5   **Three Possible Resume Formats**

CENTERED HEADINGS
This format is commonly used by college students or for academic jobs. It also has long lines so that students can prepare a one-page resume.

CENTERED

_____
_____
_____
_____

LEFT MARGIN HEADINGS
This format is commonly used by business students. It allows you to put much information on one page.

LEFT MARGINS

_____
_____
_____

CAPTION HEADINGS
This format is very popular because it is easy to read.

CAPTION

_____   _____
         _____
_____   _____
         _____
         _____

Your goal in picking out graphic features is to pick a style that matches both your personal style and the industry you are applying for. Look at the sample resumes in this chapter to see what design features and resume layouts you prefer. Plan to experiment with some graphic features with your resume. Because word-processing programs offer so many choices, readers now expect polished formats.

Writing an Objective.   Many writers do not want to limit their resumes by including an objective. However, resumes with an objective can help the reader focus on how your skills and experiences prepare you for a specific job.

An *objective* states the job you are applying for and appears immediately after the address block. Look at these possibilities and decide which style you like best:

- Electronics Technician, Level II
- Licensed Vocational Nurse at Pinehurst Nursing Home
- Accounting Clerk II, specializing in computerized accounts payable

Notice that each of these objectives seems tailored for a particular job. Because so many people use word processing today, most employers have come to expect a tailored resume, one that is prepared just for them. Therefore, a blanket resume and a blanket objective, which are prepared for many different readers, have less of a positive impact.

Sample Resume   Exhibit 10.6 ■

BARBARA ROSENFELD
9243 S.W. Green River Place • Corvallis, OR 97333 • 541/758-1944

OBJECTIVE
Seek Registered Nurse position at Good Samaritan Hospital in Corvallis, Oregon

QUALIFICATIONS
- Use the nursing process in clinical and theoretical situations.
- Possess firm working knowledge of contemporary nursing theory.
- Perform proficiently basic nursing skills: assessment, sterile technique, medication administration, intravenous therapy, and assistance with activities of daily living.
- Provide for psychosocial and cognitive needs of patients.
- Communicate skillfully with patients, colleagues, and supervisors.
- Care for patients with respect and compassion.
- Exhibit leadership and organization skills developed through participation in community, school, and political activities.
- Speak fluent Spanish and German.
- Work comfortably with computerized charting systems.

ACTIVITIES
TRAINING: Completed workshops on Oncology, AIDS, Communication, Therapeutic Touch, Interpretation of Laboratory Values, and Assessment Skills, 1992-1996.

COMMUNITY: Organized and participated in summer camp physicals for Boy Scouts and Linn-Benton Community College, June 1994 and June 1995. Took part in community health screenings at Heritage Mall, Albany, Oregon, May 1996.

EDUCATION
LINFIELD–GOOD SAMARITAN SCHOOL OF NURSING, McMinnville, Oregon
Working toward Bachelor of Science degree in Nursing, plan to graduate May 1998

LINN-BENTON COMMUNITY COLLEGE, Albany, Oregon
Associate of Science degree in Nursing, June 1996
Cumulative grade point average 3.91 out of 4.0

LEWIS & CLARK COLLEGE, Portland, Oregon
Bachelor of Arts degree in Foreign Languages, June 1988.

REFERENCES
Excellent references are available on request.

Part IV: Writing About Your Career

■ Exhibit 10.7   Sample Resume

VERONICA SCHAUP

33030 Vista del Mar Drive ■ *Lauderdale, FL 33308* ■ *305-776-4421*

OBJECTIVE

To work in the travel industry using my computer and organizational skills.

SKILLS

- Detail oriented, organized, and enthusiastic
- Skilled in human relations
- Enjoy fast-paced office environment
- High-level telephone skills
- Proficient in using travel database for booking and checking fares
- Have traveled to Canada and speak some French
- Able to prioritize work and meet deadlines

EMPLOYMENT

HORIZON AIRLINES, Detroit, Michigan
Reservation Clerk, June 1992-Present

EDUCATION

DISANTO SCHOOL OF TRAVEL, Fort Lauderdale, Florida
Certified Travel Agent, June 1992
HARPER HIGH SCHOOL, Harper Woods, Michigan
Graduated, June 1990

REFERENCES

Excellent references available on request.

Chapter 10: Writing to Get the Job You Want 335

Sample Resume   Exhibit 10.8 ■

Thomas Martinelli
1913 Green Lane
Middletown, VA 22645
703/869-7122

MECHANICAL SKILLS

- Brake systems
- Front ends
- Engine tune-up
- Heating
- Engine performance
- Electrical systems
- Tire rotation
- Electrical fuel injection
- Air conditioning
- Shock absorbers

CUSTOMER RELATIONS AND SKILLS

- Excellent rapport with customers
- Designed successful customer satisfaction form to improve service and feedback
- Motivated and hardworking (increased sales by 25% in one year)
- Improved repeat customers (over 15% increase)

EMPLOYMENT

Mechanic   **1992-Present**
MELVIN'S AUTO SHOP, FAIRFAX, VIRGINIA

- Diagnose and repair various vehicles.
- Responsible for body work, tune-ups, ignition repair, transmission repair, and maintenance.
- Develop a good working relationship with customers.
- Prepare billings and follow-up on customer concerns.

EDUCATION

Associate of Science Degree   **1992-1994**
LORD FAIRFAX COMMUNITY COLLEGE, FAIRFAX, VIRGINIA

Major: Automotive Technology

Minor: Diesel Mechanics

Activities: GPA 3.6/4.0

REFERENCES

Melvin Clarke, Melvin's Auto Shop, 2319 19th Street,
Middleville, VA 97321   703/869-1473

Susan Howells, Instructor, Diesel Program, Lord Fairfax
Community College, P.O. Box 47, Middletown, VA 22645   703/869-1120

Marshall Robinson, T & R Auto Repair, 1742 3rd Street,
Middleville, VA 97321   703/869-7193

Exercise 10.4 Writing Your Objective

Follow these steps to prepare an objective for your resume.

Step 1. Draft a list of personal goals that describe what you want to gain from working. Most people include ideas like a good salary, recognition, a secure job, or rewarding work. It's helpful to know what you want out of a job, even if this information does *not* go into your objective.

Step 2. List the fields, industries, possible employers, and job categories you want to work in. For example:

Field:	Refrigeration/Heating/Air conditioning
Company:	Service and sales/air conditioning
Job Title:	Installer, customer service, repair specialist

Step 3. List the job and the entry levels, for example, Electronics Technician Level II or Automotive Mechanic. Try to identify the job title for a promotion from this position. This will help you to consider what position you might want in three to five years.

Step 4. Pull together this information into a concise job objective that identifies the position you want, the skills you may use, and the level you will enter. Show this to your discussion group for feedback and plan to use your objective in your resume.

Filling Out the Parts of the Resume.   As you draft your resume, work hard to present supporting information as specifically and as concisely as possible. Notice how the students' sample resumes in this chapter are both specific and concise. Are there any resumes shown that you feel should be more specific? More concise?

It's a good idea to present as much information as possible. Remember that the reader will not be able to "fill in the blanks." You need to describe very specifically what work you have done, where it was done, and when it was done. Look again at the sample resumes in this chapter. Use parallel form and fragments to ensure completeness and to make your resume easier to read.

Arranging Information in Persuasive Order.   After setting up the address block and your objective on the new resume, decide which is the most compelling background you have for this job. Is it education? Previous work experience? Some specific skill? Put that section next. Recheck to make sure that the resume is arranged in order of importance, that is, to the reader.

Checking Final Format and Layout Features.   Take a final look at the format for your resume. Is it attractive, easy to read and follow, and appropriate for this field? Are sections and wording parallel? Have you taken advantage of word-processing capabilities? Is the resume balanced? Have you used enough "white" space? Are the margins equal? The reader should want to take a second look at your resume.

What's Wrong With This Resume? Exercise 10.5

Read the following resume and prepare a memo to the writer that lists the revisions you want her to make. Use the checklist on editing a resume to help you review all aspects of this resume. Highlight any strengths you think will help her. Bring your memo to your discussion group for review.

RESUME

Yolanda Bergstrom
173556 S.W. Sandy Boulevrd
Phoenix AZ 85345
503/928-2311
503/752-1718

Height: 5'11"
Weight: 140 lbs.
Age: 27
Marital status: M
Chidlren: 4

OBJECTIVE: I want a satisfying salary, rewarding work, and to be the President of my own company in 3-5 years.

EDUCATION

Maricopa Tech CommunityColg., June 1997 (currently planning to)
East Benson High School, June 1985.
Major: Electronics engineering with courses completed in DC Theory and Application, AC Theory and Application, Semiconductors, Analog Circuits I, Analog Circuits II, Integrated systems (all of which includethe fundamental base of training I need through multistage amplifiers, oscillators, communication circuits, instrumentation concepts, and some work on integrated systems (robotics). I am also good with numbers.

WORK History   I first worked at Palm Paper Company as a truck driver (1993). This was a good job, but I left beause I had a better job in a warehouse for five summers while I went to school and took care of my kids. I ws also two years self-employed in the trucking industry. Other jobs held include secretary, bookkeeper and gas station attendant. As you can tell, I'm a hard worker. I wear glasses.

                      REFERENCES   Available no request.

Part IV: Writing About Your Career

Editing for Readability, Clarity, and Correctness. Check for wordiness. Can any description be shorter? Clearer? Edit ruthlessly for any errors in punctuation or grammar and proofread carefully.

You will also have to decide how long your resume will be. Many people strongly recommend a one-page resume, especially for an entry-level job. If you are changing careers or have several years of work experience to highlight, consider using more than one page. If you work hard to compress all of the information you want to include on to one page, the result can be cluttered, difficult to read, and hard for the reader to understand.

Some companies are starting to ask for resumes that are "filed" electronically. If your school allows student WEB pages and has Internet resources, use a search engine to look for student employment and look at some student resumes. Many high tech companies are also recruiting for students by using Internet's World Wide Web.

■ CHECKLIST FOR EDITING A RESUME

Looking at Content:

_____ 1. Have I included an objective?

_____ 2. Is all information complete?

_____ 3. Have I provided enough detail about my jobs (job title, tasks, accomplishments, tools or equipment used)?

_____ 4. Have I highlighted a skills section?

_____ 5. Have I eliminated any personal information that could be discriminatory (age, religion, or marital status)?

_____ 6. Have I included references?

_____ 7. Does my resume need to include any special information?

Looking at Organization:

_____ 1. Are my sections organized in a way that emphasizes my most important background or skills?

_____ 2. Is the information in each section presented "most recent first"?

Looking at Style:

_____ 1. Can I cut any extra words to avoid wordiness?

_____ 2. Have I used action verbs and parallel form?

_____ 3. Can I edit any wording for better clarity?

Chapter 10: Writing to Get the Job You Want **339**

Looking at Layout:

___ 1. Have I proofread carefully for any typing errors or spelling, grammar, or punctuation problems?

___ 2. Does my resume format look uncluttered?

___ 3. Does my resume appear balanced from top to bottom and from left to right, with margins equal in size?

___ 4. Is there enough "white" space between sections?

___ 5. Does the color of my paper match industry expectations? (It's OK to use hot pink stationery—if you're applying to an advertising agency.)

Drafting Your Resume Exercise 10.6

Follow these steps to prepare your resume.

Step 1. *Decide on the layout of your resume.* Start with the name and address block. Put your objective next. Be sure to show the entry level of the position you want, if you know it.

Step 2. *Arrange the sections in your resume* so the most important sections (for both you and your reader) are placed first. You need to decide which comes first: The skills section? The education section? The experience section?

Step 3. *Decide on the format of your resume.* Do you want to use centered headings? Left margin headings? Caption headings?

Step 4. *Edit your resume for readability, clarity, completeness, conciseness, and correctness.*

Bring your draft resume to your discussion group for review. Revise as needed. Give your completed resume to your instructor with a brief list that summarizes strengths and weaknesses of your resume.

WRITING AN APPLICATION LETTER

Because you don't know if your application letter or your resume will be read first, your application letter needs to be crafted to introduce you to your potential employer in the most positive and informative way.

An application letter:

- Lets the reader know what job you are applying for (and, if helpful, how you found out about the job).

- Summarizes your most important reason for applying for this particular job.

- Highlights information about your skills, experiences, or education that is directly related to this position and that may not appear on your resume.

340   Part IV: Writing About Your Career

- Emphasizes ways that your skills or experience will meet the employer's needs.
- Requests an interview.

Notice how these ideas are used to organize the contents of the application letters presented in Exhibits 10.9 and 10.10.

Drafting an Application Letter

As you begin to write your application letter, two additional strategies will be useful to you:

- Writing to emphasize the reader's benefits.
- Writing courteously and clearly, avoiding business cliches.

Writing to Emphasize the Reader's Benefits.   If you were hiring someone, which statement would appeal most to you?

> "I want to work for you because this job gives me important experiences and I like the salary and benefits package you offer."

OR

> "My labwork in the electronics program prepares me to work efficiently in the clean room at Ryerson Electronics. Your procedures and processes are very familiar to me."

When you are writing about your own skills and experience, it's difficult to avoid beginning every sentence with "I." However, analyzing how your skills and experiences meet your reader's needs can help you to stand out from every other applicant. It can also help you show the reader how well you are prepared to work at a particular job.

How do you find out what knowledge, skills, and so on, are really required for a particular job? One way is to research the specific requirements for the job. You can ask people who are already working in this profession about the requirements before you write your application letter. Often, the newspaper ad or job description you are responding to has important clues about what skills are needed and why.

Writing Courteously and Clearly, Avoiding Business Cliches.   When we write formal letters, we sometimes use cliches without realizing it. Wording like "Please do not hesitate to contact the undersigned" sounds just right for ending a letter. The problem is that business cliches have been used so often that they no longer have any real meaning or impact on the reader.

Chapter 10: Writing to Get the Job You Want   **341**

Application Letter   Exhibit 10.9 ■

Brian S. Dooley, 3028 22nd Avenue SW, Seattle, WA 94302
206/622-3098

April 28, 1996

Mr. Marvin Kleinschmidt
Owner-Engineer
Kleinschmidt and Associates
1493 Greenleaf Lane
Chico CA 97302

Dear Mr. Kleinschmidt,

Please consider my application for a summer internship with your engineering firm.

For the last three years, I have worked full-time at Custom Alloy, Inc., an industrial metal corporation while attending school. I progressed from Forklift Operator to Plant Supervisor and Head of Security. The knowledge and experience I have gained about engineering designs, operations and uses would be a useful tool for Kleinschmidt and Associates.

While attending American River College, I also worked as a counseling assistant, serving students with academic advising. This experience, combined with later office work, has given me a clear and practical understanding of clients' needs and expectations.

The opportunity to work as a summer intern at Kleinschmidt and Associates means a great deal to me because I will graduate with my B.A. degree in civil engineering in June of 1997. Your summer intern program is very respected in the field, and I would be thrilled to be a part of it.

Please telephone me at (206) 622-3098 to set up an appointment. I'd really like a chance to talk with you about your summer internship program and how I might fit in your office. Thank you for reviewing my resume.

Sincerely,

Brian S. Dooley

Part IV: Writing About Your Career

■ Exhibit 10.10 **Application Letter**

Veronica Shaup

33030 Vista del Mar Drive
Lauderdale FL 33308
305/776-4421

April 28, 1996

Oscar Selivonchek
Director, Peel Travel
1357 Port Townsend St.
Lauderdale, FL 33308

RE: Application for Travel Agent position

Dear Mr. Selivonchek:

Your ad in the Florida State University Career Bulletin recruiting for a travel agent matches my skills and interests. I have worked as a reservation clerk at Horizon Airlines for the last four years and believe this experience would be valuable for helping your clients with their travel plans.

My experience with the computerized booking system at Horizon Airlines as well as day-to-day contact with a variety of people travelling all over the world would be useful resources for your office. I would need little training as one of my specialties was working with travel agents all over the Detroit region in confirming reservations and adjusting tickets.

While completing my training at the Disanto School of Travel, I coordinated a three-week trip to Toronto for 150 students, arranging all hotels, ground travel and side-excursions. I enjoyed this special booking experience and was able to travel with the group as an Assistant Tour Guide. I have a real flair for helping people solve their problems and enjoy planning vacations.

Thank you for reading my enclosed resume. Please call me to set up an interview. You can reach me directly at 305/797-3244, Horizon Airlines, from Tuesday through Saturday, anytime from 11:00 am to 9:00 pm. If you prefer, you may call me at home on Mondays at 305/776-4421. I'm looking forward to talking with you about your travel agent position.

Sincerely,

Editing Out the "I" Focus   Exercise 10.7

The following draft shows some very effective strategies, but your task is to revise the following letter for James so that the "I" pronoun is used only when absolutely necessary. You may work with a partner to complete this exercise.

When you are finished with your revision, compare your final draft to others generated by your discussion group. Try to decide which letter is "best" overall, balancing the "you" of the reader with the "I" of the writer. If asked by your instructor, turn in your drafts and notes.

This letter is addressed to a civil engineering consulting firm:

> Dear Mr. Sneed,
>
> As a student entering Oregon State University, I am seeking part-time employment with a consulting engineer while I am in school. Since I will be finishing my civil engineering degree in Corvallis, I wish to find career-related experiences close to campus. I found out that your firm was highly recommended by John Hendrix of Hendrix Construction.
>
> I worked as an engineering aide in Pulaski, Virginia, for a year before moving to Oregon. I used surveying skills extensively and converted field notes into presentable working drawings. I adapted my knowledge and skills to a variety of engineering situations. I also had many experiences in working directly with the public in a courteous and professional manner.
>
> I have also used AutoCad as a draftsman extensively. I feel I can easily adapt my computer-aided drafting skills to produce civil engineering drawings.
>
> I appreciate your time taken to review my request. I would like to speak with you further at your convenience. I can be reached any weekday after 5PM at 928-0751. I look forward to hearing from you, and I look forward to meeting you.
>
> Respectfully,
>
> James Adamson

Remember that the potential employer may be reading several hundred resumes and application letters at a time. You really don't want to sound like everyone else. Try to write your application letter so it has a positive and personal tone and avoids cliches or wording problems. Exhibit 10.11 highlights reader reactions to some of the common wording problems.

Polishing the tone of your application letter can be challenging. You want to be assertive but not aggressive! You want to be considered in the most positive way, and editing out any negatives will help create that positive first impression.

CHECKLIST FOR EDITING AN APPLICATION LETTER

Considering Audience and Purpose:

___ 1. Who is my audience? What are their specific needs?

___ 2. Have I used all resources to find out my audience's needs (newspaper, interviews)?

344 Part IV: Writing About Your Career

■ Exhibit 10.11 Common Wording Problems and Reactions to Them

Dear Sir or Madam:

Audience reaction: This person doesn't care enough about my company to find out who to write to.

REVISED:
Dear Paul,
Dear Ms. Roberts:

Thank you for your time.

Audience reaction: I shouldn't waste my time right now reading this; I have more important things to do.

REVISED:
Thank you for reviewing my resume.

I am anxious to meet with you.

Audience reaction: How anxious? I don't want to work with someone who is anxious.

REVISED:
I'm looking forward to talking with you.

If possible, I hope we could meet to talk about the likelihood of a position.

Audience reaction: This person probably can't make decisions easily because the wording is so tentative.

REVISED:
Please call me to set up an appointment to talk about this position.

I know we can benefit from meeting.

Audience reaction: How does this person know this? I don't know this person!

REVISED:
Delete this statement or say: Thank you for considering my request.

I would like an interview on May 7 between 9:30 and 10:30 AM. I will call your secretary next week to confirm the time as I will be in Los Angeles and expect to meet you on May 7.

Audience reaction: This person doesn't know I'm out of town and doesn't seem to care that my office staff has other deadlines and priorities.

REVISED:
If possible, I'd like an interview with you on May 7th, as I'll be in Los Angeles on that day. Could you have your secretary call me to confirm that an appointment has been scheduled?

___ 3. Can I list the skills, abilities, and experiences needed to be successful at this job?

___ 4. Do I know my strongest qualifications? Which skills or experiences relate to this job but don't appear in detail on my resume?

Considering Content and Organization:
___ 1. Does my first paragraph introduce the purpose of the letter and the name of the position?
___ 2. Do I interpret my skills as they apply to this job?
___ 3. Do I have at least two paragraphs that interpret my education and my experiences?
___ 4. Do I minimize my use of "I" and emphasize "you"?
___ 5. Have I asked for an interview? Included information on how to reach me and when?
___ 6. Do I thank the person for reading my resume?

Checking Style:
___ 1. Do I present all information positively?
___ 2. Am I courteous? Do I avoid the "hard sell" letter?
___ 3. Are there any business cliches? Does my letter read like it was copied out of a book?

Looking at Letter Format and Proofreading:
___ 1. Have I followed the correct letter format, considering balance of the letter on the page, use of "white" space around the text, and position of the parts of the letter?
___ 2. Are my paragraphs short and readable? No longer than six to eight lines?
___ 3. Have I proofread for any errors in grammar, punctuation, spelling, and typing?

Drafting Your Application Letter   Exercise 10.8

Get started on this assignment by finding a particular job (either in a newspaper ad or a job description) you would like to apply for. Study the newspaper ad or job description to discover exactly what the employer's requirements are. List two or three important requirements and then draft sentences that explain how your experiences meet these needs. Then,

1. Draft your application letter, writing directly to the needs of your potential employer.
2. Use the editing checklist to make your first revisions.
3. Bring your draft application letter to your discussion group for suggestions.
4. Make final revisions and proofread carefully before preparing your resume package.
5. Create a packet for your instructor, which includes your application letter, your resume, the newspaper ad or job description, your rough drafts, and your notes on the employer's requirements.

Work with your discussion group to review rough drafts and to suggest additional benefits to the potential employer. Turn in this packet according to your instructor's directions.

DRAFTING YOUR APPLICATION LETTER

Use the following form to draft your application letter by answering the questions in each section. This application letter will help you highlight the skills you may have gained through past jobs, college courses, or volunteer work. Consider using "modified block" rather than "block" style for your letter format.

As you develop your application letter, keep the following key writing strategies in mind:

1. Remember to emphasize a "you" focus.
2. Do extra work to interpret how your skills will work for this specific job.
3. Be assertive and courteous in asking for an interview at the end of your letter!
4. Proofread carefully after you're finished drafting!

Your street address
City, State ZIP

Today's date

Complete name of person to whom you are writing
Title
Department
Company
Street address
City, State ZIP

Dear Susan, or Dear Ms. Campbell, but never Dear Susan Campbell!

What is the purpose of your letter? What position are you applying for? Where or how did you find out about this position? What is the most compelling reason you are applying for this position? Note: The first paragraph can be very short!

What are your best qualifications for this particular job? Please highlight key work experiences here—consider projects you've completed, types of management or leadership experiences you've had in the workplace, at school, or in the community. Have you provided specific information as concisely as possible? Generally, use one paragraph to highlight educational experiences and a second paragraph to highlight workplace experiences.

Read over your draft. Have you put the reader's needs first? This means controlling the length of the paragraph (six to eight lines) as well as focusing the information to meet the reader's needs. Have you emphasized the "you" all the way through your letter? Have you worked to use a conversational tone? Are you friendly and professional throughout your letter?

Your goal with this letter is to sell yourself! Be positive in describing your skills, abilities, and potential. Avoid creating a "letter by the numbers" or using cliches. Have you specifically requested an interview? Have you restated how the reader can get in touch with you—by phone or e-mail? Have you cut out any cliches? Have you ended courteously by adding a sincere thank you?

Sincerely,

Your name

Chapter 10: Writing to Get the Job You Want **347**

PREPARING FOR THE EMPLOYMENT INTERVIEW

You may go on *information* interviews—a term invented by Richard Bolles, internationally recognized author and employment consultant, to describe an interview you set up to learn about a company, its operations, hiring processes, and possible job openings.

Once you have been invited to an interview, plan to spend about an hour (or longer) to prepare for the interview. Preparing for the interview can involve:

- Gathering information about your potential employer.

- Reviewing your notes, looking through company materials, and writing a list of questions to ask at the interview.

- Anticipating the kind of interview and the types of questions you will be asked.

- Thinking about how you will answer these questions. Consider concrete examples from your work and school experiences that will best illustrate your commitment, skills, ability, and potential.

This next section will summarize the kinds of interviews and types of questions you will be asked so that you can prepare for your employment interviews.

DIFFERENT KINDS OF INTERVIEWS

The kind of interview you will experience depends on the size and style of the company you are applying to and the level of the job itself.

If you are applying for an *entry-level* job, you'll most likely meet with a personnel manager face to face, and talk with your immediate supervisor after the personnel manager has "screened" out people who do not have the right mix of skills, experience, and positive motivation. Interviews for entry-level jobs tend to be informative and informal or friendly in tone. The smaller the company is, the more likely you will be talking to the owner very informally.

If you are applying for a *mid-level* job, especially one that has promotion potential, you may face more than one interview, more than one type of interview, and a more rigorous interview process.

Some larger companies use "stress" interviews that test your ability to do what you say you can do or that let the interviewer know how you respond to pressure. Sometimes a stress interview can be unexpectedly humorous; for example, the interviewer may invite you to sit down, but no chair has been provided. However, if you must demonstrate a skill, or the interviewer is purposefully hostile to test out your reactions and flexibility, you may want to be equally assertive in your questioning and negotiating with the company.

Part IV: Writing About Your Career

How you are treated in the interview process is an important clue that can tell you how you will be treated *after* you are hired. This should be a factor in your final decision of whether or not to accept the job.

Types of Interviews

You may experience any or all of these situations as you move through a series of interviews. Generally the higher the salary and the more responsibility involved in a job, the more likely you will be interviewed more than once. Part of preparing for your interview is to anticipate the type of interview. Consider your strengths and weaknesses as you read through the following types of interviews.

Telephone Interview. It's too expensive to fly you to the interview, so the interviewer asks you a list of questions either one-on-one or in a conference call. This kind of interview can be unsettling, because you can't see the interviewer's face for important cues that suggest when the interviewer needs more information. Research suggests that 70% of all communication is nonverbal, so this lack of cues about your audience's reactions is an important limitation.

Office Interview. You meet one-on-one with one person to discuss the job. The one-on-one interview typically lasts for an hour in one office.

Office "Chain" Interview. If you successfully complete the office interview, sometimes you may be unexpectedly asked to visit several other people since you "happen" to be at the company. In this type of "chain" interview, you move from person to person and office to office. You can meet with three or four people over the next three or four hours, with each person reacting as if you just arrived. Pace yourself!

Panel Interview. You arrive for your first interview and are unexpectedly introduced to a panel of interviewers. You may be talking with a small panel of 2 or 3 people, or you may be talking with a larger panel of 10 to 15 people. Your primary strategy is to stay relaxed and poised, establishing good eye contact with each person on the panel.

Company Tour. You are taken on a tour of the company's facilities. This is a wonderful opportunity to ask questions and notice what kinds of work people do, what their attitudes are, and how they are treated.

Star Performance. Occasionally, you will be asked to demonstrate that you have the skills for a particular job. You may be asked to analyze a broken machine and suggest a solution. You may be asked to write something or to make a presentation.

Chapter 10: Writing to Get the Job You Want   **349**

Lunch or Dinner Interview.   A lunch or dinner where you are also being interviewed for a job is often seen by the potential employer as an informal way for you to get to meet your co-workers. This kind of a "breaking of the bread" together is an important screening tool. Will you fit in socially with this company?

Regardless of the type of interview you experience, the interviewer's job is to evaluate a pool of applicants and to find the best person for the position, considering skills, experience, education, and motivation. Three key questions are hidden in each interview:

1.   Do you match the requirements for this job?

2.   How will you contribute to our work team?

3.   Do you really "fit" our company?

Sometimes the employment decision will be made after one interview; sometimes two, three, or five interviews must be completed before the company makes a decision. Usually, the more important the job, the more interviews you can anticipate. Look over the next box to see how these three key questions change at each interview.

INTERVIEWS: WHAT YOU CAN EXPECT

First Screen	**Second Screen**	**Third Screen**
You meet with personnel managers	You meet with supervisors and potential colleagues	You meet with senior management
Type of interview: Telephone One-on-one Office chain Tour	*Type of interview:* Panel Office chain Performance Lunch or dinner meeting	*Type of interview:* One-on-one Lunch or dinner meeting
Tone of interview: Friendly Evaluative	*Tone of interview:* Evaluative Performance Stress	*Tone of interview:* Friendly Evaluative
Personnel manager's key question: Do you match the requirements for this job?	*Supervisor's key question:* How will you contribute to our work team?	*Senior management's key question:* Do you really "fit" our company?

Your advance planning can help you prepare for the interview by anticipating the questions you will be asked (and thinking through your responses). You'll also consider how you dress to make a positive impression on your interviewers. Unless your skills are very much in demand, your goal will be to persuade the potential employer that you are the best candidate for this position.

Once the interview starts, you'll want to be poised and professional, yet relaxed. Have confidence in yourself. Know why you are there and with whom you are talking. Try to establish a good rapport with each person you are talking with. Notice the "body language" of the people you are talking to so you can adjust your answer to meet their reactions. Answer questions clearly and ask them clearly. Don't rush and don't dawdle!

Recognize ahead of time that some questions are difficult. If you can't answer a question adequately, just say so. It may help to pretend you are "on camera" so that you can present yourself positively and professionally. At the end of the interview, summarize what you think is the next step in the interview process.

What Kinds of Questions Will Be Asked?

We may wonder how best to prepare for an interview when we don't know what questions will be asked. One strategy to help you anticipate the kinds of questions that will be asked on your interview is to remember that the interviewer's job is to find the best person for this particular position. You can anticipate questions that clarify your qualifications for this particular job, considering your skills, your experiences, your education, and your motivation.

Although it may be tempting to memorize responses to the following questions, such answers can appear memorized. Your interviewer is more likely to react favorably to thoughtful, unrehearsed responses. Try to answer questions directly. Use examples to show what you mean. For example, if you are asked how you feel about working overtime, just say directly how you feel and briefly talk about your previous experience with overtime.

Background Questions. The interviewer can use background questions to clarify your interest in the job and your work experience or skills. For example:

- What type of position are you most interested in?
- Are you looking for a permanent or part-time job?
- What jobs have you held, how did you obtain them, and why did you leave your last job?
- What computer experiences do you have?

THOSE FIRST FEW MINUTES CAN GET YOU HIRED

Research shows that in the first four minutes of an interview, the person you are talking to will make a "green light" or "red light" decision about whether you should be seriously considered for the job.[1]

Even within the first ten seconds, that person is already making judgments about who you are based on *what can be seen* (your clothing, eye contact, and body language) and *what can be heard* (your style of talking, which includes volume, speed, and tone of voice).

Ask yourself the following questions:

- Does your clothing set a professional tone? Is it appropriate for the job you are applying for?

- Does your body language reinforce an open and positive attitude?
  Friendly handshake at the beginning of the interview?
  Steady eye contact?
  Poised and attentive facial expression?
  Sit next to the person rather than behind or in front of a desk?
  Project confidence by how you speak?

- Do you begin the interview positively?
  Show friendliness with appropriate "small talk"?
  Use the person's name to build rapport?

- Do you discus your qualifications effectively?
  Get to the point efficiently?
  Summarize why the employer should pick you?

- Do you enter negotiations for salary calmly?
  Stay poised, not too eager for the job?
  Avoid smiling at the wrong moment?
  Keep a confident tone and outlook throughout?

A little advance planning—and practice—can help you turn those first minutes into a positive first impression. Also, try a "mock interview" with a friend to practice fine-tuning your body language and to get some feedback on your personal style. Your goal is to be yourself but at your positive best!

Probing Questions. The interviewer uses probing questions to try to investigate issues that could affect your ability to do the job. These "red-flag" questions can sometimes be quite difficult to answer honestly. For example:

---

[1]Connie Brown Glaser and Barbara Steinberg Smalley, "Four Minutes That Get You Hired," *Reader's Digest* (August 1993), pp. 129–132.

- Have you had your driver's license revoked?
- Do you have a serious illness or injury?
- Do you have a prison record?
- How long would you expect to stay at this position?

Company-Specific Questions. Many interviewers use company-specific questions to look closely at the match between you and the company or department and to clarify the information you need to make a decision. For example:

- Do you have any questions about the company or the job?
- Why do you think you would like to work for our company?
- How do you feel about working with an older (or younger) supervisor?
- Our company is committed to total quality management and teamwork. What are your experiences in these areas?
- Are you willing to relocate if the company expects this?

Open-Ended Questions. Finally, the interviewer can use open-ended questions to find out more about you, your plans, your ability to think under pressure, or your communication skills. For example:

- What are your ideas on salary?
- Tell me something about yourself.
- What are your weaknesses and what are your strengths?
- What would you do if . . . ?
- Do you have any recommendations for us based on what you've learned today?
- We have many qualified applicants. Why should we hire you?

You may also be asked illegal questions. Questions about your age, your race, your intentions to have children, or your religion are examples of illegal questions. Besides being illegal, these kinds of questions can be used to trick you. The interviewer may be testing your tactfulness. Always respond courteously, even when you're saying no!

Exercise 10.9 Preparing for Interview Questions

Since the previously described types of questions appear in nearly all job interviews, this exercise will help you practice how to prepare for the job interview. Start by reading through the examples of questions, noting which ones might be difficult for you to answer.

Chapter 10: Writing to Get the Job You Want   **353**

- Before class, pick three questions you feel would be difficult to answer and think about how you might respond to them. Do not memorize or write out a response.

- In class, work with a partner to practice answering two of the three questions you have prepared. Then have your partner ask you a question that you did not prepare for.

- After both of you have had a chance to respond to the questions, talk about which questions were the hardest to answer and why. Come up with a few suggestions that could help you be more successful in responding to interview questions. Discuss your suggestions with the class.

What Kinds of Questions Do I Ask?

You will also need to prepare questions to bring with you. These questions can show your interest in the job *and* help you gather valuable information in making the decision to accept the job. Some of the following questions may be useful to you.

Information-Gathering Questions.

You might ask information-gathering questions, similar to the following:

- What will my responsibilities be?

- Who would I be working with? How many people will I be directly responsible to? For what kinds of tasks?

- What are the hours and days I'll be working? Is there overtime involved? Weekend work?

- What type of equipment or machines will I be using?

- How much travel is involved in this job?

- What kinds of projects will I be working on? Can you describe a typical week of activities?

Probing Questions.

You might want to include some probing questions to find out more about your potential employer. These can also be good ice-breaker questions if the interview seems to slow down. For example:

- How would you describe management's philosophy?

- What is the biggest challenge facing this company? This industry?

- In what ways do you provide professional development or training for your employees?

- What is the typical career path for someone entering a position like this one?

- What kind of a person do you think would be most successful in this job?

Part IV: Writing About Your Career

You can prepare a list of questions you would like to ask, and you can make a few notes to look at while you are in the interview. What is most important is to connect personally with your interviewers by being relaxed, establishing good eye contact, and responding thoughtfully to all questions.

RESPONDING TO A JOB OFFER

Once the job offer is made, you may need to negotiate the start date, moving costs, benefits, or salary. It is helpful if during your preparation stage, you found out exactly what range of salary and fringe benefits are the industry norm for this particular job and for someone with your skills and experience. This information prepares you for the negotiation stage.

Sometimes your new employer will have more respect for you *if* you negotiate salary and benefits. The more important the job is to you, the more difficult this process can become.

The actual job offer and salary negotiations can occur over the phone or in a face-to-face meeting. If the salary offer is not quite what you expected, try a little negotiating. You might say, "I understand the salary range for this position is . . ." and then pause. Your interviewer may need to consult others, but the result of a little delay now can lead to a higher starting salary.

Once the final offer is made, ask for a few days to consider the offer, and then follow through with your decision. Waiting too long to make a decision can create negative repercussions if you accept the job. Try to be prompt and courteous—just the way you would like to be treated.

Sometimes you must say no to a job offer. Try to do so as soon as possible, explaining your reasons tactfully. Now is not the time to say you felt the immediate supervisor was too difficult or the company's financial situation too precarious. Because the company may be holding its second choice candidates on standby, your promptness will be helpful.

Once the job offer is made, ask for confirmation in writing. This is extraordinarily important if you are moving from one city or state to another. A professional organization will not hesitate to give you this reassurance. Occasionally, employers withdraw their employment offers.

Once again, ask yourself if you want to send a thank-you letter after you have accepted the job. This may be particularly appropriate if you interviewed with many people outside the primary area you will be working in. Again, just as in the thank-you letter for the interview, be courteous, personal, and *concise*. Consider also, that because so few people *do* send thank-you letters, your written thank-you letter could be seen as intrusive.

Do I Need To Write a Thank-You Letter? Exercise 10.10

Not everyone says "thank you." In fact, very few people who are interviewed say thank you after an interview is complete. You can stand out from the competition by sending a formal thank-you letter immediately after your interview. This will place your name back on the decision maker's desk in a positive way, perhaps right at the moment of the hiring decision.

The following are some suggestions for writing a positive thank-you letter.

1. Use the standard letter format.
2. Thank the reader for the interview.
3. Highlight something positive about the company, something you learned, or something the reader said during the interview, if possible.
4. Be courteous and brief.

Notice how these guidelines are followed in this thank-you letter.

THOMAS MARTINELLI
1913 GREEN LANE
MIDDLETOWN, VA 22645
703/869-7122

April 28, 1997

Clayton Woods
Clayton's Car Service
2785 Madison Street
Middletown VA 22645

Dear Clayton,

    Thank you for taking me around your shop Friday afternoon. I enjoyed visiting with you and I liked your shop, especially seeing your computerized diagnostic system, which is superior to the one we use at Melvin's Auto. As I mentioned, Melvin is retiring in four weeks. I was very excited to learn about your need for an experienced mechanic.

    I've enjoyed working at Melvin's during the past two years, and I can tell from having seen you that it would be fun to work in your shop. Thanks for considering me, and I hope to hear from you soon.

                                                 Sincerely,

                                                 Tom Martinelli

---

Writing a Thank-You Letter Exercise 10.11

Write a thank-you letter for a job you have applied for. Follow conventional letter format. Bring your typed letter to class for discussion.

356   Part IV: Writing About Your Career

WHAT'S NEXT?

We can expect continued change in every career—but one thing remains constant. People with good technical skills and good communication skills will be in demand.

You can decide what you want to do, where you want to do it, and why. You can find opportunities at entry level, you can return to school for more training and change careers at any point, and you can start your own business. Any of these options will require you to think about your priorities, skills, and interests. Use the concepts presented in this chapter to prepare yourself for finding the job you want and for interviewing successfully.

■ SUMMARY EXERCISE

What's Wrong with This Application Letter?

Prepare a memo to the writer that lists the revisions you want the writer to make in his application letter. Consider content, organization, style, and format changes that could improve this letter.

Share your list with your discussion group. If your instructor wants you to revise the letter, please do so, turning in the final copy with all notes to your instructor.

DRAFT APPLICATION LETTER

To: Beth Miller
From: Brant McLannahan

2195 SE Geary #31
Corvallis, OR 97330

February 11, 1996

Dear Columbia Logging,

My attention has been caught by your ad for timber cutting in the *Gazette-Times* on February 1. This job is very interesting to me because of the good salary, the possibility for year round employment. I would also enjoy traveling to different states to work.

My work experinece includes three summers and one winter job for a total of around 12 months of cutting Douglas fir, Grand fir and cedar, maple and oak. I learned falling techniecs from my boss and I have read some real informative books on safe timber falling.

I would be more than happy to answer any additional questions if you have any. Please do not hesitate to contact the undersigned if I may be of any further assistance.

Sincerely, Brant McLannahan

Chapter 10: Writing to Get the Job You Want **357**

CONCEPT REVIEW ■

Can you define the following key concepts from Chapter 10 without referring to the chapter? Write your definitions in a journal entry or review them out loud.

resume	stress interview
performance interview	networking
application	application letter
document design	layout
readability	chronological format
"you" focus	"skills" resume

ASSIGNMENTS ■

Assignment 1: Investigate Your Field.

Working alone or with another person, find two or three articles about the employment trends in your field. Look for articles that explain changing job demands, the new skills that potential employees need, salary ranges, or the degrees students will need for different types of work.

Summarize each article and draw conclusions from the information you have gathered. Reconsider the training you are getting in school. Do you feel you will be adequately prepared for this changing workplace? Prepare a memo that presents your findings and your analysis for the class.

Assignment 2: Develop Interview Skills.

Working with two other classmates, summarize the most useful interviewing suggestions from this chapter. Talk to three or four students about their interviewing experiences. Do their experiences validate the suggestions in this chapter? What additional suggestions can your group make?

Set up five-minute "practice employment interviews" with one person in your group playing the role of the interviewer, another being the interviewee, and the third person being the recorder to take notes during the interview and the discussion that follows. Take time after the practice interview to discuss what you learned from each point of view. Switch roles and repeat this exercise. When you are through with the role-playing, try to summarize what you learned from this experience.

What advice would you now give to someone preparing for an employment interview? Prepare a presentation for the class (and a memo for your instructor) based on your reading of the chapter, your interviews of other students' experiences, and your role-playing.

Assignment 3: Conduct an Information Interview.

List the questions you have about your future career, considering not only the first entry-level job, but the next step. For example, *what* is that next step and *what* kind of responsibilities will it

358 Part IV: Writing About Your Career

involve? *What* training or skills will your job require in five years? Make sure you have at least nine questions, using *what, how,* and *why* as your start.

Find someone who has worked in your field for several years and use your questions to interview that person about the career opportunities in your field. Stay flexible in the interview to ask both planned and spontaneous questions. Record your impressions immediately after the interview. Be sure to analyze the implications of this information. Did your understanding of the opportunities or requirements of work in your field change? Summarize your interview results in a memo for your class.

Assignment 4: Prepare a Resume Package.   Review the concepts and guidelines in this chapter. Find a newspaper ad that describes a job you would like to apply for. Prepare an application letter and resume for this particular job, describing your skills and experiences as they are today. It's important to be able to write persuasively about what you can do today, because we rarely have the right mix of skills or experience for any job. Use your discussion group for additional suggestions on how to improve your application letter and resume. Turn in your completed resume and application letter to your instructor with earlier drafts.

■ **PROJECT: WHAT'S OUT THERE FOR ME?**

Your assignment: Work alone or with a partner to gather information about the jobs that are available at three or four employers in your field in your region. Consider what you want to do and where you want to work as well as who are the major employers.

- *List the types of jobs that you want to investigate.* You will need to find out three or four job titles. List the kinds of information that you want to find out, such as salary ranges, education requirements, or different types of entry-level jobs. Work with your career center. Use the yellow pages in your phone book. It may help you to work with a partner to draft questions you would like answered.

- *Prepare a grid or chart to help you organize the information* you will gather. List the factors that you will need information about and that are most important to you.

- After you gather the information, *write up a memo to your class and instructor* that highlights what you found and what you learned. After you are satisfied with your rough draft, share it with another group for feedback. Revise it, turning the final copy in to your instructor. Remember to turn in all drafts and notes with your final report.

Chapter 10: Writing to Get the Job You Want   **359**

JOURNAL ENTRY

Consider how you can improve your interviewing skills. Describe (perhaps in list format) the steps you went through recently in an interview for a job you wanted. What went well? What could you improve? Are there any changes you would make in preparing for a job interview after reading this chapter and working through the exercises? List them.

11

Writing for Promotion

Chapter PREVIEW

This chapter helps you to think about the future and to develop a career plan. It also summarizes what the requirements will most likely be for your early on-the-job writing. Finally, it will help you to practice a very difficult writing task often assigned to new employees—writing performance appraisals.

Journal ENTRIES

Begin thinking about your job-hunting experiences and the career you'd like to have in the next three to five years. Write a short one-paragraph response to any of these questions in your class journal, or prepare answers for class discussion.

1. What makes a "fair" supervisor? Think about "fair" and "poor" supervisors you have worked with. How did these people contribute to your morale? Your productivity?
2. How would you describe your career plan to a friend? To your supervisor?
3. Many times, new employees are asked to carry out performance evaluations on their co-workers. If you were asked to do this, what performance criteria would you use?
4. What are the effects of a negative performance evaluation?

Chapter READINGS

Sometimes we accidentally find the "perfect" job that fits our interests, education, and skills. Sometimes we have to prepare ourselves for the job we want.

This first article profiles changes in the high-technology field of semiconductors. Not only does the article point out very specifically what workers will need in education and training, but it also notes that employers will most likely have to go outside their region to find the skilled workers they want.

The second article examines the dreams of a couple who are hoping for entry-level production jobs in the semiconductor industry. It explains what future training and education will be needed for them to advance in this industry.

BLUE COLLARED: THE HIGH-TECH INFLUX MAY DEEPEN THE DIVIDE BETWEEN RICH AND POOR

Roger O. Crockett

Semiconductor companies planning to locate or expand in the Portland area will start low-level production workers at about $6 an hour.

The vast majority of those workers will come from Oregon.

The same high-tech companies will start engineers at $16 an hour.

The vast majority of those engineers—along with a large percentage of the technicians and some of the upper-echelon administrators—will come from out of state.

That's right, most of the higher-paying jobs will go to people from out of state.

Although any jobs that offer benefits and possible advancement are good news for Oregon's unemployed and underemployed, some experts say the burgeoning semiconductor industry is not a panacea for the local workforce. Among their concerns:

- The gap between upper and lower incomes will be widened because more than half of the new jobs will pay less than the average Portlander's wage of $27,068.

- Low-level workers who do not obtain advanced technical training probably will be replaced by automation within five years.

- A shortage of technical training programs will prevent Oregonians from competing with skilled workers from out of state.

Such concerns are causing some to look askance at the second-generation high-tech boom that is expected to bring from 5,200 to 7,500 new jobs to the Portland metropolitan area by the year 2000.

"Adding thousands of low-wage jobs on the lower end of the pay scale is going to depress wage scales," says Margaret Hallock, the director of the University of Oregon's Labor Education and Research Center.

But Oregon doesn't have enough skilled workers to fill the more lucrative jobs being created.

In addition to the 11,500 Oregonians already working in the semiconductor industry, 11 chip- and wafer-making companies plan to build, expand, or add employees in Multnomah, Washington, and Clark counties. Another three companies are exploring the possibility of moving here. Additional projects outside the three-county area also will compete for workers.

"Certain jobs we can fill from within the local area," says Jim Harper, Wacker Siltronic Corp.'s director of human resources. His factory expects to add 300 jobs by mid-1995.

"But as the skill level and requirements get bigger, we have to enlarge the circle that we hire from. As we need programmers or engineers—people that have more advanced degrees—then we start looking beyond Oregon."

Does that mean Oregon workers lose out?

Not necessarily.

The Value of a Job.   More than 3,700 of the new jobs in the three-county area are expected to go to Oregonians, according to industry officials and government documents. More would be added if the three companies currently scouting the area decide to settle here.

Although most of those jobs are at the low end, they are all a cut above the minimum-wage service jobs that have marked economic growth in Oregon in recent years.

"I've been adamant on the decline in wages in Oregon, but these jobs are better than the average jobs we've created in the state," says Joseph Cortright, a local economist who has studied wage distribution for the Legislature.

Semiconductor manufacturers, like many high-tech companies, provide employer-paid health plans, life insurance and pensions. They give bonuses, opportunity for overtime pay and regularly scheduled raises.

They also provide training—at company expense, but on personal time—if the worker is motivated to move ahead and avoid being replaced by a machine.

Such training is crucial.

At Intel Corp., for example, advances in technology already have eliminated the need for operators. Now technicians with associate's degrees are the lowest-ranking production workers.

What if an employee doesn't have a degree or is unwilling to pursue one?

"They may have to look for a different career or be at the bottom of the heap," says Cheryl Hinerman, Intel's manager of workforce development. "Time marches on without them."

Darcy J. Williams, 27, doesn't plan to be left behind by advances in automation.

Since quitting her part-time job at Jiffy Lube five years ago and coming to work at Wacker's wafer plant in Northwest Portland, Williams has seen her pay go from $6 to $10 an hour.

Hired as an operator in the lapping area—where workers smooth and flatten wafers about the size and shape of a compact disc—she's taken advantage of company training almost from the beginning. Today she's a computer-support specialist.

"Getting a job here pretty much turned my life around," says Williams, who plans to continue training for further advancement.

Hers is the type of success story that high-tech recruiters like to tell.

Working with JobNet, a publicly funded consortium involved in recruiting and training workers for the industry, the high-tech firms say they'll hire as much as 90 percent of their floor operators from among the thousands of workers like Williams in the Portland area. That labor pool includes 40,000 unemployed, 24,000 temporary workers and thousands more who are working part time.

Paycheck to Paycheck.   The entry-level operator jobs, which require basic work skills and a high school diploma, start at $6 an hour and top out at $17. With an average wage of $10 an hour, their annual income is $20,800—below the Portland area's average wage of $27,068.

"Those that don't have a lot of skills are efficient working people, but the labor market is not paying those people as they used to," say Paul Warner, a state economist.

Already in Oregon the top 20 percent of households take home 54 percent of the pretax income, according to state tax returns. Warner and others say the semiconductor industry's pay structure could exacerbate the disparity. And it will add mostly out-of-staters to the top tier of wage earners.

Based on data from five companies, Oregonians can expect to earn an average of $25,794 a year between 1994 and 2008, while the more skilled workers recruited from out of state would earn $35,529—a difference of $9,735. But the demand for workers at all levels could push wages up across the board, Warner says.

"We think it's going to push wages up because it will be increasing demand for labor at a time when there's not a lot of labor available," he says.

For Joyce E. Perry, any job is better than what she was doing.

"I was basically doing nothing," says the 41-year-old, who is going through Wacker's new operator-training program. "I expect to have a stable situation now."

Her aspirations are modest. With her children grown, Perry expects to rent a larger apartment if she gets hired full time at Wacker. That's good for Perry, but Gerald Kissler, a UO vice provost and an economist who studies wage distribution, warns that the state will not be well served if production jobs are the end point.

High-end professional and managerial jobs come from creating an environment in which research-intensive start-ups can flourish. With that kind of employment landscape, "there will be fewer jobs, but they will be higher-wage jobs to complement the 21st-century blue-collar jobs," he says.

For example, the state of Washington has an economy based on agriculture, lumber, and fishing; but its annual per-capita income is more than $2,000 higher than Oregon's. The reason: a bigger community of high-tech companies doing research and development, Kissler says.

Companies planning to locate here will emphasize production over research.

Training a Workforce. Feeding a chip-hungry world, the semiconductor industry should experience about 18 percent growth in revenues through 1999, according to Mark A. Giudici, a semiconductor analyst for Dataquest, a division of Dunn & Bradstreet in San Jose, Calif.

With that growth will come increased automation, eliminating many low-skill jobs on the plant floor.

The jobs of the future, Guidici says, will require education beyond high school: An associate's or even a bachelor's degree in electrical physics could be the requirement for entry-level jobs.

Intel offers in-house college courses. Richard Crossley, a 36-year-old technician, enrolled so he eventually can become an engineer.

"Right now, where I'm at, there is little room for growth," he says. "To go anywhere, you have to get to the next level."

But Oregon schools do not have sufficient training programs to meet the demand. Consequently, many Oregonians don't qualify for many of the jobs that pay solid family wages: $30,000 to $60,000 a year.

Concerns about the lack of qualified workers have swelled with announcements of each company's plans to locate or expand in the Portland area.

"You can't stick a sign in your yard—that's for sure," says Intel's Hinerman, when asked how the chip company plans to recruit 1,755 skilled employees for its expansion.

To get the best talent, Intel goes out of state. "If you hire all there is, and you still have more openings, you have to go where you can," Hinerman says.

Fujitsu's Gresham general manager, Jun Nakano, whose company needs 445 workers for its $1 billion expansion, also has voiced concern about the competition for workers. "I worry about it," he says.

"What was a serious challenge becomes an urgent, even more serious challenge," says Lisa Nisenfeld, the Portland Development Commission's director of workforce and target industries. "We're trying to stay ahead of the wave of high-tech expansion."

That means developing more high-tech training programs so residents can enhance their skills.

Regional workforce development task forces are meeting throughout the valley to discuss ways to build technical training centers and improve the link between businesses and schools.

"It's not a situation of saying that the river is too wide," Nisenfeld says, refusing to let the challenges defeat her. "We do believe there is a way to get across."

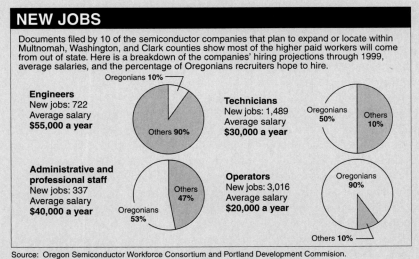

Roger O. Crockett, "Blue Collared: The High-Tech Influx May Deepen the Divide between Rich and Poor," *The Oregonian* (August 6, 1995), pp. A1, A14. Used by permission.

ENTRY-LEVEL WORKERS CAN BUY ONLY A WAFER-THIN SLICE OF AMERICAN PIE

Roger O. Crockett

After wiggling her 5-foot, 7-inch frame into a clean-room bunny suit for the first time, Candi Declues strikes a triumphant pose.

"Heeey," she hollers to her classmates at Wacker Siltronic Corp.'s technical training center. "I did it."

For her, slipping into the high-tech, sky-blue suit signals a dramatic change in her life.

But is it a change that will allow her to realize her dreams of financial security? It depends.

She and her husband, Donnell, have been unemployed for seven months. With a 6-year-old and a newborn, they are hungry for opportunity. And the Portland area's burgeoning semiconductor industry is giving them one.

The Declueses are both vying for entry-level production jobs with Wacker Siltronic Corp., a wafer manufacturer building a $230 million plant next to its existing plant in Northwest Portland.

The prospect of a job with a future is exhilarating for the couple. She left her $6-an-hour job at the post office because of complications with her pregnancy. He was laid off in January from a similar job in the same department.

Training is just the first step toward a future at Wacker.

The two must complete a probation as temporary employees. Wacker and other semiconductor companies commonly use temps during peak production periods. They also use temporary agencies to bring trainees, such as the Declueses, into the plant.

As such, their future is tentative.

"After about 90 days, if they want to come to us, and we want them to come to work, then they can come on board," Jim Harper, Wacker's human resources director, says of temporary employees in general.

If hired, their pay will go from $6 to $7.60 an hour, putting their combined annual income at $31,616 plus whatever overtime pay they earn.

Although the Declueses hope the jobs will allow them to save for a car, a house, and their children's education, that may not be realistic.

Living with Candi's mother since Donnell lost his job, the Declueses have monthly expenses of only $988. But once they get their own apartment, complete with utilities and other incidentals, there won't be much left.

Furthermore, their work schedules will be rigorous.

Wacker employees work 12-hour stints, usually 36 hours one week and 48 hours the next.

They also must obtain advanced training to move into higher-paying jobs.

Donnell is convinced he'll move up to a technician's job, where he can earn as much as $50,000 a year. To do that, he will have to earn an associate's degree on his own time.

"I have no problem with that. Wacker is just like the Air Force," says the former senior airman. "All you had to do was study to make the grade."

Candi hopes to eventually move into a job in human resources, which also would require a degree. She knows that working 12-hour shifts, going to school and caring for her children will be exhausting. "The hardest thing is the hours," she says with a sigh.

But then she thinks of her kids: "I really care how my children look at me. I want them to say, 'My mom's not lazy. She works.'"

HIGH-TECH JOBS

Here's what you need to know about jobs in the semiconductor industry:

Temporary Operator

- **Skills:** Basic reading, writing, math and oral communication; keen hand-eye coordination; ability to perform multiple tasks; desire to learn.
- **Experience and Education:** High school diploma or GED certificate, one year of production experience preferred.
- **Pay:** $12,500 to $27,000 a year, depending on experience.

Operator

- **Skills:** Technical reading and some computer literacy; oral and written communication; basic math including decimals, fractions, percentages and averages; calculator and computer keyboarding; keen hand-eye coordination; ability to perform multiple tasks.
- **Experience and Education:** High school diploma or GED; one year of production experience preferred; some higher-level operators need some college technical course work or previous semiconductor experience.
- **Pay:** $13,500 to $35,500 a year, depending on training and experience.

Technician

- **Skills:** Technical, verbal and written communication; understanding of semiconductor equipment maintenance; ability to handle measuring devices; command of algebra and geometry.
- **Experience and Education:** Associate's degree in electronic engineering, microelectronics, manufacturing, chemistry, physics or a similar technology degree; or the equivalent in technical on-the-job experience.
- **Pay:** $20,500 to $50,000 a year.

368   Part IV: Writing About Your Career

Engineer

- **Skills:** High-level mathematics and statistical-process control techniques; technical troubleshooting; communication and presentation; knowledge of chip, wafer or semiconductor methods and theory.
- **Experience and Education:** Bachelor's degree in physical science, electrical or mechanical engineering, physics or related science; or the equivalent in technical on-the-job experience.
- **Pay:** $33,600 to $90,000 a year.

Roger O. Crockett, "Entry-Level Workers Can Buy Only a Wafer-Thin Slice of American Pie," *The Oregonian* (August 6, 1995), pp. A1, A14. Used by permission.

Analyzing THE READING

Use the following questions to analyze the content of these two articles and to develop your reactions to them. Write a one-page journal entry in your class journal or prepare your answers for class discussion.

Reading for Content

1. In the "Blue Collared" article, where does the introduction end? Find a sentence in the introduction that introduces the main idea. List three or four main supporting ideas that Crockett wants you to remember.

2. In the "Blue Collared" article, where does the conclusion begin? After you finish reading the entire article, what is the main idea that Crockett wants you to remember? How does this idea compare to the introductory main idea?

3. Now read "Entry-Level Workers." Find the introduction and conclusion to this article. List the three or four main ideas.

4. Does Crockett provide enough facts, examples, and anecdotes in each of these articles so that his points are believable? Are they credible? Why or why not?

5. Who was the audience for the "Blue Collared" article? Is the audience different for the "Entry-Level Workers" article? What kinds of audiences would read these articles? How would they react? What do you think was Crockett's purpose in writing these two articles?

6. Describe the writer's style. What specific writing strategies does Crockett use to make his points? Is his tone argumentative? Why or why not?

Reading for Reaction

1. What is the message in these two articles for college students? What particular skills should schools emphasize so that students will have the necessary skills to be successful in a high-tech environment?

2. What are the issues that the Decluses are dealing with? Are their expectations realistic? What further complications can the Decluses expect as they try to raise their children with both parents working 12-hour shifts and continuing their education?

3. What are the advantages and disadvantages for the employer who uses temporary agencies to hire part-time employees? What are the advantages and disadvantages for temporary workers like the Decluses?

4. How does this article "Entry-Level Workers" clarify the "American dream" in the 1990s? What are your personal reactions to this article?

WRITING FOR PROMOTION

After you get the job you want, you will be using your technical skills and communication skills every day. You will also receive useful feedback about your performance from your supervisors and your co-workers, either formally in planned evaluations or informally in casual conversation.

Your daily performance is the most important factor for your future success. Your commitment to consistent high-quality work will open opportunities for you.

Only you can decide how hard you will work and what position you might want next. To help you think about that next job, you will need some kind of career plan that will help you achieve your personal goals. To start on this task, begin by analyzing your work experience.

Analyzing Your Work Experience   Exercise 11.1

To identify what aspects of work are most *satisfying* and *unsatisfying* to you, draft a 30-minute response to the following question.

> Compare the most productive and least productive work situations you've ever been in, whether at home or outside the home. What components made these situations successful or unsuccessful? Why?

After you've written your response, reread it, underlining key features. Then make a list that shows the parts you most like and dislike in a job. Write one sentence that defines your current career plans, then ask yourself how this list matches up with your current career plans. Bring your list to your discussion group and use it to identify additional "likes" and "dislikes" to use in drafting your career plan later. If your instructor asks you to do so, turn in a copy of your response to your instructor.

CLARIFYING PERSONAL GOALS

When you prepared your resume, you probably prepared an objective to help you identify exactly what position you wanted to apply for.

When you are thinking about your overall career, you need a clear sense of your personal and professional goals *and* a career plan (how you will achieve your personal goals).

The first step is to identify your personal goals. These may have little to do with your job or career plans, but for you to feel fulfilled with your work, there should be some connection between the two. You might want to consider two kinds of personal goals—short-term (those goals you'll try to achieve in about three years) and long-term personal goals (those goals you'll be working toward over the next five to seven years, or longer).

Personal goals, whether they are written down or not, often spell out:

- The kind of income you would like to earn.
- The education you would like to have.
- Where you would like to live.
- How many hours you want to work—or how much leisure time you want.
- Whether you want to work in a shop, in the field, or in an office.
- What kind of work or activities you want to do.

It may be useful to think about your personal goals for your family, your community, or your spiritual commitments.

Exercise 11.2 Drafting Personal Goals

Draft a list of eight to ten personal goals, answering these prompts:

After graduation, I want to . . .

In the next three years, I want to . . .

In five years, I want to . . .

After you have completed this list, look over the suggested list of personal goals. Have you left out any aspect of your future personal life that you think is important? If so, put it down on paper.

You'll be using this list to prepare your career plan. If your instructor asks you to do so, share your list with your discussion group or turn it in to your instructor.

Chapter 11: Writing for Promotion   **371**

CLARIFYING PROFESSIONAL GOALS

Professional goals are very different from your personal goals. Professional goals are directly related to your future career. They spell out the kind of company you want to work for and what kind of positions you want to have, first at an entry level and then after several years of work.

Professional goals can be very brief; however, they should help you to develop the "big picture" about your career, that is, both your short-term and long-term goals. Many people never put these goals down on paper, but they know where they are going and why. Even informal professional goals can be helpful. Many people change their careers because of external factors (for example, downsizing or closing of companies). Others change their careers because they are following a dream.

Start by thinking about what you want to achieve in your profession. Think about both short-term goals (those you'll try to achieve in about three years) and long-term goals (those goals you'll be working toward during the next five to seven years, or longer).

Professional goals often identify:

- The specific field you want to work in.
- A sequence of positions you want to hold (for example, construction worker to project manager to owner of construction engineering firm).
- The kind of people, equipment, or projects you want to work with.
- The professional associations you would like to join.
- The level of responsibility you want to assume over time.
- The number of people you may want to supervise.

The following list includes professional goals for Steve, a student who's making a mid-career change from working in a steel mill that closed down to starting his own business.

- I want to open my own business to prepare tax returns and provide accounting services for small businesses in Albany.
- For the short term, I'll work at a small accounting firm that does tax consulting.
- I want to join the Association of Tax Preparers locally to build a network.
- I like to work with automated accounting and tax reporting software.
- I'll need to work long hours during each tax season.
- I'll need constant training to keep up with changes in tax law.
- I'll need to understand how to run a small business.

Notice how Steve started to list the kinds of skills he needed to be successful in his new business. He has already begun to work on his career plan!

Exercise 11.3 Drafting Professional Goals

To prepare for work on your career plan, draft a list of five to six professional goals, answering these prompts:

After graduation, I want to . . .

In the next three years, I want to . . .

By five years, I want to . . .

After you have completed making this list, look over the suggested list of professional goals. Have you left out any aspect of your future professional life that you think is important? If so, put it down on paper.

You'll be using this list to prepare your career plan. If your instructor asks you to do so, share your list with your discussion group or turn it in to your instructor.

DEVELOPING A CAREER PLAN

At the beginning of a working life, we may not see clearly where we are going. We can only see the first step. Researchers suggest that workers today will change their careers three to five times over their working life. These career changes are partly driven by external factors—changes in demand for a particular kind of worker, for example.

A few students begin their studies at college and know exactly what they will do after graduation. Often students who are returning to school after having worked for several years have very definite ideas about what they want to achieve. A career plan written by Robert, an older-than-average student who is trying to move out of the carpentry industry into a retail sales position for building materials, is presented in Exhibit 11.1.

Robert began by thinking about the job description for the entry-level position he wanted as well as his long-term goal. He then identified the skills and experiences he had that matched this position as well as the training he needed for his short-term and long-term goals.

You can see that Robert's career plan will help him know where he is going, what kind of training he needs, and what specific kinds of jobs he needs to complete to broaden his experiences.

Sometimes we can't make up our minds about what we really want to do. This problem calls for a major investigation into what careers are possible—given your aptitudes, interests, and skills. Most colleges have a career center with

Robert's Career Plan   Exhibit 11.1 ■

Position: Sales clerk, retail sales of building materials
Within 3-5 years: Store manager of chain retailer of building materials
Reports to: Store manager

What Will I Do in This Position?

1. Work closely with customers. Need ability to build customer loyalty.
2. Complete customer orders, follow up on back orders, provide customer service, to retail and wholesale customers.
3. Explain how to complete major and minor home repair and building projects.
4. Help customers select the right materials and tools for the repair job.
5. Maintain inventory records, complete sales records accurately.

What Skills Do I Already Have for This Job?

1. I already know about sales and service of building materials, paints, installation procedures for hardware fixtures, insulation, etc. because of my six years experience as a carpenter.
2. I work well with people, whether they are novices or experts. I'm service oriented.
3. I have three years of experience in working with customers on wholesale catalog orders.
4. I'm familiar with installation and repair of the exterior of the house (including roofing, exterior siding, windows, doors, garage doors, etc.).
5. I've worked on many projects to repair, upgrade, and replace plumbing, interior doors, cabinets, and specialty wood trim.
6. I can do simple electrical repairs and read blueprints.
7. I'm very skilled with a variety of power tools (skillsaw, router, table saw, radial arm saw, power miter, planer, band saw) and pipe threading/cutting equipment.

What Skills Do I Need?

1. Need to upgrade my computer skills (especially using databases).
2. Need some experience in using bar-code-based inventory management.
3. Need to upgrade my knowledge of new paint products.
4. Need a refresher course in current building codes in my county.
5. Need to refresh my accounting knowledge (long-term goal).
6. Need a class in managing skills (long-term goal).

---

counselors to help you with career testing and advising. If you are undecided, a good first step is to talk to people who are already working. Observe what people do when they are working and think about what you would like to do. Consider:

- Completing an aptitude test at your career center.
- Scheduling a "careers options" conference with a career counselor.

Part IV: Writing About Your Career

- Attending any job fairs that may be held at your school and talking to the participating managers about what kinds of jobs are available.

- Noticing employment trends and new careers highlighted in the newspaper.

- Investigating summer internships at companies you are interested in.

- Thinking seriously about what you really enjoy doing and would like to do.

If you have identified your personal and professional goals, this next writing exercise will help you draft your career plan.

Exercise 11.4 Analyzing Jobs for Your Career Plan

Having a career plan can help you to prioritize how to use your time and resources wisely. Follow these steps to begin the analysis that will support your career plan.

Step 1. *Where are you now?* Describe the job you now have (listing responsibilities and tasks, skills needed, education needed). List any special projects you have worked on. Use this form to list the information as specifically as possible. Even if you feel your current job is not related to your future career, please do this analysis. You may be surprised by what you learn about yourself from analyzing your responsibilities, tasks, and skills.

Position: _____

Reports to: _____

Summary of job: _____

_____

Responsibilities	Tasks/Projects	Skills Used

Step 2. *Where do you want to go?* Describe the job you want to have in three to five years (listing responsibilities and tasks of the job, skills needed, education needed). Consult your personal and professional goals as needed. Prepare a position summary for entry level and (if possible) for three years after you have held the position.

Position: _____

Reports to: _____

Summary of job: _____

_____

Chapter 11: Writing for Promotion **375**

What will I do in this position? _____

_____

What skills do I already have for this job? _____

_____

Step 3. *What skills do I need?* Read over your notes for the two previous steps. Look for places where you need more specific experience. For example, the job you want to work toward might require working more with customers or monitoring or planning the budget. Draft a list of action steps to acquire the experience or skills needed to prepare you for the new position. Once you set up a schedule, you have the beginnings of your career plan.

Step 4. *Read over your personal and professional goals.* Ask yourself how the personal and professional goals in your earlier list are met by any of the action steps or the position you want to attain. Try to integrate your personal and professional goals into each stage of your career.

Step 5. *Share your job analysis* with your discussion group. Getting feedback now can be very useful in spotting places where you may need more experience or skills. Make any revisions needed and turn in your job analysis to your instructor.

Drafting a workable career plan can be challenging. Sometimes people don't want to talk about their future plans with their supervisors because the information might be used against them. Trusted co-workers or a person outside the company you work for can give you some very useful feedback about your career plan.

Finding someone who is experienced in your field to listen to your plans and to give you suggestions is a very important step. This person can serve as a *mentor* or guide as you develop a realistic career plan. If you can't talk with your immediate supervisor, you can get very useful information from him or her about your strengths and weaknesses in a performance review that you can then use in your career planning. Good friends, colleagues, relatives, and teachers can also be excellent resources.

AFTER ACCEPTING THE JOB, ANTICIPATE EVALUATION

You've got the job. After the settling-in process, a key concern many students have is: How will I be evaluated after I start to work? Another perhaps more important concern is: What are the skills and performances that will help me get a raise or lead to a promotion?

Your performance evaluation starts with your very first day. Your co-workers and immediate supervisor have very clearly defined performance expectations.

A supervisor might comment, "I like Max. He's always here on time, and I can count on him to do the job right the first time." This is an example of the kind of informal work appraisal that happens every day and that supports your reputation on the job.

You can make the settling-in process a little easier by observing those around you and asking questions to clarify the performance standards for your job.

- How do I know if I'm doing a "good" job?
- What do I need to do a "good" job?
- In what ways can I improve my performance?
- What resources are available to me?

You may need to meet a dress code or follow company policy on work hours, customer interaction, or training. You'll need to know more precisely what the goals of your work unit are and how they fit in with the goals of the company. You'll notice how your co-workers are dressed, what their attitudes are, and how they are rewarded. You may need to know why the previous worker left or why you were selected for this job.

Exercise 11.5   Clarify Work Expectations

Describe the work expectations for a job you have been responsible for, either at a company, in the community, or for your family. Answer these questions in your draft:

- What was the job you did and how were you hired?
- What skills do you think got you the job? What skills did you actually need to be successful once you had the job?
- Were there deadlines or a schedule you needed to meet?
- Was there any kind of dress code?
- Was a customer-oriented attitude required?
- How was your performance evaluated or rewarded?
- Was there a performance evaluation? How did it work? Do you think the evaluation was given fairly?
- What did you gain from this experience?
- If you were the supervisor, would you change anything?

Bring your draft to class and review it with your discussion group. Prepare a list in your small group that identifies the most important factors a person should expect in a performance evaluation. Present your list to your class or turn it in to your instructor, along with your preliminary draft about your job expectations.

UNDERSTANDING PERFORMANCE EVALUATIONS

Performance evaluations (sometimes called appraisals) can help you know in a formal way if you are meeting your employer's expectations.

For new employees, performance evaluations usually occur after three or six months of employment. After a positive performance evaluation, the employee moves from "probationary" status, and performance reviews occur semiannually or annually from then on, depending on company policy. If an employee does not meet performance standards, he or she may be placed on probationary status for a specific time.

The larger the company, the more likely you will find very formal performance evaluation forms and schedules. In larger companies, if an employee is returned to probationary status, there may be very specific steps to follow, especially if the company is unionized. The smaller the company, the more likely that performance evaluations are informal and sometimes given at the whim of the employer.

Many companies ask employees to complete a self-appraisal as part of their performance evaluation. In this document, you "grade" yourself on key components of your job. Other times, your colleagues may be asked to appraise you, using a peer evaluation form. Exhibit 11.2 is an example of a self-evaluation form used as part of a faculty performance review that is scheduled every three years. As you read through Exhibit 11.3, an annual appraisal for hourly employees, notice how the categories reinforce the main areas of the job and the goals of the company for quality customer service or product quality.

UNDERSTANDING APPRAISAL FORMS

Most appraisal forms include the following elements:

- *Standardized performance criteria* for the quality and quantity of work, for the skill and knowledge of the worker, and for knowledge of special operating conditions, especially safety.

- *Standardized way to evaluate intangibles* like motivation, attitude, initiative, cooperation, or adaptability to change.

- *A specific time period* the evaluation covers or is completed.

- *A ranking for each performance criteria* (for example, from "outstanding" to "needs improvement," to "unsatisfactory").

- *A supervisor's summary comment.*

- *An overall rating* that may be tied to either a pay increase or a promotion.

- *An employee section* that may require a signature or a written reaction to the performance appraisal.

When a company uses standardized performance appraisal forms and a schedule, employees know in general what criteria will be used for evaluation and when or how often they will be evaluated.

The evaluation process gives each employee a formal way to receive feedback on his or her work. Performance appraisals also can support a supervisor's recommendation for a promotion, a pay increase or, a reprimand, a demotion, or even a dismissal.

378 Part IV: Writing About Your Career

■ Exhibit 11.2 Faculty Self-Appraisal Form

Faculty Member _____ Date _____

Supervisor _____

Please mark your responses according to the following scale:
 A = I strongly agree
 B = I agree
 C = I disagree
 D = I strongly disagree
 E = Does not apply or not enough information to comment

Please attach a paragraph of discussion for *each* statement.

_____ 1. Contributes to department and division. Contributions I have made include:

_____ 2. Develops appropriate instructional materials. Materials I have developed include:

_____ 3. Develops new course and/or improves curriculum. Course(s) I have developed and/or curriculum improvements I have made include:

_____ 4. Stays professionally current. Activities to help grow professionally include:

_____ 5. Maintains openness to innovation in instructional methods. Opportunities I have taken for innovation in teaching methods include:

_____ 6. Meets responsibilities to college committees as assigned. The committees I have served on and my contributions to them include:

_____ 7. Communicates student opportunities for employment in the following ways:

_____ 8. Communicates course goals and objectives clearly to students in these ways:

_____ 9 Demonstrates a professional attitude conducive to discussion and inquiry. Ways I promote discussion and inquiry in the classroom include:

_____ 10. Communicates in a variety of ways to express and explain information. Ways I communicate with students and colleagues include:

_____ 11. Is aware of campus resources in the following ways:

ADDITIONAL COMMENTS: _____

_____

_____

However, the best performance appraisal system may not work if it is not used fairly. Standards and ratings may vary widely from supervisor to supervisor. A performance that earns an "outstanding" rating from one supervisor may only gain a "meets expectations" from another supervisor. When personal feelings enter into the evaluation process, morale can be affected.

Chapter 11: Writing for Promotion **379**

Annual Appraisal for Hourly Employees   Exhibit 11.3 ■

Date _____

Employee _____   Job Assignment _____

Supervisor _____   Department _____

Date Hired _____   Prior Appraisal Date _____

PERFORMANCE AREAS	Outstanding	Above Average	Average	Below Average	Unsatisfactory	Not Applicable
JOB KNOWLEDGE Comment:						
QUALITY OF WORK Ability to effectively use time and materials. Ability to meet deadlines/production schedules. Comment:						
SAFETY: Result of work under safe/unsafe practices. Comment:						
ATTENDANCE _____ Days not at work since last appraisal: 　　Personal: _____ Illness: _____ _____ Days late to work since last appraisal. Comment:						
PERSONAL FACTORS						
INITIATIVE/LEADERSHIP Takes on new tools, skills, or projects. Comment:						
FLEXIBILITY Adjusts to new situations. Comment:						
COOPERATION Works well with assigned teams and supervisor. Comment:						
DEPENDABILITY Completes the job reliably. Comment.						

Part IV: Writing About Your Career

■ Exhibit 11.3 (concluded)

EMPLOYEE'S PRESENT JOB DESCRIPTION (Attach current job description and note any changes in job assignments here.)

EMPLOYEE'S STRENGTHS (What should this employee be commended for?)
1.
2.
3.

AREAS TO IMPROVE (What should this employee improve?)
1.
2.
3.

NOTE: What has been accomplished toward areas to improve set by the previous appraisal? If the employee has not taken any action, indicate why.

SUMMARY RATING:

_____ Performance always exceeds job requirements.
_____ Performance is above average in meeting job requirements.
_____ Performance consistently meets job requirements.
_____ Performance does not always meet job requirements.
_____ Performance frequently does not meet job requirements.

SUMMARY COMMENT ON RATING:

Contributions to company-related activities (Safety Committee, annual holiday party, etc.)

EMPLOYEE'S COMMENT:

Employee's signature/date   Supervisor's signature/date

COMMENTS by next level of supervision:

Manager's signature/date

CORRECT THE PAST OR DEVELOP FOR THE FUTURE?

Is your supervisor interested in evaluating your *past* performance or is your supervisor asking you how you can improve your *future* performance? Most supervisors emphasize either the past or the future in the questions they ask in performance reviews.

Evaluations that emphasize the past tend to be reactions to problems after they occur. Think about a movie you have seen that made you wish you had stayed home. This reaction (or experience) may cause you to stop going to movies. Your supervisor, for example, may react to your being late to work by lowering your overtime hours.

Evaluations that emphasize the future tend to be proactive, in which supervisors anticipate problems before they occur. You could, for example, decide if a movie is not very good, that you could walk out after 15 minutes. You are now ready with a solution if a problem occurs.

In the same way, evaluating your own performance can help you to identify areas in which you may need more training, before your supervisor identifies a problem.

When appraisals are oriented to the future and based on clear performance criteria, you can get some good ideas about what you need to do to meet your employer's needs. Sometimes you may need help in gaining skills by taking classes or attending workshops. Sometimes your training may be on the job. However, a good evaluation program will match rewards to your progress and encourage your supervisor to work with you to set specific goals so that your overall productivity—and value to your employer—will increase.

People can change their behavior with consistent, clear, and positive supervision. If you don't have such a supervisor, knowing about these key elements in a performance appraisal should help you to find the resources you need to earn your promotion.

Writing an Evaluation of Yourself.   The challenge in writing a self-evaluation is to present your performance positively and honestly, neither overstating nor understating your achievements. Also, you are looking at your performance over a longer period of time—usually about a year.

You'll want to document what you have done as specifically as possible. Sometimes it's difficult to report fairly on your performance because it's hard to remember what happened last month, let alone last year. Do you keep a calendar of key events to help you remember all of your accomplishments?

Exercise 11.6 What Could Be Improved in This Appraisal Form?

Consider the following performance evaluation form developed to evaluate interns serving as pharmacy technicians at a small (125-bed) hospital. Can you suggest any improvements for these two excerpts, from the first and last pages of the four-page form? Bring your suggestions to class.

Pharmacy Department
MERCY HOSPITAL

Pharmacy Technician and Intern Performance Evaluation

The following criteria are used in evaluation of performance related to the provision of excellent service based on Mercy Hospital key values.

Employee Name   Date

Core Competencies   Meets   Does Not Meet
                                                    Standards   Standards

DEFINITIONS:

- **Exceeds Standards:** Employee may exhibit one or more of the following qualities; including, but not limited to: serves as a mentor or role-model, inservices staff or other customers regarding a department or job-specific function, functions with little or no supervision.
- **Meets Standards:** Employee meets the standard of excellence for the hospital/department.
- **Does Not Meet Standards:** Employee does not reflect the standard of excellence established by the hospital/department. (Prior work plan must be in place.)

A. Interpersonal Relationships

1. Consistently maintains and promotes excellence in service through open and honest interpersonal relationships with employees, visitors, patients, and medical staff.   _____   _____

2. Consistently demonstrates effective communication and conflict resolution skills.   _____   _____

3. Consistently promotes positive teamwork in the department and between departments.   _____   _____

4. Consistently compensates for personal issues without disrupting the work environment.   _____   _____

5. Consistently utilizes constructive feedback to improve own work performance or behavior.   _____   _____

6. Consistently demonstrates flexibility to meet fluctuating department demands.   _____   _____

COMMENTS: _____

_____

_____

Chapter 11: Writing for Promotion **383**

Performance Evaluation/Pharmacy Technician and Intern
Page 4

E. Professional Practice Performance (continued)	ES	MS	DNMS

Competencies: (continued)

9. IV admixture and chemotherapy preparation is completed
accurately and efficiently. _____ _____ _____

Comments: _____

_____

_____

10. Employee has completed aseptic technique evaluation form. _____ _____ _____

Additional Comments: _____

_____

_____

_____

Revised: 11-30-95

Employee Signature: _____ Date: _____

Use the following questions to start your own performance appraisal:

- What are my intermediate and long-range goals?

- What are my goals for my department and employer?

- How will I know I have met these goals?

- What deadline should I use to meet these goals?

- How often should I check my progress?

- What criteria should I use to evaluate my progress?

Complete a Self-Appraisal   Exercise 11.7

Use the previous performance questions to "appraise" your performance on a recent task you completed, either at work or at school.

Use any of the forms shown in this chapter if they are useful, or make up one of your own. If you design your own appraisal form, plan to review all appraisal forms with your discussion group before completing the form. Have the group comment on the strengths and weaknesses of each appraisal form, using a tape recorder to list the group's main ideas. Identify which areas are most

Part IV: Writing About Your Career

difficult for the supervisor to complete (and why) and which areas are most difficult for the employee to complete (and why).

Revise your appraisal form (if needed), then complete it. Bring your completed appraisal form to your discussion group for review. Which areas were, in fact, most difficult to complete? Do you think the form accurately and fairly reports your performance? Review the appraisal forms to identify which seem most useful to a supervisor and to an employee. Summarize your comments in a short memo (use a list format) for your instructor and turn the entire project (with all group notes) in to your instructor.

***

Writing an Evaluation About Others.   Every supervisor is responsible for evaluating the performance of all employees on some sort of a regular schedule set by the employer, typically after the first three months when the employee is new, and then every six months or annually.

New employees also can expect that the first three or six months on the job will be a probationary period. If you do not meet your employer's expectations during the probationary period, you will simply be dismissed.

Each company has a different system for evaluating employees' performance. There are many reporting forms, and the performance appraisal may be called different names, such as an evaluation or a performance review.

Some companies use a *summative evaluation* that covers the last performance period and is used to assess what went right and what went wrong.

Other companies use a *proactive evaluation* that covers the strengths and weaknesses of a person's performance as they relate to the future. Through the proactive evaluation the company attempts to reward strengths and to encourage the person to receive training in weaker areas. A few companies use a mix of both systems. Notice from the forms provided in this chapter how important the summary of previous work performance is.

As a technical specialist, you may be asked to evaluate a co-worker's or a supervisor's performance, especially if your employer emphasizes a "team" approach to how the work is done.

Evaluating a co-worker's or a supervisor's performance is more exacting than checking the "right" boxes on another form. Sometimes people are tempted to give an "outstanding" rating for every category of the performance evaluation, but you will frequently be asked to write a narrative paragraph that explains any unusually high or low ratings.

In writing any evaluation, whether for yourself or for a co-worker, you will need to be fair and objective, to support your findings with facts, and to provide only information that is relevant to evaluating this particular job and its requirements.

Chapter 11: Writing for Promotion **385**

In other words, you will need to have a clear picture of the job requirements and an acceptable range of performance standards for this job before you can begin the evaluation.

The following checklist offers guidelines that may help you complete either a self-evaluation or a performance evaluation for someone else—a co-worker or a supervisor.

CHECKLIST FOR WRITING PERFORMANCE EVALUATIONS ■

Before Preparing the Evaluation:

___ 1. Who is the audience for this performance evaluation? Will the person read my comments? (In some companies the employee will sign the evaluations of their peers as part of the review process.)

___ 2. What type of personnel evaluation is this? For a new employee? For an annual review? For someone who has been placed on probation or is being considered for a promotion?

___ 3. Do I understand the criteria being used in this evaluation? Do I know the acceptable ranges of performance?

___ 4. Will feedback from other co-workers be obtained? From my supervisor? How important is my evaluation?

___ 5. Can I complete this evaluation fairly? Is there any reason I should not prepare this evaluation? If you have any questions at all about the relevance or fairness of your evaluation, you should clarify them with your supervisor or with the personnel manager.

Preparing the Evaluation:

___ 1. Photocopy the evaluation form so you can mark it up.

___ 2. Review the job description for this position if one is available.

___ 3. Read over your calendar or any other document that will give you a clear picture of the major accomplishments of this individual during the last year (or the period being evaluated).

___ 4. Read over the evaluation form several times, highlighting areas of strengths and weaknesses, making notes of especially useful or troublesome areas.

___ 5. List facts you can use as examples to illustrate the person's strengths and weaknesses. Try to be as balanced as possible, neither overstating nor understating your summary. If suggestions for improvement are required as part of the evaluation, draft this list as well.

Completing the Evaluation:

___ 1. Fill out every section on the form. Draft any comments and the summary narrative, if that is required.

___ 2. Let the draft "rest" for at least 24 hours before revising it. Then, review it for thoughtful content, logical organization, clear style, and a readable format.

___ 3. Complete the form carefully, proofreading ruthlessly. Then, date it, sign it, and deliver it personally to the person requesting the evaluation.

___ 4. Review the completed evaluation with the person being evaluated. Listen carefully to the person's discussion and focus on future performance. Be ready to revise your appraisal for completeness and fairness.

WHAT'S NEXT?

Not everyone wants a job that has potential. Not everyone wants to "move up the ladder." Not every field has clear directions for promotions, although jobs with more responsibility generally offer more recognition and pay.

What do you want to do over your working career? Do you have the education or training needed to shift from one career to another? Most research suggests that people will shift careers at least three times during their working life. More formal education may be required. For example, more than formal training is required for shift work in an automated factory.

Exercise 11.8   Evaluating an Appraisal

Read over the following staff evaluation recently completed for an office assistant in a small tax office. Use the guidelines from this chapter and the following questions to analyze this performance evaluation. You may find it efficient to use a list format to take notes.

- Does this supervisor evaluate the performance of the employee fairly?
- What criteria does the supervisor use to evaluate performance?
- What reaction do you think the employee would have to this evaluation?
- Does the evaluation meet the guidelines discussed here? Why or why not?
- Is any information missing?
- Which areas of this performance evaluation appear to be the most difficult to prepare and why?
- Do you have any suggestions for the supervisor who wrote this evaluation? For the employee?

Bring your list of notes to your discussion group for review. Share experiences your group has had with performance evaluations. As a group, discuss and agree on the four major revisions Alex Barry, the supervisor, needs to make. Prepare a group memo for your instructor listing your revision suggestions. Finally, if Alex Barry were completing an evaluation for the first time, what additional advice would you give him? Turn in this informal memo (with your analysis of Rhonda Thomas' performance evaluation attached) to your instructor.

Chapter 11: Writing for Promotion 387

PERFORMANCE EVALUATION   Date   *5/1/96*

Employee: *Rhonda Thomas*   Job Assignment: *Office Assistant*

Supervisor: *Alex Barry*

Date Hired: *1/2/95*   Last Appraisal Date: *5/1/95*

JOB DESCRIPTION: *Computer input of tax returns, interface with clients, answer telephones, schedule appointments, return completed returns to clients and collect payments, file client records and tax forms, keep office clean. Other areas: do limited payroll and bookkeeping work.*

OVERALL PERFORMANCE: *Fair. Performs simple tasks well and efficiently but gets flustered and confused when dealing with more complex tasks or when trying to handle more than one problem at the same time. Rhonda tends to be afraid to learn new skills and tends to focus excessively on details rather than overall business goals. She tends to require close supervision rather than taking initiative to develop new skills. Her shyness tends to make her hesitant to call clients to inform them that their returns are ready. This shyness leads her to occasionally talk to clients unprofessionally. In general, she meets deadlines dependably, particularly when given a very precise job to do and told exactly how to do it. She is usually punctual and rarely misses a day's work. She has improved her knowledge. Last year, we talked about her need to improve her knowledge of tax law, and we sent her to school. She did well in her classes but still seems afraid to apply the knowledge she gained. We've given her additional training in payroll and bookkeeping, but she needs more skills and we're going to send her to a payroll class this summer. She really needs to build her confidence but I don't know how to accomplish this. She still does not clean the office and I, as supervisor, usually end up doing it myself.*

PERFORMANCE RATING:

_____ Always exceeds job requirements.

_____ Is above average in meeting job requirements.

_____ Consistently meets job requirements.

✓ Does not always meet job requirements.

_____ Frequently does not meet job requirements.

COMMENT: *Although I've marked that Rhonda does not always meet job requirements, she does many of her tasks, especially the carefully supervised ones, very well, very efficiently and with few mistakes. I believe that with a little more training and with a lot more self-confidence, she can learn to do her job extremely well. If she can improve her self-confidence and get some more education, she has potential for promotion in a couple of years.*

EMPLOYEE'S COMMENT: *I have always tried to do my best. I didn't realize that I was such a poor employee and I will try to improve.*

Rhonda Thomas

Employee's signature/date *5-1-96*

Alex Barry

Supervisor's signature/date *5-1-96*

Many companies are looking for a different kind of worker: people who can work in a team setting or independently, as needed; people who can use a variety of skills; people who are comfortable with a variety of computer applications; and people who are responsive to change. Have you identified what kind of education or training you will need for the next three years? For the next 10 years?

At the same time, jobs at higher levels can challenge you in new ways. The key to your future success is your ability to keep anticipating what's needed for the next level ahead of you and your willingness to acquire the people skills and the technical skills that you'll need. For example, if you're a shop supervisor, what's required to become a plant manager? You may need more training and more skills. You may have to become a different kind of person, that is, more outgoing or more reserved.

You have to know what you are willing to give up and where you want to go. You have to know what skills you need. Will you need more background or skills in accounting? Budgeting? People-managing skills? Planning? Running schedules?

Make your own career plans and support them by getting the training you need. Sometimes you can get the training you need right on the job by volunteering for special projects. Sometimes you can gain it by attending workshops or by volunteering for community jobs. You can ask your friends and co-workers for advice; but in the case of your own career, you are in charge.

Commit to Continuing Your Education. Are you a life-long learner? Most companies are looking for people who can be flexible with new situations and skills. Once you finish your formal education, you will have additional opportunities for training. Sometimes companies will pay for more formal education. Even smaller companies will pay for workshops to help build your skills. When these programs are available, they are an excellent resource for helping you improve your pay scale or to prepare for your next promotion.

For example, George, an auto mechanics major, took a writing class at the community college he was attending. After preparing a resume, he got a job in his field instead of continuing to work at a minimum wage job. The resume opened some opportunities for him because it highlighted his skills as they related to being a mechanic. He had a career plan and followed it through.

Consider carefully the following question: How will you use your skills to achieve your goals? The small decisions you make every day can open new doors for you. Think, plan, and act. Think about what's happening around you and decide what you really want to do. Then, in the words of a major tennis shoe manufacturer, "Do it!"

Your education, motivation, and skills will be matched by a wide range of opportunities. Use the concepts presented in this book to develop your skills. Your dreams and your tenacity will help you create the career opportunities you want.

Chapter 11: Writing for Promotion   **389**

CONCEPT REVIEW ■

Can you define the following key concepts from Chapter 11 without referring to the chapter? Write your own definitions in a journal entry or review them out loud.

appraisal	performance evaluation
probation	performance standards
entry level	middle level
goal setting	attendance
termination	personal goals
career plan	mentor
reprimand	

SUMMARY EXERCISE ■

Preparing a Career Plan for Ted

As part of developing a career plan, Ted has written a paper comparing and contrasting two work situations. Read his paper (printed below) then take the following steps to prepare a career plan for Ted.

Step 1. What is the problem facing Ted? Read through his paper, thinking about what kinds of work Ted seems to prefer and what kinds of skills he might need to be more successful.

Step 2. What recommendations do you have for Ted? Prepare a memo with an introduction, a discussion, and training recommendations for Ted. Discuss your memo with your discussion group and present a summary career plan and recommendations to the class.

Here's the question Ted answered: Compare the most productive and least productive work situation you've ever been in, whether at home or outside the home. What components made these situations successful or unsuccessful? Why?

Ted's response:

Alive vs. Dead

Luckily, my least productive work situation is in the past and unfortunately my most productive work situation lies there, too. I really felt strong and fulfilled as an assistant varsity boy's basketball coach. Sadly, I felt empty and lost as a floor supervisor for Whitmore's Drugs.

Being a varsity basketball coach made me feel alive. I was constantly charged up when I held this position. I was in control; I knew this stuff. Every decision and idea I had or made was simple and clear. All my years of playing basketball and studying

Part IV: Writing About Your Career

its finer points paid off. I enjoyed teaching the game to the kids. Working at their level and showing them a certain technique satisfied me, but best of all, seeing that they understood was the icing on the cake.

Results. I could see results! What a joy it was to see an individual progress as the season progressed. Seeing that happen made me feel successful. I had focus during my stint as an assistant. My charts and diagrams were well thought out and were easily understood by the kids. I was really feeling my job. To top it off, coaching made me feel young and inspired. I was excited and filled with anticipation every day I woke up.

Being in charge as a supervisor was not fun. Although I was in charge, I really felt I had no control. I was always running around, rushing to do or get something. I couldn't grasp it. It was too much! I was alone, isolated as *the* supervisor. I treated everyone fairly, but I always felt judged by my fellow employees.

I couldn't see any tangible results. Customers that I really helped rarely made me feel I was doing them a service or helping them out. I had no previous knowledge to fall back on when it came to schedules, ordering, making displays, or dealing with unhappy employees.

I also did not have any focus. I didn't see the big picture and it showed in my evaluations. I got mediocre and poor marks as a supervisor. This was no job for me, and I left feeling that I had wasted my time and the company's. Being unsuccessful as a floor supervisor and being successful as an assistant varsity basketball coach taught me one thing. Choose a career that makes you feel good and alive.

■ ASSIGNMENTS

Assignment 1: Preparing a Career Plan for Yourself.
Working alone or with another person, write a career plan for yourself that identifies what kind of education or training you will need to find a good job in your field once you are finished with school. You may need to use resources at your career planning center. Plan to photocopy useful background information to support your career plan.

In Your Career Plan, Include:

1. An introduction that describes and defines the positions you want. Include entry and middle-level jobs.

2. A summary of education, skills, and work experiences required for this position.

3. A list of the salary ranges for each position.

4. A conclusion that recommends what actions you should be taking now to prepare for this position.

Find at least one article in a trade journal that profiles workers in your field. Summarize information about the type of job, its working conditions, responsibilities, skills, and level of commitment from the worker. Include a citation on the article.

Chapter 11: Writing for Promotion **391**

If possible, contact someone who is already working in this field and interview that person about the entry requirements and promotion opportunities. Include the information in your career plan. If your instructor asks you to, include a comment that summarizes what you learned from this assignment.

Assignment 2: Conduct an Information Interview.
List the questions you have about your future career, considering not only the first entry-level job, but the next step. For example, *what* is that next step and *what* kind of responsibilities will it involve? *What* training or skills will your job require in five years? Make sure you have at least nine questions, using *what, how,* and *why* to begin.

Find someone who has worked in your field for several years and use your questions to interview that person about the career opportunities in your field. Stay flexible in the interview to ask both planned and spontaneous questions. Include questions about the technical and communication skills that are required for success. Ask if there are any advantages or drawbacks in this position.

Record your impressions immediately after the interview. Be sure to analyze the implications of this information. Did your understanding of the opportunities or requirements of work in your field change? Share your findings with your team as you gather information. Prepare a memo that summarizes your findings and includes recommendations for the class (take into consideration people who want to enter this field).

Assignment 3: Evaluate Performance.
You have been working for several weeks now in a discussion group for review and revision of work done for this class. Discuss as a team the criteria needed for effective group work. Decide whether your evaluation will be *summative* or *proactive.* Your team may decide to design a form to be used for this evaluation.

As an individual, prepare performance evaluations for each member of your discussion group and a self-evaluation for yourself. Discuss the review process with your group, noting any discrepancies between the "ideal" and "actual" performance desired. Conduct a formal debriefing with your group, giving feedback to each group member as constructively as possible. Focus on ways to improve future performance, even if your group elected to prepare a summative evaluation.

Write a summary of your own impressions of the review process. Turn in your completed and confidential evaluations with your summary to your instructor.

PROJECT: WRITING A PERFORMANCE APPRAISAL ■

As a class, discuss some performance criteria that are important to both success in the class and success later in the workplace, such as motivation, timeliness, or quality of work. After the class has decided on the five most important criteria, work in small groups to evaluate the performance of each person in your discussion group.

Part IV: Writing About Your Career

Step 1. Prepare a form evaluation on another student in your group. Consider that person's contributions during the last several weeks to your group work. Your form can include:

- A description of the tasks this person completed.

- A list of key performance factors (for example, *tangible factors* such as quality and quantity of work, usefulness of work, timeliness of work; and *intangible factors* such as motivation, attitude, initiative, flexibility); cooperation or willingness to work as a team member.

- A ranking system (outstanding, good, needs improvement) for each factor.

- A summary comment from you about this person that includes an overall rating with a list of two recommendations this person could use to improve his or her performance.

Step 2. After you have written the evaluation, discuss it with the student. Have the student write a reaction to the evaluation that becomes a part of the evaluation.

Step 3. Make any final revisions and turn the evaluation in to your instructor, adding a brief note about what you learned from this assignment. Consider the process you used to gather information, to evaluate the information, and to write the final evaluation.

■ JOURNAL ENTRY

We can anticipate continued change in the job market. However, for people beginning or changing their careers, it can be very difficult to "see" where new jobs are being developed. Spend about 10 minutes listing as quickly as you can all the new jobs you can think of. Think as creatively as possible. You can get started on your list by answering some of the following prompts.

- The most nontraditional jobs I've heard of were . . .

- The most interesting jobs I've heard of are . . .

- I'd really like a job that lets me . . . These jobs are . . .

- Some people get to work at "real" jobs that are like hobbies. These are . . .

- New technology is opening up more jobs. These are . . .

Remember that your goal is to come up with a new understanding of where the "good" jobs are for you—and where these jobs will be in the future.

After you have completed writing your lists, reread them and decide which jobs have the most real potential and why. Write a brief paragraph summarizing what you learned from this exercise.

Appendix A
Letters and Memos

Use this appendix to check the format you are using for letters and memos. Brief comments have been provided to highlight the most common features of letters and memos.

Most workplace writers spend much of their time preparing memos and letters. More often today, memos are sent by voice-mail or e-mail, in some instances, drastically changing how we deliver information to our co-workers.

Our writing goal is always to produce documents that give our readers:

- *Thoughtful content* that solves the reader's problem.

- *Logical organization* that presents the information in an understandable order.

- *A writing style* that presents information clearly, concisely, correctly, courteously, and completely.

- *Document design* that is easy to read.

- *Careful editing and proofreading.*

Notice in this appendix how writers have used conventional formats to meet these guidelines.

QUICK FORMAT: LETTERS

"Please send me a letter!" You will hear this statement many times in your career as your customers, suppliers, and sometimes government agencies ask for information or have a problem that needs to be solved. Putting your information or proposal in writing can be more efficient that relaying the information over the phone.

A letter helps many others see exactly what you mean and often establishes an informal contract between the reader and the writer, with the writer legally representing the company. A letter also provides formal confirmation of decisions made or important guidelines. For example, you might receive a letter confirming

Appendix A: Letters and Memos   **395**

you've been hired, or you might send a letter requesting information about a product or service.

Because letters represent your employer, it's important that they are written professionally. The same guidelines used throughout this book are useful in writing letters. Your letters need to be thoughtful, well organized, clearly written, concise, and courteous. They must also follow conventional letter format.

Here are the main parts of a letter, which generally appear in this order:

Letterhead	Shows the name and logo of the company sending the letter or the return address of the sender. People often make up a personal letterhead with a logo using word-processing features.
Date	Shows the date the letter was written.
Inside Address	Lists the name, title, department, company name, street address, city, state and zip code of the person the letter is being sent to.
Greeting	Says hello to the person the letter is written to. Also called a salutation. The greetings most commonly used are: Dear Ms. Smith: Dear Mr. Smith, Dear Henry, Dear Manager: *Note:* Using a comma after the greeting is considered a more personal element, and using a colon after the greeting is considered formal. When selecting a format style, check to make sure all your elements match!
Body	Presents the main part of the letter organized in an introduction, discussion, and conclusion pattern. *Note:* The letterhead introduces the company or the person writing the letter. Therefore, the first sentences do not introduce the writer but concentrate on the purpose and contents of the letter.
Closing	Ends the letter with a conventional expression. This is sometimes difficult to write, especially when you are writing to someone you don't know. However, the closing is expected and adds to your courteous and professional tone. Leaving out the closing can result in a letter with a cold tone. The most commonly used expressions today are: Sincerely, Sincerely yours, Cordially, Respectfully, Expect changes in letter parts and formats. The most common closing 150 years ago was: Your humble and obedient servant.

Appendix A: Letters and Memos

Signature Block   Shows your name and title. If the letter is a formal contract, the company's name can appear in the signature block.

References   Shows initials of person who prepared the letter (if not the sender) and lists materials enclosed with the letter. This line is less frequently used but can be useful.

Companies often have very clear internal guidelines about the letter format that should be used for their outgoing letters. This appendix shows the three most commonly used formats, but expect variations!

- **Block:** Shows all parts of the letter lined up tightly to the left margin. This format is considered formal and emphasizes efficiency.

- **Modified Block:** Puts date and signature block slightly right of the middle of the letter. Paragraphs are not indented. This format is most commonly used as a balance between formal and personal formats.

- **Indent:** Indents paragraphs five spaces. Also puts date and signature block slightly right of the middle of the letter. This format is personal and often is used for writing to customers, personal letters on company stationery, or when applying for jobs.

Notice how the writer's choice of format affects the reader as you read three versions of the same letter (Exhibits A.1, A.2, and A.3).

Appendix A: Letters and Memos   **397**

Block Letter   Exhibit A.1 ∎

Allen Turgenyev
Real Estate and Property Management
Serving You Since 1984

290 King's Boulevard, Spokane, WA 29567
206/754-8012 FAX 206/754-8011

May 29, 1996

Michael Moss
1675 NW Alpaca Drive
Spokane, WA 29567

Dear Mr. Moss:

Our maintenance man, John Fitzgibbons, stopped by your property to assess the siding. He recommends:

1. Caulking the entire duplex along the horizontal siding, and

2. Painting the T.111 on the south side of the building with one coat of paint.

He estimates costs and materials will be $275.

We recommend that you go along with this suggestion; however, if you have any questions or concerns, please call John at 754-9614, or call our office. In the meantime, unless we hear otherwise from you by June 14, we will go ahead and have the work done.

Sincerely,

Allen Turgenyev

kw

Start your block letter format about 1 inch from the top of the page.

The center of the letter needs to appear at about the center of the page. Adjust the number of blank lines (typically three to eight blank lines) between the date line and the inside address.

Despite the formal block style, notice the personal greeting.

The first paragraph states the purpose of the letter.

Allow one blank line between each paragraph.

Use lists to highlight key information.

Notice how the writer creates a warm yet professional tone with just a few words.

Three blank lines allow room for the letter writer's signature.

Initials show who prepared the letter.

398   Appendix A: Letters and Memos

■ Exhibit A.2   Modified Block Letter

In the modified block letter, the date moves to just a little past the middle of the letter.

Allen Turgenyev
Real Estate and Property Management
Serving You Since 1984

290 King's Boulevard, Spokane, WA 29567
206/754-8012 FAX 206/754-8011

May 29, 1996

Michael Moss
1675 NW Alpaca Drive
Spokane, WA 29567

Dear Michael,

Our maintenance man, John Fitzgibbons, stopped by your property to assess the siding. He recommends:

1. Caulking the entire duplex along the horizontal siding, and

2. Painting the T.111 on the south side of the building with one coat of paint.

He estimates costs and materials will be $275.

We recommend that you go along with this suggestion; however, if you have any questions or concerns, please call John at 754-9614, or call our office. In the meantime, unless we hear otherwise from you by June 14, we will go ahead and have the work done.

Sincerely,

Allen Turgenyev

kw

No indenting of paragraphs with the modified block format!

Always allow one blank line between each paragraph.

Notice how the date line, the closing, and the signature block are all lined up exactly.

The position of the signature block changes, but the position showing the initials of the person who prepared the letter does not change.

Appendix A: Letters and Memos **399**

Indented Letter   Exhibit A.3 ∎

Allen Turgenyev
Real Estate and Property Management
Serving You Since 1984

290 King's Boulevard, Spokane, WA 29567
206/754-8012 FAX 206/754-8011

May 29, 1996

Michael Moss
1675 NW Alpaca Drive
Spokane, WA 29567

Dear Michael,

Our maintenance man, John Fitzgibbons, stopped by your property to assess the siding. He recommends:

1. Caulking the entire duplex along the horizontal siding, and

2. Painting the T.111 on the south side of the building with one coat of paint.

He estimates costs and materials will be $275.

We recommend that you go along with this suggestion; however, if you have any questions or concerns, please call John at 754-9614, or call our office. In the meantime, unless we hear otherwise from you by June 14, we will go ahead and have the work done.

Sincerely,

Allen Turgenyev

kw

In the indented letter format the date moves to just a little past the middle of the letter.

Indent paragraphs five spaces with the indented letter format!

Always allow one blank line between each paragraph.

Notice how the date line, the closing, and the signature block are all lined up exactly.

The position of the signature block changes, but the position showing the initials of the person who prepared the letter does not change.

400   Appendix A: Letters and Memos

QUICK FORMAT: MEMOS

The conventional memo format is shown below. Start using it now for as many of your assignments as you can to give you more practice in writing memos.

Please note that companies use a standard memo format in different ways and often include a letterhead in their memos. As companies switch from paper trails to paperless offices, the memo may change its medium, but you'll still see these main parts to a memo, generally in this order:

Date: Shows when the memo was written. May document the exact date an event took place.

To: Shows name, title, and department of person who is receiving the memo.

cc: Used to show names of people receiving copies.

From: Shows name, title, and department of person writing the memo.

Subject: States *topic* and *purpose* of memo in five to seven key words. May show *urgency* of memo.

Body: Organized into introduction, discussion, and conclusion pattern. May include graphics or tables of data. May use bullets or a numbered list.

Exhibit A.4 shows how these elements are used in a conventional memo.

Appendix A: Letters and Memos   **401**

Memo   Exhibit A.4 ∎

Today's date

TO: Name and title of person receiving memo

FROM: Name and title of person sending memo

SUBJECT: Topic and purpose of memo in 5–7 key words

Memos typically begin with a brief introduction that explains the purpose of the memo. Summaries or sentences that preview what's in this memo are very helpful for longer memos.

Format. The discussion section of the memo presents supporting information. Memo writers frequently use:
- short paragraphs (typically 5 to 7 lines)
- lists of key information or ideas (can use numbers or "bullets" to show priority).
- headings to separate the sections

Editing. Memo writers are ruthless about writing clearly and concisely.

Courteous tone. Readers react strongly to negative memos. Always check your writing style to see if you have written courteously and have stated ideas positively.

Conclusions. Most memos end with a conclusion that reinforces the most important points and tells why these are important for this situation and this reader. Usually readers look more closely at the first and last paragraph of a memo, so here is where you request your reader to take action.

Start your memo format about 1 inch from the top of the page.

Readers need to know right away what this memo is about. Make sure a person who has no knowledge of the situation could file the memo in the right place by looking at the subject line.

Use headings and short paragraphs to help *cluster* the information into easy-to-read blocks.

Notice how "white" space (line spacing between each paragraph) is used to make the memo easy to read.

When your memo ends, just stop! Don't worry about the bottom margin. Memos typically do not fill an entire page and are not centered from top to bottom like letters.

Appendix B
Reports

If a memo is very long, over eight or nine pages, it becomes a *formal report*— whether the document is intended for an audience inside or outside your employer. Readers appreciate a report format that helps them find information quickly. For example, a table of contents is useful in a 25-page report, and a novice reader will appreciate a glossary that lists key terms with definitions.

The major difference between a memo report and a formal report is the number of pages and how prefacing and supplementing sections are used.

Prefacing sections go before the main body of the report and usually include:

- Title page.
- Cover or transmittal letter or memo.
- Table of contents.
- Glossary (if needed).
- Executive abstract.

Supplementing sections go after the conclusion and recommendations sections and may include:

- Bibliography.
- Blank survey form.
- Tally sheet.
- Supporting maps (if needed).
- Transcripts of interviews (if needed).
- Copies of supplementing materials.

You must decide which of these formal report elements are needed based on your employer's guidelines, the needs of your audience, the type of material you are presenting, and the length of your report.

Appendix B: Reports **403**

A formal report prepared by four students for a technical writing class is included in this appendix. As you read it, you will notice comments that highlight the writing strategies and report elements that will help you understand the choices the student authors—Tim, Joel, Sang, and Tony—made as they worked on the five-week project to prepare a problem-solving formal report.

Following the students' report, another section of this appendix, Footnotes and Bibliographies (see pages 442–450) discusses when to use footnotes, how to write them, and how to prepare different styles of bibliographies.

COMMENT ON THE COVER MEMO

CONTENT: The authors' purpose for writing the report is stated very clearly in the first and second paragraphs. Notice also how this team limits the scope of the report in the third paragraph. The reader knows what to expect in this report.

The last paragraph lets the reader know that this team is ready for further discussion. How will the reader get in touch with Tim, Joel, Sang, and Tony? It's always a good idea to add that information.

The cover memo or letter is also the best place to connect on a personal level with your reader. You may want to thank the reader for being helpful or for making resources available.

ORGANIZATION: The team organizes the cover memo into a conventional pattern of introduction, discussion, and conclusion.

STYLE: Although this team uses a formal voice to present its findings, notice how the "you" focus creates a courteous tone.

DOCUMENT DESIGN: Tim, Joel, Sang, and Tony chose a cover memo because the reader works for the same organization they work for. They could have chosen a cover letter if they wanted to create a more formal tone. Some design elements that work well:

- Short, readable paragraphs.
- Standard memo format for date/to/from/subject lines (no surprises).
- Standard memo format for paragraphs (no indentations, single-spaced text).

MEMORANDUM

DATE: December 5, 1995

TO: Beth Camp; Chair, Distance Education & Instructional Technology Committee,
Linn-Benton Community College (LBCC)

FROM: Tim Haag, Joel Ewing, Sang Lim, Tony Falso

RE: REPORT ON STUDENT ACCESS TO LBCC INFORMATION TECHNOLOGY

The enclosed report represents a group project required for your Technical Writing 227 class. We hope that you will find the recommendations useful enough to share with your policy subcommittee members.

As you know, this report is not simply the academic exercise of four LBCC students. It is our intent that this report be put to practical use. Furthermore, this report is intended to guide LBCC staff and faculty in the development of fair and useful policies in dealing with student use of LBCC's informational technology.

The subject of student access to information technology at Linn-Benton Community College, and development of policies to deal with such, is both complex and controversial. Nevertheless, with limited time at our disposal considering the broad scope of issues involved, our team has worked hard in an effort to provide the policy subcommittee with what we hope are valuable insights and recommendations.

If you, or any other interested party, have questions about this report or would like further assistance from any of us, please feel free to contact us at any time.

Source: Tim Haag, Joel Ewing, Sang Lim, and Tony Falso, class project, Decmeber 5, 1995. Used with permission.

406 Appendix B: Reports

COMMENT ON THE TITLE PAGE

CONTENT: The clearly written title for this report identifies the topic of the report, the scope of the report, and the student point of view—all in ten key words.

That's the writer's goal when choosing the title of the report. Can the reader understand what this report is about? Who should read it and how important is this report? You should be able to answer these questions just by reading this title.

Notice how specifically both the reader and the writers are identified. If possible, include the name, title, department, and name of company for the reader and the writer.

ORGANIZATION: The title page is organized into four main parts:

- A clear, descriptive title.

- A "prepared for" section to show the primary reader.

- A "prepared by" section to show the writer of the report.

- A date.

STYLE: Word choice is an important consideration of writing style. Will every reader who picks up this report understand what *information technology* means?

DOCUMENT DESIGN: Notice how the four key elements of a title page are balanced from the top of the page to the bottom of the page. "White space" adds emphasis to each element.

Student Use of Information Technology
at
Linn-Benton Community College

for

Beth Camp
Technical Writing Instructor
Chair, Distance Education & Instructional Technology Committee
Linn-Benton Community College
Albany, Oregon

by

Tim Haag
Joel Ewing
Sang Lim
Tony Falso

(Technical Writing 227 Students)

December 5, 1995

408   Appendix B: Reports

COMMENT ON THE TABLE OF CONTENTS

CONTENT: Notice the descriptive headings throughout the report, which organize the discussion. These descriptive headings show up in the table of contents and give an excellent preview of report contents. Often readers of very technical reports will study the table of contents to identify the topics covered and their relationships to each other.

Three types of headings appear in this table of contents.

1. *Functional headings* describe the purpose of a section. These include, for example, Introduction, Conclusion, and Bibliography.

2. *Topic headings* introduce the topics that will be discussed in a particular section. For example, Problems, Goals, Guidelines, Penalties, and Future Growth state the topic for those sections.

3. *Descriptive headings* combine both the topic and the purpose of the section. These section headings are longer and are often presented in question format. One example: What instructional technology is available to LBCC students?

ORGANIZATION: You can tell at a glance which topics make up the main sections and which topics are supporting sections. This organizing strategy is called *subordination.*

Information that would be useful to the reader but distracting if it were included in the main body of the report is organized into two supporting Appendixes.

STYLE: Parallel form is used pretty consistently throughout this table of contents within each section. Notice how the background section uses key words for its topic headings, and the Access section uses questions for its more descriptive headings.

DOCUMENT DESIGN: Several design elements show up on this title page. They include:

- Capital letters to emphasize main sections.

- Indenting to show supporting sections.

When you number the pages, note that pages before page 1 of the main body of the report use Roman numerals (i, ii, iii, iv, v, etc.), and the report pages use Arabic numbers (1, 2, 3, etc.). Notice that page 1 is not numbered at all; this follows widely accepted standards for report page numbering.

CONTENTS

	page
REPORT SUMMARY	iii
INTRODUCTION	1
BACKGROUND	1
Problems	1
Goals	3
Guidelines	3
Future Growth	3
ACCESS	4
What instructional technology is available to LBCC students?	4
Does LBCC offer its students reasonable access to technology?	4
What possible issues spring from access to the Internet?	5
Access: Conclusions	9
RELATED ISSUES	9
Privacy	9
Fair Use	10
Penalties	10
Informing Students	11
Related Issues: Conclusions	11
SURVEY SUMMARY	12
CONCLUSION	13
RECOMMENDATIONS	14
BIBLIOGRAPHY	16
APPENDIX A: Sample Survey Form and Tally Sheet (Raw Data)	17
APPENDIX B: Survey Form Written Comments	19

410   Appendix B: Reports

....

COMMENT ON THE SUMMARY

CONTENT: The summary is called by many different names, such as *executive summary, synopsis, abstract,* or simply *summary.* This one-page summary is written to "replace" the entire report for busy readers. Thus, the summary must include the essence of the report. You will need to:

- State problem and purpose of the report.

- Provide just enough background so the casual reader understands the extent of the problem. (You may need to use statistics or percentages.)

- Summarize very briefly the methods you used to collect data.

- Highlight key findings.

- Repeat the recommendations.

- Conclude with a compelling reason that motivates the reader to act on your recommendations.

Not every reader who picks up your report will read through the entire report body, but most readers will read your summary.

ORGANIZATION: Notice how the previous writing goals or *key ideas* are organized into an introduction, discussion, and conclusion format.

STYLE: Since nearly all readers of a report will read the summary, writers need to use words that communicate ideas clearly to all levels of readers, whether novice, intermediate, or expert. Some other strategies the writing team used to improve style:

- Transitional phrases that lead the reader through the summary of new information being presented.

- Clear "pointers" to specific resources or pages in the report.

Tim, Joel, Sang, and Tony chose to use a "pointer" to later recommendations; many summaries repeat the recommendations section in its entirety.

DOCUMENT DESIGN: Once the reader begins to read the report seriously, the document design should reinforce the main ideas to make it easy for the reader to understand and retain the ideas being presented. The following design elements help to achieve this goal in this summary:

- Short, readable paragraphs.

- *Italics* and underlining to highlight section headings.

- Bulleted lists to highlight key concepts.

SUMMARY

The Challenge: New information technology brings with it many new challenges for Linn-Benton Community College. LBCC must determine what technology it will be able to offer, and it must decide what information it will allow students to have access to and what it will restrict. Along with this responsibility, LBCC must develop guidelines for student use of this technology and determine how to deal with violations of these guidelines. Furthermore, LBCC must develop strategies to educate students on what resources are available as well as inform them of the guidelines for use and penalties for misuse of those resources.

Policy Problems? LBCC is faced with many problems related to the development of an effective and comprehensive instructional technology use policy. These problems are not insurmountable and LBCC is in a position to take heed of the old cliche about reinventing the wheel. Many other institutions are faced with similar problems and concerns and have already developed approaches for dealing with the issue. Using these institutional guidelines as well as those of the American Library Association (when combined with LBCC staff and student input and participation) should result in the development of an effective policy.

Some Simple Solutions: Although the issues are broad, many of the solutions are very simple. For instance, the Computer Ethics Institute's "Ten Commandments for Computer Use" are simple yet effective guidelines that will help to inform most any student (or any other system user) of their expected behavior when using any of LBCC's information technologies. These simple guidelines (see report page 7) can be an effective tool in avoiding controversy over issues of censorship and restriction.

How students are informed of LBCC's policies is of critical importance. Students all need to be well informed about what resources are available to them and how they are expected to use them. Three specific recommendations worthy of mention are:

- The use of college publications.
- The use of bulletin boards and fliers for information dissemination.
- The use of an on-line screen that identifies user responsibilities.

Student Viewpoints: An important tool of this report was its survey (see Appendix A in this report). Seeking student input was imperative in determining their needs, views, and insights. A wide range of responses was anticipated and received. Somewhat surprisingly, however, students expressed some very clear views about some issues. Additionally, they were not afraid to share them, and we received many written comments along with the survey. A comprehensive listing of their comments may be found in Appendix B.

These survey responses and comments can be used as an effective tool in guiding LBCC's policy decisions. Although our initial survey obtained some very interesting results, it is our recommendation that a more comprehensive follow-up survey would benefit LBCC even further.

Additional specific recommendations can be found in the recommendations section of this report.

412   Appendix B: Reports

COMMENT ON THE INTRODUCTION AND BACKGROUND

CONTENT: The introduction and background sections are key parts of your report. These sections will

- Introduce the problem that the report will address.
- Clarify the purpose of the entire report.
- Define key terms, if needed.
- Provide background, if needed.
- Preview exactly what's in this report.

Although these elements are repeated in the cover letter or memo and in the report summary, the writer's main purpose in writing an *introduction* is to prepare the reader for the detailed discussion that follows. The writer's goal in preparing a *background* is to give the reader just enough information so that the discussion that follows makes sense. Not all readers will be well informed about the report problem. That's why a background is needed. Most writers will forget to add the preview; however, the *preview* helps the reader to understand the breadth of the report and the order of the sections that are presented in it.

Tim, Joel, Sang, and Tony wrote a very brief introduction section. They moved directly to a background section. Notice how carefully this writing team previews the exact order of the report in the first paragraph of the background and then defines the *context* (the setting or circumstances that affect this problem) throughout the background.

ORGANIZATION: The introduction, discussion, and conclusion pattern is used consistently to organize each section throughout the introduction and body of the report.

STYLE: In the introduction, the authors used short paragraphs to give the reader time to absorb the new information being presented. In the background section, this writing team has used words like *preview* and *overview* to emphasize the purpose of this section. As the amount of information being presented increases, writers will need to revise for

- Short, readable paragraphs.
- Clear transitions between ideas and sections.
- Precise and easy-to-understand wording.

INTRODUCTION

We have entered the information age, an age where the world is connected in ways never known before. The way we access, process, and use information has radically changed with advances in computer technology.

This new technology has brought with it new challenges for organizations that use it. Particularly challenged are educational organizations. How much information technology is adequate and just what information should be made available to students? Should all information be made accessible or should access to certain kinds of information be restricted?

This report will attempt to provide some assistance in developing a policy that will establish what is adequate access and what are student rights and responsibilities as users of this access. It is hoped that the material contained in this report will aid in the development of meaningful and comprehensive guidelines for users of this technology here at LBCC.

BACKGROUND

In order to better understand the nature and scope of the issues regarding student access to information technology at LBCC, an overview of some current issues will be presented. First, we will look at some examples of potential problems related to putting LBCC's system "on-line." We will also look at some sample goals and guidelines for educators providing students with access to new information technologies. In addition, we will briefly discuss the situation LBCC administration is faced with regarding future growth of its information technology.

Problems

As with all informational technologies, there are problems that arise in terms of maintaining these technologies and keeping them up-to-date. These problems require sufficient funding and staffing to ensure that the systems operate smoothly. Setting goals for system development, implementation, and maintenance is of paramount importance if efficient and useful technologies are to be offered.

Another problem that arises when Internet access is brought into the academic arena is determining what is acceptable use and how to deal with what is determined to be unacceptable.

The following are some scenarios of actual incidents and their outcomes.

Scenario 1.

March 12, 1994 (Saturday)

You are a computer science professor at Denver University and creator of NYX, a free public access service affiliated with the University. The early edition of the Sunday *Denver Post* includes a story about NYX carrying alt.sex.intergen, an electronic free-speech forum about inter-generational sexual relations. The newspaper didn't interview you and their story leaves the impression that the news group contains illegal material when, in fact, it discusses the question: "Is pedophilia ok/evil?" Do you kill the news group? (Kadie)

1

414   Appendix B: Reports

DOCUMENT DESIGN: If you are working on a team, several key design decisions need to be made. Some examples follow.

- What style of headings should be used throughout the report?
- What kind of sections and subsections will we use?
- What kind of in-text citation format will we follow?
- How will graphics be used in each section?
- Are there any special features of page layout we should use throughout the report?

The challenge facing your writing team is to produce a document that has been written by more than one writer to look as if it were prepared by only one writer!

Appendix B: Reports **415**

The **"Library Bill of Rights"** says:
"Materials should not be proscribed or removed because of partisan or doctrinal disapproval."
(Kadie)

Comment:
"If you believe that the principles of intellectual freedom developed for libraries apply to public access computer sites, you should not remove the news group." (Kadie)

The Rest of the Story
NYX Sys Op Andrew Burt has declined to kill the news group. NYX is still on-line. The *Denver Post* has received faxes critical of its one-sided reporting.

Scenario 2.

February 1992–present

You are a faculty member at Iowa State University. Your computer administrators have set up a system in which members of the university community can access sex-related news groups (such as alt.sex) only by asking for a computer account on a special computer and by "acknowledging their responsibility in accessing, using, and distributing material from it" by signing a statement to that effect.

Some people say the university's action is OK because anyone who wants access can get it. What do you think?

Library Policy Says:
American Library Association's "Books/Materials Challenge Terminology" statement says "censorship" is:

The change in the access status of material, made by a governing authority or its representatives. Such changes include: exclusion, restriction, removal, or age/grade level changes.

The ALA "Restricted Access to Library Materials" says:

Attempts to restrict access to library materials violate the basic tenets of the LIBRARY BILL OF RIGHTS. [. . .] In any situation which restricts access to certain materials, a barrier is placed between the patron and those materials. [. . .] Because restricted collections often are composed of materials which some library patrons consider objectionable, the potential user may be predisposed to think of the materials as objectionable and, therefore, are reluctant to ask for them.

The Rest of the Story
The restrictions at Iowa State University continue. (Kadie)

2

COMMENT ON THE BODY

CONTENT: Each section begins with a general idea, which is supported by examples, facts, or quotations. The body of the report should answer the key questions of your audience. In this report, the authors asked: What are the key issues involved when students use computers at the school? What resources are currently available to students? How have other schools written their policies? and What do students think should be included?

How the question is worded affects what kind of information is gathered: *what* questions define the problem and key factors, *how* questions clarify process, *why* questions emphasize analysis, and *what if* questions anticipate and explore coming changes.

ORGANIZATION: Each section should begin with an introduction that states the purpose and previews what's in the section. If the report is long or complex, use an outline to organize it.

There are two commonly used outline patterns: The feasibility outline can be used to answer the question, "Given these criteria and this performance information, should we do this?" The report outline can answer the question, "Given the results of these findings, what should we do?" Using an outline can help the writer to organize a large "body" of information so that it flows logically. It can help the writer to check for completeness and clarity of connections between ideas, and can become the basis for section headings and the table of contents.

STYLE: A straightforward objective reporting style avoids editorial comments or bias and helps build the credibility of the report.

DOCUMENT DESIGN: Use *lead-ins* to introduce the information and *in-text citations* to show the source of the information.

> Example of *lead-in:* Some of the guidelines were, in part, *suggested by the American Library Association.*
>
> Example of *in-text citation:* To make the Department's computer systems readily available to our students (*Brock University*).

The authors (Tim, Joel, Sang, and Tony) chose an informal documentation system commonly used in the workplace, but they carefully distinguished between what they are saying as a writing team and what the research is saying. This minimizes plagiarism, which is, very simply, using someone else's ideas or information as your own.

At the end of this Appendix you will find an abbreviated guide on how to use three styles of documentation: the Modern Language Association, the American Psychological Association, and the numbered system.

Goals

The Department of Computer Science and Information Processing at Brock University of Ontario, Canada, states its key goal for providing information technology this way: "The computing systems . . . are intended to support its educational and research purposes and to enhance the educational environment" (Brock University).

The Department then identifies these more specific goals:

- To ensure that the computing systems are properly maintained.
- To ensure that computer system up-time is at a maximum.
- To announce in advance any downtime that is scheduled.
- To make available state-of-the-art computer systems and software, subject to available funds and space.
- To make available suitable stations for those with special needs, subject to available funds and space.
- To make the Department's computer systems readily available to our students (Brock University).

Guidelines

As stewards of the information age, educators who provide students with access to technology, particularly the Internet, take on the role of librarians. They must guard against restricting information available on the Internet and from judging the users of that information and what they choose to access. What better guidelines than the Library Bill of Rights, which says in part:

Libraries and librarians exist to facilitate freedom of speech and freedom to read by providing access to, identifying, retrieving, organizing, and preserving recorded expression regardless of the formats or technologies in which that expression is recorded (ALA).

Some of the guidelines, in part, suggested by the American Library Association:

- Follow the Library Bill of Rights.
- Have legal counsel check policy.
- Train staff. Include empathy training.
- Consider any policy that limits access very carefully. Provide a clear description of the behavior that is prohibited so that a reasonably intelligent person will have fair warning.

Future Growth

LBCC administrators have made plans for future expansion of information technology in the way of more Internet access stations, teleconferencing and telecourses, and access to some of the on-campus resources through the Internet.

In regard to future expansion and addition of new technologies, there is the problem of implementation in terms of staff resources. Rick Barker, LBCC's media technician, revealed in an interview that staffing is one of the problems in implementation and maintenance of the current and future systems on campus. Barker has a seemingly impossible list of responsibilities and projects that he is currently undertaking with the help of only one part-time work-study student.

There is little doubt that LBCC's current resources are well used and that the usage level will increase with time. Additionally, the need for these resources will continue to rise as these kinds of technologies become integral components of the educational process.

3

COMMENTS ON THE SURVEY SECTION AND GRAPHIC ELEMENTS

CONTENT: Here the writing team highlights findings based on its primary research. As they introduce the survey, notice how simply they report the methodology of the survey; the reader learns that 41 students from LBCC were surveyed. In more formal reports, more attention to the methods used to collect the information would be required.

ORGANIZATION: The writing team continues to use *problem, purpose,* and *preview* checkpoints to prepare the introductions to each section and supporting section.

STYLE: The style continues to be objective in tone.

DOCUMENT DESIGN: Notice also by comparing the survey form itself in the appendix that the writing team did not highlight every single finding from the survey data. They choose to

- List key findings.
- Select two to show visually in a standard bar chart.
- Report interview highlights in a paragraph form.

The following suggestions may help when you use graphics in your report:

- Place the graphic as closely as possible to the idea it supports.
- Use a descriptive title to introduce the graphic.
- Conclude with an interpretation of the graphic.

ACCESS

Before reaching conclusions about workable policies on instructional technology, it was important to survey what resources are currently available, how accessible those resources are to students, and what concerns might arise from the use of those resources, especially the Internet. The following sections will address these questions:
- What instructional technology is available to LBCC students?
- Does LBCC offer its students reasonable access to this technology?
- What possible issues spring from access to the Internet?

What instructional technology is available to LBCC students?

In order to assess LBCC's ability to adequately meet its students' needs for instructional technology, an informal inventory of these resources was conducted. Currently, LBCC's instructional technology resources include the following:
- The Forum Lab (over 30 IBM-compatible and Macintosh computers).
- The Learning Resource Center (25 IBM-compatibles offering Internet access).
- The campus VAX system (offering communication with campus library, faculty, and users of Forum Lab computers).
- Two library terminals connected to the Internet.
- Two IBM-compatible labs at its satellite campus in Corvallis (Benton Center).
- An IBM-compatible lab at its satellite campus in Lebanon.
- An IBM-compatible lab at its satellite campus in Sweet Home source.

Does LBCC offer its students reasonable access to this technology?

Those who actually use LBCC's instructional resources (students and staff members) are most qualified to provide insight into the subjective issue of reasonable access.

Among the 41 students surveyed:
- 83% of the respondents felt LBCC should offer more computer training. (Figure A)
- 90% felt LBCC should better inform students of the available resources. (Figure B)

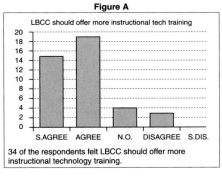

Figure A

34 of the respondents felt LBCC should offer more instructional technology training.

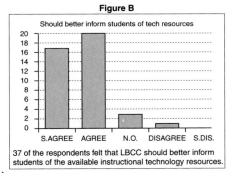

Figure B

37 of the respondents felt that LBCC should better inform students of the available instructional technology resources.

420 Appendix B: Reports

COMMENTS ON THE BODY (CONTINUED)

CONTENT: In the rest of the body of the report, Tim, Joel, Sang, and Tony summarized the findings of their interviews. Sometimes, attributing the comments directly to the person interviewed may create problems for that person or his or her employer. In these cases, you may decide to use pseudonyms or simply delete the person's name. You can, when introducing the interview material, note that no names are being used to protect the confidentiality of the sources.

The next sections further develop the materials needed so that the readers can decide which elements should be included in a campus computer use policy. You will see additional comments by the conclusion section on page 13 of the report.

Appendix B: Reports   **421**

One staff member agreed with this latter finding. "We do a pretty good job of keeping it a secret, though not intentionally," the staffer (Smith) said.

Another LBCC instructor, however, felt that LBCC offered sufficient access. "Everyone who takes a class at LBCC has the chance to use the computer labs available at the main campus, as well as the outlying centers. This access is free of any extra fees and is only limited by the space available and the time that the labs are available" (Fay).

There are concerns, however, about some of the technology available. "One continuing frustration for students has been the very diverse platforms (Windows, DOS, and Macintosh) we use in different areas of the college," stated one instructor (Camp). Said another, "Some of the labs have very outdated equipment which are [sic] not capable of handling much of the new software on the market today" (Fay).

A third staffer notes that there is different software at the different campuses. This lack of standardization "makes it difficult for students who begin a project at LBCC and want to continue it at Benton Center" (Smith).

LBCC has made progress in offering students access to the Internet. To its two connected terminals in the library, LBCC has added Internet access in the Learning Resource Center. Internet classes will now be offered there as well as at the lab at Oregon State University.

What possible issues spring from access to the Internet?

While hardware and software issues are often related to budget concerns, student access to the Internet raises key issues centering on intellectual freedom, censorship, and "unacceptable use." This section will examine these two issues:

- Should LBCC restrict what is accessed on the Internet?
- What might be considered "unacceptable use" of the various Internet tools?

Should LBCC restrict what is accessed on the Internet?

When discussing restrictions, the concept of *freedoms* cannot be excluded. Thus, to provide background to this discussion, consider the following excerpt from the American Library Association's Library Bill of Rights:

> Libraries and librarians exist to facilitate [freedom of speech and freedom to read] by providing access to, identifying, retrieving, organizing, and preserving recorded expression regardless of the formats or technologies in which that expression is recorded.

One staff member clearly subscribes to the same philosophy. "In my opinion, it is NOT up to LBCC to determine what is acceptable and unacceptable content," said the instructor (Fay).

An LBCC student echoes that opinion: "The Internet and access to it must be kept free from regulation lest personal rights be sacrificed. It is not so important that students may browse pornographic material. What is important is that they have the RIGHT to. Censorship is by far a greater evil than any we as citizens can imagine" (survey comment).

5

422 Appendix B: Reports

Another surveyed student added: "I don't care how people use the Internet as long as there is equal opportunity, they don't break the law, or violate my rights of privacy (i.e., personal information or private e-mail)." Note: See Appendix B for a comprehensive list of survey comments.

What might be considered "unacceptable use" of the various Internet tools?

Concerns arise over substantial student use of instructional technology as well as student access to the vast amounts of information available through these resources. The issues of "unacceptable use" of computer resources, the approval of restrictions, and possible censorship to avert perceived abuses are controversial ones. This section will

1. Offer general guidelines for "acceptable use" of campus technology.
2. Explore the subject of "unacceptable use."
3. Present divergent views on restriction and censorship.

An Inclusive Policy of Acceptable Use of Campus Technology

Any institution intending to set its own policies would do well to research similar existing policies. The following guidelines may be helpful in establishing instructional technology policy.

The Ten Commandments for Computer Use (from the Computer Ethics Institute)

1. Thou shalt not use a computer to harm other people.
2. Thou shalt not interfere with other people's computer work.
3. Thou shalt not snoop around in other people's files.
4. Thou shalt not use a computer to steal.
5. Thou shalt not use a computer to bear false witness.
6. Thou shalt not use or copy software for which you have not paid.
7. Thou shalt not use other people's computer resources without authorization.
8. Thou shalt not appropriate other people's intellectual output.
9. Thou shalt think about the social consequences of the program you write.
10. Thou shalt use a computer in ways that show consideration and respect.

LBCC might also consider combining an institution's insights with those of its users. An LBCC student offers his own guideline, saying, "Instructional technology should not be used as a toy. If work is LBCC related, fine. If not, buy your own computer. No public subsidized toys!" The concept of *public subsidy* adds a dimension of responsibility to users of instructional technology. It is the responsibility of the students to use these resources in an "acceptable manner" or, conversely, to steer away from "unacceptable uses." But what is an unacceptable use?

What constitutes unacceptable use of computer resources?

To set the stage for a discussion of possible restrictions, this section will introduce actions often viewed as unacceptable uses of computer resources. Included in the list of unacceptable uses of instructional technology (as outlined in Brock University's Department of Computer Science policies) are the following:

6

Appendix B: Reports

a. *"Modifying the computer system's files, such as modifying the autoexec.bat, config.sys, and system folders."* (The student survey bore out strong agreement with this item. Ninety-eight percent of the respondents felt that attempting to introduce a computer virus into the system should be considered a violation of a campus computer policy.)

b. *"Displaying, playing, or transmitting materials that may be considered offensive by others."*

The students provided a more specific look at unacceptable material available electronically. Our survey listed five different categories of material that can be found on the Internet. The students were asked to indicate any of those five items they considered unacceptable. The categories of material and percent of students who designated that material as unacceptable follow:

Child pornography: 93%

Adult pornography: 71%

Information that could lead to human or animal injury: 61%

Sexually explicit discussion: 59%

Profane language: 46%

Brock University's further discussion of offensive materials is germane to this section:

"Offensive material takes many forms in the world of computing and networking. Originators (those who send) have a responsibility to monitor their material so that it is not obscene, vulgar or harassing. Should you receive material such as a message, picture, or suggestion that offends you, tell the originator that the material is unwelcome and offensive to you." The term *offensive,* then, is relative. If the receiver deems it 'offensive', then the originator faces possible penalties. If not, then much wider latitude is granted."

LBCC policy makers might also consider another kind of electronic harassment, such as the kind experienced at Eastman-Kodak, Inc. Kodak's Information Services Security Manager, Robert L. Mirguet found persistent e-mail messages about users' personal affairs to be a major problem within its communication system. That company, and others, decided it needed to clearly define what is unacceptable content in employees' e-mail (*Computerworld*).

At this point, it is important to differentiate between sending unwanted offensive material and calling it up for your own purposes. This is an important line to draw when discussing restrictions and censorship, which comprise our next section.

Restriction and censorship: Do they have a place in LBCC's instructional technology policies?

The controversy over issues such as restriction, censorship, and intellectual freedom is ingrained in American schools. This report will not even attempt to settle the issue. Instead, this section will present some views of LBCC students and staff members on these issues and offer an opposing view by a college's computer policy committee.

The opinions on the issues of restriction and censorship vary from one staff member to another. One instructor felt LBCC was not in a position to be a gatekeeper. "In my opinion, it is NOT up to LBCC to determine what is acceptable and unacceptable content." Another staffer felt it was definitely not the job of the instructors, saying, "I want someone to tell me what is unacceptable. I don't want to be in the position of being a censor." That same individual does see the need for at least one kind of limitation, saying, "What needs to be restricted is what they print."

7

Where do the students stand on restrictions? Of the students surveyed (Figure C) 53% felt LBCC should restrict access to certain material on the Internet, while 83% (Figure D) felt the school should impose penalties on students who violated policies on acceptable use.

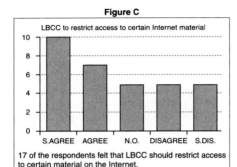

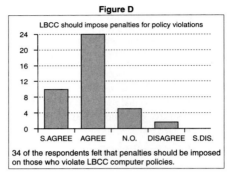

Additionally, students were asked to consider seven different categories as violations of a campus policy on acceptable use of instructional technology. Results were as follows:

1. Attempt to introduce a computer virus into system (98%).
2. Attempt to access other students' accounts without their consent (95%).
3. Downloading pornographic materials (63%).
4. Overloading system resources with your own data (61%).
5. Use of profanity in any written communication (32%).
6. No action (2%).
7. Other (0%).

Students showed strong reaction to introducing a computer virus to the system, perhaps due to the destructive nature with which it could affect all the system users at the same time.

The Computer Policy Committee at the University of Wisconsin-Milwaukee, on the other hand, allows much wider latitude. To paraphrase that university's campus administrative policy:

- Only after extensive research and consultation should restrictions be placed.
- The principles of intellectual freedom apply to computer resources as well as traditional resources.

It is necessary to continuously define, modify, and adapt to any new issues on this subject due to the rapid developments in technology. Therefore, LBCC must pay close attention to its restrictive policies and carefully monitor the related issues.

Appendix B: Reports **425**

Access: Conclusions

This section has presented a well-conceived list of proscriptions that can guide users of instructional technology. Adhering to guidelines from the Computer Ethics Institute could easily prevent much controversy in the issues related to responsible use. The list places considerable accountability on the individual. It is when the individual's own interpretation of "offensive" and "unacceptable" contradicts wider interpretations of the same that issues of restriction and censorship arise.

LBCC students and staff, and one computer policy committee, showed an expected range of views in the areas of restriction and censorship. Resolution of these issues is not the focus of this report.

RELATED ISSUES

Various issues are related to student access of instructional technology and some of these will be discussed in this section. Among those are privacy; fair use; restrictions; student violations; penalties; and the issue of informing and educating students about available resources, as well as LBCC's technology use policy.

Privacy

Privacy is a very sensitive issue related to student access of instructional technology. Each user has a right to privacy in his or her electronic work space. Users must respect other users' computer files, as these are their own private property. "Such actions as destroying, altering, copying or simply snooping through the data of other users is a serious breach of ethics" (Brock University). In order to prevent computer hackers from invading users' privacy, certain safeguards must be developed. There are a few possible safeguarding methods which need further exploration. They include the following:

A. Use of technology that forces users to use different passwords each time they turn on a computer: This helps to deter any invader from monitoring a user's file (Chronicle of Higher Education).

B. Use of technology that eliminates the ability of computer users to masquerade as someone else: This makes invaders unable to disguise their identity and can create in them a sense of insecurity that might prevent violations (Chronicle of Higher Education).

C. Use of software that would encrypt (scramble) the information: This makes it difficult and improbable for invaders to be able to break into a user's file without unscrambling the code (Chronicle of Higher Education).

LBCC's network system administrators should be "expected to treat the contents of electronic files as private and confidential" (Univ. of Illinois). According to our student survey, a majority of the students (56%) said that they would not approve of any kind of inspection of their private e-mail files even when the inspection was called for by law-enforcement authorities (see Figure E).

9

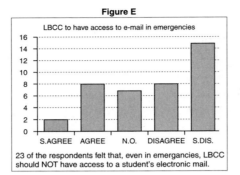

Figure E — 23 of the respondents felt that, even in emergancies, LBCC should NOT have access to a student's electronic mail.

Any inspection of students' files and any action based upon such inspection should be governed by applicable U.S. and Oregon laws and by LBCC policies. All students need to be fully informed of the restrictions placed upon them, and these restrictions should be included in the contents of any published policy statement or written agreement between LBCC and student users (University of Illinois).

Fair Use

Defining "Fair Use" of copyrighted materials in various communication media could get fairly complex.

Various fair-use exemptions have been available in education for copying portions of books, periodicals, and other protected works, such as TV programs. But, with new electronic versions of copyrighted works, it could become a completely different story. Since regulations for the use of electronic versions of copyrighted works have not been fully established, it is possible that academic institutions may not get the same fair-use exemptions.

Without these exemptions, students and faculty members would have to seek permission or pay a fee to use electronic versions of copyrighted works. If we don't preserve our fair-use right in an electronic environment, the usage will be prohibitively expensive or not available at all.

Penalties

After LBCC establishes certain restrictions on the use of its computer systems, penalties for those who violate the policy need to be set accordingly. Our survey shows that a majority of the students (83%—34 of 41 respondents) agreed on imposing penalties for violation of LBCC's policy. The degree of penalty imposed should be closely related to the importance of the issues that students have identified as unacceptable use, as discussed in the previous section on "restrictions."

Serious offenses like death or bomb threats should be turned over to law-enforcement agencies for investigation and prosecution. More typical violations like harassment or forged e-mail will be best dealt with through college judicial boards. Therefore, it is necessary to incorporate those relevant instructional technology policies with the campus judicial system.

Two full pages of "Proposed LBCC Students' Rights, Responsibilities & Conduct" were published in *The Commuter*—a weekly student publication in LBCC on October 4, 1995. This is a good baseline to see how a campus judicial system would operate and relate it to the policy on instructional technology.

Due to new and fast developing technologies that could make existing penalties meaningless, it is imperative that LBCC update the penalty list in the policy on a regular and frequent basis.

Informing Students

The students at LBCC have the right to be properly informed of available college resources and of any policies related to the use and misuse of those resources. There are many ways that this can be achieved. Among them, the following examples seem to be excellent steps toward properly informing students:

A. Use of college publication: By publishing in the college catalog, student handbook, and other periodicals like the campus newspaper.

B. Use of bulletin board: By posting information at various college locations or distributing fliers.

C. Use of on-line screen: By developing computer screens that students will see every time they use the system.

Use of an on-line screen seems to be a very effective way to inform students because it can provide students with selected information that is helpful in understanding the particular work they are doing and related precautions. In this way, students will be able to access resources without getting overloaded with an enormous amount of information at once.

Informing students of their violation should not be done publicly. They should be informed through private conversation or mail, postal or electronic, and in a manner which encourages students to resolve the problem.

Related Issues: Conclusion

All the issues mentioned in this section are important and further study of these issues is recommended. It would be in LBCC's best interest to encourage student participation through further research in their college classes such as writing, computer, engineering, and other technical or communication-related classes. This will allow LBCC to be better positioned in solving any problems that may arise from these issues.

Appendix B: Reports

SURVEY SUMMARY

LBCC is drafting a policy on instructional technology on the campus currently and we believe it is important to reflect various students' opinions in policy making. Our survey was designed to help determine the importance of instructional technology to LBCC students and to enable students to express their views on issues surrounding access to the Campus Computer System, including access to the Internet.

The survey was conducted among 41 students at both the LBCC main campus and at Benton Center classes during the period November 11, 1995, through November 12, 1995. In order to avoid sampling error, survey locations were carefully selected. Seventy percent of the survey was conducted in the LBCC cafeteria, where students with broad backgrounds were exposed to the survey. The remaining surveys were conducted at several different Benton Center classes.

A summary of our findings is as follows:
Note: (%) indicated represents the percentage of students who agreed with the statement given unless otherwise noted. (See Appendix A for detailed survey results.)

1. Students agreed that LBCC should better inform the students of available resources on campus (90%—37 of 41 respondents) and offer more classes that train them to utilize computer technology (83%—34 of 41 respondents).

2. Students agreed they would utilize Internet accounts on LBCC campus if they were available (56%—23/41) with a majority (86%—32 of 35 respondents) indicating they would be willing to pay $20.00 or less per quarter school term.

3. Among "unacceptable" materials for access in publicly funded institutions students strongly agreed to the following:
   Child pornography (93%).
   Adult pornography (71%).

4. Students agreed that LBCC should restrict access to certain material on the Internet (53%) and impose penalties on students who violate policy on acceptable use (83%).

5. Students showed strong rejection to invasion of their privacy. They disagreed with any inspection of their electronic mail file even by law-enforcement officials (70% disagree).

6. Six different categories were given to the students and they were asked to choose the violations of a campus policy on acceptable use of instructional technology. The following shows highest agreement.
   Attempt to introduce a computer virus into the system (98%).
   Attempt to access other students' accounts without their consent (95%).

Many valuable comments from students were obtained through this survey. (A complete list of respondents' comments can be found in Appendix B.) Overall, this survey gave us great confidence in making our recommendations from the viewpoint of students. We were tempted to conduct a second survey for more in-depth research, but we were not able to do so with our limited time and resources. It is our belief, and recommendation, that a more comprehensive and thorough (and perhaps more valid!) survey should be attempted by LBCC. Such a survey could result in findings that would be essential in developing a successful information technology policy for LBCC.

12

430 Appendix B: Reports

COMMENTS ON THE CONCLUSION

CONTENT: The main purpose of the conclusion is to prepare the reader for the recommendations that follow. To do this, the conclusion must restate the problem that the overall report is addressing, restate the report's purpose, and then highlight and repeat the key findings presented in the body of the report.

Restating the main ideas may seem repetitive, but this technique is used because many readers do not simply pick up the report and read it straight through. They may read and reread the report. They may be interrupted many times before they finish reading the report. The job of the conclusion is to support the reader with summaries that help reinforce the main findings of the report.

STYLE: Notice how carefully transitions (key words like *currently, further, in addition, an important next step*) are used to hold the conclusion together.

DOCUMENT DESIGN: Although there are no lists or descriptive headings to break up the text on this page, the writers have used short paragraphs to improve readability.

Appendix B: Reports **431**

CONCLUSION

In the case of Linn-Benton Community College, the Internet has arrived and, as is often the case, with new technology, come new concerns. This step forward has raised questions about the quality and quantity of instructional technology in the school's system, including the amount of student access to it, how to govern its use and how to curb its misuse.

Currently in its system, LBCC offers several locations where information technology serves the student body both on and off campus. To some, this is an adequate amount of hardware, while others feel that even though the number of computers may be adequate, their availability through classes and open lab hours does not necessarily conform to students' work and class schedules.

Further, there are concerns that the current technology may be underpowered in terms of storage and speed to meet the demands of effective access to the Internet, as well as new advances in software. Staff members are aware of this perceived shortcoming, yet school finances will most likely determine the school's ability to move forward technologically. In other words, the resolve to upgrade is there, but the money may not be.

In addition, once networked technology is in the students' hands, workable policy must be set on issues such as acceptable use, user privacy, intellectual freedom, and restriction and censorship. Compatibility with current LBCC policies is the first step toward development of any new policies. In the case of violations, for example, there is no need to establish a different system of due process.

An important next step is to research and adapt other institutions' policies on these issues. Using the experience and guidelines of other colleges as a foundation for one's own policies has proven successful for others. It would be logical to follow that same course. In the area of intellectual freedom, many universities have adopted, or at least adapted, the definition quoted earlier in this report used by the American Library Association.

Linn-Benton Community College is making a concerted effort to advance technologically and to establish workable computer use policies. This suggests that it is also an institution that will continue to pursue progress and set workable guidelines to match that progress. That effort strongly suggests that it is an institution that will meet the challenges of the future. The list of recommendations that follows should assist all who work toward that endeavor.

432 Appendix B: Reports

COMMENTS ON THE RECOMMENDATIONS SECTION

CONTENT: The most important sections in the report are the executive summary and the recommendations. They highlight solutions to the problem, and the recommendations section supports each group of recommendations. Note the following five questions to ask as you write your recommendations:

✓ *Can we solve the problem?* Restate the problem; give your reader a clear answer. You can say, "Yes, we should do this because. . ." or "No, we should not do this because . . ." Adding the "because" includes the key reason to act or not act. You can also say, "Yes, we should do this, but before we take action, we need to . . . ," or "No, we should not do this, but before we make a final decision, we should . . ."

Tim, Joel, Sang, and Tony say that yes, the college should have a computer use policy and that students should participate in setting that policy. This recommendation is placed first, and it answers the primary question that the whole report faces.

✓ *Can we implement the solution?* Tell the reader where, when, and how these changes will be made. Be specific about *how* the recommendations will be carried out, and who is responsible for carrying them out.

✓ *Can we evaluate the solution?* Set deadlines and performance criteria for carrying out the solution. Without setting these kinds of guidelines, how will we know the solution worked?

✓ *Can we plan for the future?* Anticipate changes; create a way that people can report critical changes in demand or performance.

✓ *Do we delegate leftover work?* If a major task was not completed before the report was finished, this is the place to suggest who should do what.

The recommendation is the most important part in your formal report. Sometimes your audiences want just the information, but most of the time they want your recommendations.

STYLE: The authors used parallel form and action verbs for their lists. This technique makes it easier to analyze and compare the recommendations being presented.

Appendix B: Reports

RECOMMENDATIONS

The Policy
The computer system policy development guidelines provided by the University of Illinois (see Figure F on the following page) should be considered during the early stages of LBCC's policy development. If a policy has already been drafted, or completed, then it should be checked against this list of guidelines in order to ensure that all the considerations have been addressed.

Solving Student Access Problems
The following recommendations will assist LBCC in providing its students the best opportunities to access LBCC information technology.

- Offer more computer training classes for students and *proactively* encourage computer literacy.
- Expand distance learning programs and offer modem access to the campus network for students who can't access the resources located on campus.
- Offer various means of informing students of college resources and related policies. An on-line college bulletin board is recommended.
- Offer a special computer lab that will provide physically disabled students proper access by using specially designed software.

Financial Constraints
LBCC should continually explore ways to reduce costs in providing information services to its students. Regardless of long-term budget battles, LBCC will be required to provide the most it can to students with whatever resources are available from whatever sources are available. The following are some recommended strategies for dealing with future financial constraints.

- Explore ways to circumvent requiring access fees to LBCC's Internet users.
- Search for and accommodate any discount programs offered to educational institutions by network/on-line providers.
- Expand current efforts toward forming business and community partnerships, for example, through the formation of financial support groups, such as among high-tech firms in the community, that could provide equipment, technical support, and funding. In exchange, LBCC will provide them with employee education and a higher quality workforce.

Student Participation
LBCC should actively and routinely encourage student participation in all aspects of development and maintenance of information technology resources. The following projects are examples of how students can provide assistance such as through college classes like writing, computer, and other science or communication related classes.

434   Appendix B: Reports

- Research on future usage of instructional technology including wireless access.
- Research on possible problems and solutions related to the instructional technology.
- Research on methods to increase the students' access.
- Research on increasing financial supports from businesses in the community.
- Research in an effort at identifying and adopting new technologies and related policies.

Additionally, LBCC should make every effort at improving articulation of its goals among departments and staff, including between LBCC and its satellite campuses. In doing this, and following the recommendations provided for above, LBCC will be moving forward confidently while offering its students the best information technology services available.

Figure F

**INSTRUCTIONAL TECHNOLOGY
POLICY DEVELOPMENT GUIDELINES**

The policy:

1. Should reflect the current U.S., Oregon, and LBCC campus policy regulations concerned with instructional technology.

2. Should be written for the best interest of students. Any controversial issues involved in usage have to be reasonably and sensitively defined to reduce any escalation of possible conflict.

3. Should provide extensive training to the staff who are involved in enforcing the policy so they will fully understand their actions.

4. Should be fully distributed to the system users in order to eliminate any possible misinterpretation.

5. Should be reviewed and updated on a regular basis in order to adapt to fast-changing technology.

6. Should be written in such a way that any novice user could get a reasonable understanding of the contents.

7. Should provide students the right to participate in making or adapting any new policies (University of Illinois).

436 Appendix B: Reports

COMMENT ON THE BIBLIOGRAPHY

FORMAT: The writing team has chosen to use an informal bibliography for its workplace audience. You will have to match the documentation format to your audience's expectations. An informal bibliography provides an alphabetically arranged list of complete information about the resources that were used in preparing the report. This can be challenging if you are using the Internet or CD-ROM resources, because these formats are just being developed and will continue to change over the next several years. The key concept to follow is to provide enough information to help the reader find the exact same source. For example,

> Crislin, Ronald. "Community Advocates for Health Increasing." *Journal of Mental Health,* Volume 17, No. 3 (June 1996), 42–47.

Academic audiences are usually much more exacting when it comes to the type of in-text citations and bibliography formats they expect. The section following this report, Footnotes and Bibliographies (see pages 442–449), discusses formal footnoting and preparing bibliographies more completely.

BIBLIOGRAPHY

Anderson, Rhonda. "Trends in Computer Use Policies." *The Chronicle of Higher Education* (February 15, 1994), A-25.

Betts, Mitch. "IS Policies Target E-mail Harassment." *Computerworld* (February 13, 1995), 74–77.

Camp, Beth, Writing Instructor, English Department, Linn-Benton Community College, Albany, Oregon. Personal interview via e-mail (November 17, 1995).

Computer Ethics Institute. *The Ten Commandments for Computer Ethics* (undated).

Department of Computer Science and Information Processing, Brock University at Ontario, Home Page. Downloaded from http://www.cosc.brocku.ca/policies/

Fay, Mike, Microcomputer Applications Instructor, Benton Center, Linn-Benton Community College, Albany, Oregon. Personal interview (November 15, 1995).

Kadie, Carl. Summary of the American Library Association's "Guidelines for the Development of Policies and Procedures regarding User Behavior and Library Use." Downloaded from: http://www.eff.org/CAF/faq/policy.best.html

Kadie, Carl. "The Application of Library Intellectual Freedom Policies to Public and Academic Computer Facilities (or "The Sermon in the Loop"). For Computers, Freedom, and Privacy '94 Version 1 (kadie@eff.org) (corrections requested).

Linn-Benton Community College. *Schedule of Classes, Winter Quarter 1996,* and *Draft Policy on Computer Lab/Internet Use.* Albany, Oregon (1995).

"Proposed LBCC Students' Rights, Responsibilities & Conduct." *The Commuter* (October 4, 1995).

Smith, Roberta, Microcomputer Applications Instructor, Benton Center, Linn-Benton Community College, Albany, Oregon. Personal interview (November 9, 1995).

University of Illinois. "Interim E-Mail and Computer File Privacy Policy." (From *Edu— University of Illinois at U-C—Privacy, undated.)

University of Wisconsin-Milwaukee. "Network Policy Resolutions of the Computer Policy Committee." Adopted as campus administrative policy (February 23, 1993).

438　Appendix B: Reports

COMMENT ON THE SURVEY

CONTENT: The goal of a survey is to collect useful information that is directly related to solving your problem. Survey data needs to be valid (connected to your topic) and reliable (someone else should be able to run the same survey and obtain roughly the same results). These two principles guide academic research.

In this section, the writing team has placed the raw data for their survey results organized into a tally sheet that shows the reader exactly how many responses fall into the different categories for each question.

Notice that in the design of this informal survey, these elements are at work:

- An introduction that describes the purpose and gives directions on how to complete the survey.

- A definition of the key term "instructional technology" to focus the responses to questions

- Several different formats of questions that allow the report writers to translate the data from raw numbers to percentages:

  A *Likert scale* that collects more precise responses than a "yes" and "no" question.

  A *checklist* with an open line to gather ranked importance of key issues.

  A final *open-ended question* to collect student opinions. These opinions are highlighted on an additional page in the appendix.

A survey can be a quite sophisticated tool for gathering information. The survey that follows illustrates how you can experiment with different ways of collecting information, analyzing it, and then selecting highlights to present in the main sections of the report.

Appendix A   Survey Results—Raw Data

We are LBCC students completing a project for our technical writing class.

This survey is a major component of our research project, entitled, "Instructional technology on the LBCC Campus: Research and Recommendations."

There are two purposes to this survey:
- To help determine the importance of instructional technology to LBCC students' views on instructional technology on campus.
- To enable LBCC students to express their views on issues surrounding access to the Internet.

If you have any questions, please ask the student distributing the survey.

For questions 1–5 and 7–9, circle the word that best indicates your feeling on the given statements.

Example: It will rain in Oregon sometime in January.

Strongly agree   Agree   No opinion   Disagree   Strongly disagree

For the sake of this survey, the term "instructional technology" refers to computer lab resources, tentatively planned Internet access, and the campus e-mail system.

1. LBCC should better inform students of the availability of instructional technology resources on campus.

Strongly agree	Agree	No opinion	Disagree	Strongly disagree
17	20	3	1	0

2. LBCC should offer more classes that train students how to utilize computer technology.

Strongly agree	Agree	No opinion	Disagree	Strongly disagree
15	19	4	3	0

3. If it were available, I would utilize an Internet account on LBCC's campus.

Strongly agree	Agree	No opinion	Disagree	Strongly disagree
11	12	12	5	1

4. All LBCC students, no matter what their major, should have equal access to campus Internet resources.

Strongly agree	Agree	No opinion	Disagree	Strongly disagree
25	12	2	1	0

5. I would be willing to pay the following amount (**per quarter**) to use a personal Internet account on LBCC's campus.

$0	$10	$20	$30	$40	$50	Other
13	10	9	3	1	1	4

440   Appendix B: Reports

6. I consider the following material (available on the Internet) 'unacceptable' for access in publicly-funded institutions. (Check all that apply.)

<u>13</u>   Profane language

<u>17</u>   Sexually explicit discussion

<u>29</u>   Adult pornography

<u>38</u>   Child pornography

<u>25</u>   Material that provides information that could lead to injuries to humans or animals (i.e., instructions on producing a bomb)

<u>1</u>   Other (Explain)

7. LBCC should restrict access to certain material on the Internet.

Strongly agree	Agree	No opinion	Disagree	Strongly disagree
10	7	5	5	5

8. In cases of emergency or by order of law enforcement officials, LBCC should be allowed access to students' electronic mail.

Strongly agree	Agree	No opinion	Disagree	Strongly disagree
2	8	7	8	15

9. LBCC should impose penalties on students who violate campus policies on acceptable use of instructional technology.

Strongly agree	Agree	No opinion	Disagree	Strongly disagree
10	24	5	2	0

10. Put a check next to any actions below that you feel should be considered violations of a campus policy on acceptable use of instructional technology.

<u>25</u>   Overloading system resources with your own data (i.e., saving excessive data on campus disk drives)

<u>26</u>   Downloading pornographic material

<u>40</u>   Attempting to introduce a computer virus into the system

<u>39</u>   Attempting to access other students' accounts without their consent

<u>3</u>   Use of profanity in any written communication

<u>1</u>   No actions should be considered violations

___   Other _____

11. Please add any opinions on the issues discussed in the survey. _____

_____

_____

18

Appendix B: Reports

441

Some Students' Comments Written on the Survey Form

Q5: "What is the cost to run it?" "What is average use?"

"Willing to pay $5."

"Include in tuition."

"OSU is free to students."

Q6: "Except in personal E-mail."

"I probably would not use it."

(Re: "Adult pornography") "Definitely limit downloading (except perhaps for Human Sexual Behavior, etc., classes by special permission)."

Profane language—"can't monitor efficiently, is too wide opened."

Q7: "Vehemently disagree."

Q9: (Answered both Agree and Disagree) "Yes, but not $, limit access."

Q10: ("use of profanity...") "Personal communication should have no regulation."

("other") "Accessing unauthorized areas."

Q11: "If an 'open' talk should be regulated, E-mail is like personal mail not to be regulated."

"I feel that violations, whatever they may be, made on the instructional technology would be to just ban that student's access."

"Restricting access is preposterous." Re: Question #8: "Don't users deserve their privacy?"

"Overloading system is hard to regulate. How much is too much and who decides."

"While educational financial resources are being threatened, the costs to students and programs are rising; however, computers will continually take academic advances in the workplace. Computer illiteracy should be the main focus of modern educators."

"Sounds like you are looking for some negative feed back on the usage of the computer systems here at LBCC. Are you for or against this program? Are you working for the government? Are you following me?"

"I think that gaining access to the Internet would greatly increase LBCC's students interest in computers."

(Regarding question #10) "On #10, overloading system resources with your own data, who decides what is overloading the system?"

"I think you have something good going."

"We need all of the computers to be connected to the internet. Maybe the same one as OSU!"

"I don't care how people use the Internet as long as there is equal opportunity, they don't break the law, or violate my rights of privacy (i.e., personal info., or private E-mail)."

"The reasoning for my opinions is that I do not use the Internet. I based my answers on what I know of it."

"Use the library."

"1. Instructional technology should not be used as a toy. If work is LBCC related, fine. If not, buy your own computer. No public subsidized toys. 2. Censoring Internet is repulsive just like book burning or removing freedom of speech."

"The internet and access to it must be kept free from regulation lest personal rights be sacrificed. It is not so important that students may browse pornographic material. What *is* important is that they have the *RIGHT* to. Censorship is by far a greater evil than any we as citizens can imagine."

19

442   Appendix B: Reports

FOOTNOTES AND BIBLIOGRAPHIES

Workplace readers are sometimes informal in their use of proper citations and bibliographies; however, ethical use of sources can be a major issue. You can decide whether or not to use a footnote based on the answers to these two questions:

- Did you use someone else's ideas, information, or materials?

- Does your audience expect footnotes and a bibliography for this particular paper or report?

If your answer to either of these questions is yes, plan to select and use one of several widely used footnote and bibliography systems. You will need to decide which system to use. MLA is commonly used by people from the humanities, the APA system is used by scientists and psychologists, and the numbered system is used most often by technical people (e.g., engineers). To decide which system to use, consider what field you are majoring in and what your audience expects. A discussion of each—MLA, APA, and numbered—is included here.

Because documentation systems continually change, if you memorize one of the three systems discussed in this appendix, you can expect the format to be different in a few years. For example, formats for citing materials from Internet or CD-ROM resources continually change. The key concepts are knowing how to:

- Create explanatory footnotes when needed.

- Use in-text citations in the body of your paper or report to show the page and author where specific information was found.

- Use a bibliography to list all the resources consulted for your paper or report.

Each of these concepts is explored in this section.

Using MLA Documentation

Most academic writers in the humanities (social sciences, arts, literature) use the MLA documentation system. MLA stands for the Modern Language Association, which publishes guidelines for preparing in-text citations, footnotes and bibliographies.

Using MLA to Prepare In-Text Citations.   The real purpose of providing in-text citations is to help the reader know when you, as the writer, are speaking, and when you are summarizing, paraphrasing, or quoting from research. Clear *in-text citations* help the reader know exactly where to find the information

being quoted, paraphrased, or summarized. With the MLA system, the writer shows the author's last name or key words and the page number in the in-text citation. For example:

> Fortunately, 95% of community-based health clinics have community advisory boards to help set policies. (Crislin 47)

This in-text citation shows the exact page this information may be found.

Writing consistent *lead-ins* is very important so the reader can easily distinguish between your original writing and writing that is based on the work of others (by using paraphrasing, summarizing, or quoting). Lead-ins add credibility to the discussion if they include information about the expert, for example:

> According to Ronald Crislin, Director of the Institute of Mental Health, community-based health clinics must follow the following four key guidelines:

Using MLA to Prepare Explanatory Footnotes.

Occasionally, you will want to provide the reader with additional explanations or definitions that might seem to interrupt the reader if they are included right in the text.

Simply create an explanatory footnote, which is placed at the bottom of the page. Some writers use asterisks, but most prefer numbers so that the reader can match a numbered footnote in the text to the footnoted material. A few writers put all footnotes in an end notes section at the back of the paper, but most authors prefer to add the footnote at the bottom of the page where it is easiest for the reader to find.

In the following example, the body of the paper gave a definition of schizophrenia, and at the bottom of the page, this explanatory footnote appeared:

---

[1]Researchers disagree strongly about the definition of schizophrenia, but Wallace's definition is the most widely accepted.

Using MLA to Prepare a Works Cited Page.

With the MLA system, the bibliography is called Works Cited. This means that every in-text citation in the body of your paper must be linked to each resource listed in your bibliography. If you read an article, for example, but did not quote, paraphrase, or summarize it, you should not include it in your bibliography.

Notice how the Works Cited page is organized alphabetically by the last name of the author. If no author is given for a particular source, list it according to the title of the article or book. Be exacting in proofreading for correct punctuation and use of capital letters, as shown in this sample MLA Works Cited page. Notice that the first line of each entry is flush with the left margin, and all other lines are indented five spaces. Notice also that the entire bibliography is double-spaced.

444 Appendix B: Reports

To cite materials downloaded from Internet resources on your works cited page, include the http address and use the date the document was last updated as well as the total number of pages, usually shown as 1 of 3 pages or 1 of 7 pages.

Department of Computer Science and Information Processing, Brock University of Ontario, Home Page. Downloaded from http://www.cosc.brocku.ca/policies. Last updated June 1996, 1–5.

To cite materials downloaded from CD-ROM resources on your works cited page, include the item number, if available.

Henderson, Paul. "European Recovery Aids U.S. Oil Firms' Revenues." *European Chemical News* 30 Jan. 1995: 14. *F & S Index Plus Text*. CD-ROM. SilverPlatter. 1995, 1–6.

Using the correct format can be challenging if you are using Internet or CD-ROM resources, because these formats are just being developed and will continue to change over the next several years. The key concept to follow is to provide the information that will help the reader find the exact same source.

Exhibit B.1 shows the sample report's bibliography in MLA format.

Using APA Documentation

Academic writers in psychology and the sciences generally use the APA documentation system because it provides the date the research was published right in the body of the text. APA stands for the American Psychological Association, which publishes guidelines for preparing in-text citations, footnotes, and bibliographies.

Using APA to Prepare In-Text Citations.

The real purpose of providing in-text citations is to help the reader know when you, as the writer, are speaking, and when you are summarizing, paraphrasing, or quoting from research. Clear *in-text citations* help the reader know exactly where to find the information being quoted, paraphrased, or summarized. With the APA system, the writer includes the author's last name or key words, the date, and the page number as part of the in-text citation. For example:

Fortunately, 95% of community-based health clinics have community advisory boards to help set policies (Crislin, 1996, page 47).

This in-text citation shows the exact page where this information may be found as well as the date the research was published.

Writing consistent *lead-ins* is very important so the reader can distinguish easily between your original writing and writing that is based on the work of

Appendix B: Reports   **445**

Exhibit B.1   **Bibliography in MLA Format** ■

Works Cited

Anderson, Rhonda. "Trends in Computer Use Policies." *The Chronicle of Higher Education* (Feb. 15, 1994), A-25.

Betts, Mitch. "IS Policies Target E-mail Harassment." *Computerworld* (Feb. 13, 1995), 74–77.

Camp, Beth, Writing Instructor, English Department, Linn-Benton Community College, Albany, Oregon. Personal interview via e-mail (Nov. 17, 1995).

Computer Ethics Institute. *The Ten Commandments for Computer Ethics* (undated).

Department of Computer Science and Information Processing, Brock University at Ontario, Home Page. Downloaded from http://www.cosc.brocku.ca/policies/, 1–4.

Fay, Mike, Microcomputer Applications Instructor, Benton Center, Linn-Benton Community College, Albany, Oregon. Personal interview (Nov. 15, 1995).

Kadie, Carl. Summary of the American Library Association's "Guidelines for the Development of Policies and Procedures Regarding User Behavior and Library Use." Downloaded from: http://www.eff.org/CAF/faq/policy.best.html, 1–7.

Kadie, Carl. "The Application of Library Intellectual Freedom Policies to Public and Academic Computer Facilities (or "The Sermon in the Loop"). For Computers, Freedom, and Privacy '94 Version 1 (kadie@eff.org) (corrections requested), 1–9.

Linn-Benton Community College. *Schedule of Classes, Winter Quarter 1996,* and *Draft Policy on Computer Lab/Internet Use* Albany, Oregon (1995).

"Proposed LBCC Students' Rights, Responsibilities & Conduct." *The Commuter* (Oct. 4, 1995), 1–4.

Smith, Roberta, Microcomputer Applications Instructor, Benton Center, Linn-Benton Community College, Albany, Oregon. Personal interview (Nov. 9, 1995).

University of Illinois. "Interim E-Mail and Computer File Privacy Policy" From *Edu—University of Illinois at U-C—Privacy (undated), 1–3.

University of Wisconsin-Milwaukee. "Network Policy Resolutions of the Computer Policy Committee." Adopted as campus administrative policy (Feb. 23, 1993), 1–4.

446 Appendix B: Reports

others (by using paraphrasing, summarizing, or quoting). Lead-ins add credibility to the discussion if they include information about the expert, for example:

> According to Ronald Crislin, Director of the Institute of Mental Health, community-based health clinics must follow the following four key guidelines.

Using APA to Prepare Explanatory Footnotes. Occasionally, you will want to provide the reader with additional explanations or definitions that might seem to interrupt the reader if they are included right in the text.

Simply create an explanatory footnote, which is placed at the bottom of the page. Some writers use asterisks, but most prefer numbers so that the reader can match a numbered footnote in the text to the footnoted material. A few writers put all footnotes in an end notes section at the back of the paper, but most prefer to add the footnote at the bottom of the page where it is easiest for the reader to find, and to use in-text citations rather than footnotes at the back of the paper or report.

In the following example, the body of the paper gave a definition of schizophrenia, and at the bottom of the page, this explanatory footnote appeared:

---

[1]Researchers disagree strongly about the definition of schizophrenia, but Wallace's definition is the most widely accepted.

Using APA to Prepare a References Page. With the APA system, the bibliography is called References. Just as with the MLA system, you will list all resources cited in the body of the paper.

The APA bibliography is also organized alphabetically by the last name of the author. Notice that you use only the first initial of the first name. If no author is given for a particular source, list it according to key words from the title of the work. If more than one source is listed for a particular writer, list the most recent first.

Be exacting in proofreading for correct punctuation and use of capital letters, as shown in this sample APA References page. One major difference is that only the first word in the title of the source is capitalized; all other words are shown in lowercase letters. Capitalize the first word appearing after a colon.

Notice that the first line of each entry is flush with the left margin, and all other lines are indented five spaces. The entire bibliography can be double-spaced as shown in the MLA works cited page, or it can be single-spaced with one blank line between each entry as shown here.

To cite materials downloaded from Internet resources on your References page, include the http address and use the date the document was last updated as well as the total number of pages, usually shown as, for example, 1 of 5 pages.

Department of Computer Science and Information Processing, Brock University of Ontario. (June, 1996). Home page. Available from: WWW: http://www.cosc.brocku.ca/policies

To cite materials downloaded from CD-ROM resources on your references page, include the item number, if available.

Henderson, P. (1995, January 30). European recovery aids U. S. oil firms' revenues. *European Chemical News*, 14. *F & S Index Plus Text*. CD-ROM. SilverPlatter, pp. 1–6.

Using the correct format can be challenging if you are using Internet or CD-ROM resources, because these formats are just being developed and will continue to change over the next several years. The key concept to follow is to provide the information that will help the reader find the exact same source. Exhibit B.2 shows the sample report's bibliography in the APA format.

Bibliography in APA Format   Exhibit B.2 ∎

References

Anderson, R. (1994, February 15). Trends in computer use policies. *The Chronicle of Higher Education,* A-25.

Betts, M. (1995, February 13). IS policies target e-mail harassment. *Computerworld,* 74–77.

Camp, B. (1995, November 17). Writing instructor, English Department, Linn-Benton Community College, Albany, Oregon. Personal interview via e-mail.

Computer Ethics Institute. (Undated). *The Ten Commandments for Computer Ethics.*

Department of Computer Science and Information Processing, Brock University at Ontario. (Undated) Home page. 1–4. Available WWW: http://www.cosc.brocku.ca/policies/

Fay, M. (1995, November 15). Microcomputer Applications Instructor, Benton Center, Linn-Benton Community College, Albany, Oregon. Personal interview.

Kadie, C. (Undated). Summary of the American Library Association's guidelines for the development of policies and procedures regarding user behavior and library use, 1–7. Available WWW: http://www.eff.org/CAF/faq/policy.best.html

Kadie, C. (Undated). The application of library intellectual freedom policies to public and academic computer facilities (or "the sermon in the loop"). For *Computers, Freedom, and Privacy '94,* Version 1, 1–9. [kadie@eff.org]

Linn-Benton Community College (1995). *Schedule of Classes, Winter Quarter 1996,* and *Draft Policy on Computer Lab/Internet Use.* Albany, Oregon.

Proposed LBCC students' rights, responsibilities & conduct. (1995, October 4). *The Commuter,* 1–4.

Smith, R. (1995, November 9). Microcomputer Applications Instructor, Benton Center, Linn-Benton Community College, Albany, Oregon. Personal interview.

University of Illinois (undated). *Interim E-Mail and Computer File Privacy Policy.* Available WWW: *Edu—University of Illinois at U-C—Privacy.

University of Wisconsin-Milwaukee. (1993, February 23). Network policy resolutions of the computer policy committee, 1–4.

448   Appendix B: Reports

Using a Numbered System

Before word-processing programs made the preparing of footnotes and bibliographies so much easier, scientists and technical professionals wanted a way to document in-text citations that did not interrupt the reader with so much text. They also wanted a way to add new citations to the bibliography without having to retype the entire bibliography.

The result was the numbered system, which numbers the items on the bibliography and which uses these numbers and the page numbers for in-text citations.

Several different systems are used in such applied sciences as chemistry, geology, mathematics, and medicine. One widely used guide is *The CBE Manual*, published by the Council of Biology Editors, which follows many of the conventions of APA. Check guidelines used in your field. The sample shown here is a less formal format of the numbered system.

Using the Numbered System to Prepare In-Text Citations.   Use an in-text citation when you are summarizing, paraphrasing, or quoting from research. Clear *in-text citations* help the reader know exactly where to find the information being quoted, paraphrased, or summarized. With the numbered system, the writer shows the number of the citation from the bibliography and the page number in the in-text citation. For example:

> Fortunately, 95% of community-based health clinics have community advisory boards to help set policies (3: 45).

This in-text citation shows which item in the bibliography this came from (3) and the exact page on which this information may be found (45).

Writing consistent *lead-ins* is very important so the reader can easily distinguish between your original writing and writing that is based on the work of others (by using paraphrasing, summarizing, or quoting). Lead-ins add credibility to the discussion if they include information about the expert, for example:

> According to Ronald Crislin, Director of the Institute of Mental Health, community-based health clinics must follow the following four key guidelines.

Using the Numbered System to Prepare Explanatory Footnotes.
Occasionally, you will want to provide the reader with additional explanations or definitions that might seem to interrupt the reader if they are included right in the text.

Simply create an explanatory footnote, which is placed at the bottom of the page. This will not be confused with the in-text citation, because the numbered footnote appears on the same page and is not placed inside parentheses.

In the following example, the body of the paper gave a definition of schizo-phrenia, and at the bottom of the page, this explanatory footnote appeared:

[1]Researchers disagree strongly about the definition of schizophrenia, but Wallace's definition is the most widely accepted.

Using the Numbered System to Prepare a Bibliography.

With the numbered system, the bibliography may be called Bibliography, References, Literature Cited, or References Cited.

You will need to number each entry on the bibliography. Some writers insist that the bibliography must show the citations exactly in the same order as they appeared in the paper. Other writers say that the bibliography can be alphabetized and numbered, and that it really doesn't matter in what order the numbers appear in the body of the paper. Check with your instructor to see which is preferred.

In our sample bibliography, the writer has alphabetized each entry. Copy this format exactly as it appears. Be exacting in proofreading for correct punctuation and use of capital letters.

To cite materials downloaded from Internet resources on your bibliography page, include the http address and use the date the document was last updated as well as the date the document was accessed.

Department of Computer Science and Information Processing, Brock University of Ontario, Home Page. [on-line]. Last updated June 1996. Available from: http://www.cosc.brocku.ca/policies via the Internet. Accessed November 2, 1996.

To cite materials downloaded from CD-ROM resources on your bibliography page, include the item number, if available.

Henderson, Paul. "European Recovery Aids U.S. Oil Firms' Revenues." *European Chemical News* 30 Jan. 1995: 14. *F&S Index Plus Text*. CD-ROM. SilverPlatter. 1995, pages 1–6.

Note: the formats for Internet or CD-ROM resources will continue to change over the next several years. The key concept to follow is to provide the informa-tion that will help the reader find the exact same source. Exhibit B.3 shows the sample report's bibliography using the numbered system.

450 Appendix B: Reports

■ Exhibit B.3 **Numbered Bibliography**

Bibliography

1. Anderson, Rhonda. "Trends in Computer Use Policies." *The Chronicle of Higher Education* (Feb. 15, 1994), A-25.

2. Betts, Mitch. "IS Policies Target E-mail Harassment." *Computerworld* (Feb. 13, 1995), 74–77.

3. Camp, Beth, Writing Instructor, English Department, Linn-Benton Community College, Albany, Oregon. Personal interview via e-mail (Nov. 17, 1995).

4. Computer Ethics Institute. *The Ten Commandments for Computer Ethics* (undated).

5. Department of Computer Science and Information Processing, Brock University at Ontario, Home Page. Downloaded from http://www.cosc.brocku.ca/policies/, 1–4.

6. Fay, Mike, Microcomputer Applications Instructor, Benton Center, Linn-Benton Community College, Albany, Oregon. Personal interview (Nov. 15, 1995).

7. Kadie, Carl. Summary of the American Library Association's "Guidelines for the Development of Policies and Procedures Regarding User Behavior and Library Use." Downloaded from: http://www.eff.org/CAF/faq/policy.best.html, 1–7.

8. Kadie, Carl. "The Application of Library Intellectual Freedom Policies to Public and Academic Computer Facilities (or "The Sermon in the Loop"). For Computers, Freedom, and Privacy '94 Version 1 (kadie@eff.org) (corrections requested), 1–9.

9. Linn-Benton Community College. *Schedule of Classes, Winter Quarter 1996,* and *Draft Policy on Computer Lab/Internet Use.* Albany, Oregon (1995).

10. "Proposed LBCC Students' Rights, Responsibilities & Conduct." *The Commuter* (Oct. 4, 1995), 1–4.

11. Smith, Roberta, Microcomputer Applications Instructor, Benton Center, Linn-Benton Community College, Albany, Oregon. Personal interview (Nov. 9, 1995).

12. University of Illinois. "Interim E-Mail and Computer File Privacy Policy" From *Edu—University of Illinois at U-C—Privacy (undated), 1–3.

13. University of Wisconsin-Milwaukee. "Network Policy Resolutions of the Computer Policy Committee." Adopted as campus administrative policy (Feb. 23, 1993), 1–4.

Appendix C
Writing Skills Review

In all kinds of jobs, the ability to use basic English grammar correctly is a fundamental requirement. Some companies include a writing test as part of their screening process for entry-level jobs. Mastering the foundation skills that are reviewed in this section will help your reader understand what you are writing without being distracted by errors in punctuation, grammar, or spelling.

Good editing and proofreading skills can help you move out of the back office or shop, assume more responsibility with customers, or be promoted. This appendix reviews the most commonly used areas of grammar, punctuation, and spelling that you will need to use for editing and proofreading all of your workplace writing.

HOW TO USE APPENDIX C

Use Appendix C to review punctuation, grammar, and spelling skills. Each section begins with a short review, with exercises for you to complete. As you complete each exercise, check your answers at the end of this appendix. If you need additional review, ask your instructor about other resources.

If you completed the exercises in each chapter as you worked through this book, you have improved the clarity and correctness of your writing.

You can use this appendix for review and for reference. The following list will help you locate the various review sections within this Appendix.

Understanding Sentence Structures　　　　　　　　　　Go to page 452

- Reviewing parts of speech
- Combining sentences

Building Punctuation Skills　　　　　　　　　　Go to page 465

- Using periods, semicolons, and question marks
- Using commas

451

452   Appendix C: Writing Skills Review

- Changing sentence patterns
- Using apostrophes

Revising for Correct Pronouns   Go to page 473

- Using subject and object pronouns
- Proofreading for pronoun agreement
- Clarifying pronoun reference

Revising for Effective Verb Use   Go to page 480

- Understanding verbs
- Reviewing verb tenses
- Looking at irregular verb forms
- Checking for subject/verb agreement
- Using active and passive verbs

Proofreading   Go to page 488

- Building spelling skills
- Using commonly confused words

UNDERSTANDING SENTENCE STRUCTURE

Your goal in workplace writing is to write concisely, using as few words as possible, and, at the same time, to write clearly and correctly so that your reader will not be distracted by grammar, punctuation, or spelling problems.

Having a formula to answer all your questions about grammar, punctuation, and spelling could be helpful. However, although we can memorize many rules to help guide how we use grammar, nearly every rule has an exception.

To avoid having you memorize many different rules and then being uncertain when to use them, this section will help you understand how different words are used in sentences (**parts of speech**), and how they are put together into sentences (**phrases and clauses**). This section will help you review:

Parts of Speech   Go to page 453

- Using content words (nouns, adjectives,
  verbs, and adverbs)   Go to page 453

- Using connector words (pronouns, articles,
  prepositions, and conjunctions)   Go to page 456

Sentence Patterns   Go to page 461

- Finding subjects and predicates

Combining Sentences   Go to page 461

- Adding words
- Adding phrases and clauses

Summary Exercise   Go to page 465

Reviewing Parts of Speech

You will be able to recognize the basic categories of words (called *parts of speech*) when you are finished with this section. These are:

Nouns   **Adjectives**   **Verbs**   **Adverbs**

These parts of speech make up nearly all of the words you use. They are called "vocabulary" or "content" words. Your reader understands your meaning because of the vocabulary you use.

Pronouns   **Articles**   **Prepositions**   **Conjunctions**

These parts of speech are used to connect "content" words together; these "connector" words show relationships. We need both kinds of words to be perfectly clear to our readers.

Do either of these "sentences" make sense?

Come market. (Uses "content" words only)

To the with me. (Uses "connector" words only)

Combined: Come to the market with me.

This next section will help you understand what function different words have in a sentence. *Knowing what words do* in a sentence can help you decide *how to punctuate* the sentence.

Using Content Words (Nouns, Adjectives, Verbs, and Adverbs)

A **noun** is any word that describes a person, place, or thing. For example, the *cook* works in the *kitchen* and prepares *dinner*. A proper noun refers to a particular person, place, or thing. For example, *Chef Mark* works for the *Cafe Paris* and prepares *chicken Kiev*.

Writing with Nouns   **Exercise C.1**

Write one paragraph about a recent visit to a restaurant that includes at least 10 nouns and 5 proper nouns. Underline the nouns in your paragraph. Review the sample response given in the back of this appendix. You may want to work with a classmate to check your work.

Adjectives are descriptive words that add specific detail to your writing. You could write that you work in a building, or you could write, I work in an *old office* building *downtown* that has *new* computers.

An adjective describes a noun by adding detail (*red, large, happy*) and is usually placed as close as possible to the word it modifies. For example:

She is *sleepy* in the *early* morning, but once she has eaten a *very large* breakfast, Rachel is ready for *hard* work.

Exercise C.2   Using Adjectives

Add as many adjectives as possible to the following sentences.

1. _____ gas was leaking from the hot water heater.
2. The _____ odor was _____ strong.
3. I called my _____ landlord.
4. She said this was not a _____ problem.
5. The _____ smell was seeping into my _____ bathroom.
6. Every time I took even a _____ shower, the water smelled like _____ broccoli.

Verbs are very important because they describe action—past, present, or future.

Rhonda *runs* quickly, she *hits* the ball nearly every time, and she *can catch* equally well. She *has been* a valuable baseball player, and she *will be* a good addition to our team.

Each highlighted verb describes an *action* or a *state* or a *condition of being*.

Exercise C.3   Finding Verbs

Underline the VERB or the VERB PHRASE in the following sentences:

1. Open the window, please.
2. In Mexico, windows are always opened early in the morning.
3. Every window has an iron grill in front of it.
4. Nearly every house is built around a patio.
5. The patios are filled with many different plants and flowers, each one in its own earthen pot.

Verbs work either as the entire **predicate** of a sentence or as an essential part of the predicate:

Subject	+	Predicate	=	Complete sentence
Rhonda	+	*runs*.		(Complete sentence)
Rhonda	+	*can run* faster than any other player.		(Complete sentence)
Rhonda	+	*is* not *running* this morning.		(Complete sentence)

Verbs can also be used in different parts of the sentence, as the next examples will show. But, when this happens, the sentence is not complete without a **main verb.** The following sentences show verbs used in other places in the sentence and highlight the main verb.

Using an infinitive to modify the main verb:
Rhonda WANTS *to be* on our team. (main verb = wants)

Using a verb phrase (called a participle) to modify the subject:
The person *approving all new players* IS our coach. (main verb = is)

Using a verb phrase (called a participle) as a subject:
Having no time for homework IS a problem. (main verb = is)

More Practice Finding Verbs Exercise C.4

Underline the *main verb* or *main verb phrase* in each of the following sentences.

1. "To be or not to be," cried Hamlet.
2. Trying out for a play can be stressful.
3. After working hard to memorize his lines, Fernando was ready for the audition.
4. Fernando wanted to be selected for a part.
5. The director selecting the main characters chose Fernando.

An **adverb** is a word, phrase, or clause that can modify a verb, an adjective, or another adverb and is usually placed after the verb.

Janet *drank* her orange juice *slowly.*
   (*slowly* modifies the verb *drank*)

The orange juice was *very fresh.*
   (*very* modifies the adjective *fresh*)

He *is always ready* to travel.
   (*always* modifies the verb phrase *is ready*)

Most of the time, adverbs show *how*. For example, in the sentence He ran *quickly,* the word quickly shows how he ran. Many times, adverbs will end with *-ly: anxiously, happily, thoroughly, perfectly, correctly, poorly, or completely,* for example. Adverbs can also show when, how, where, and to what extent.

Using Adverbs Exercise C.5

To describe the action more precisely, write in an adverb for each blank line.

1. Jorge said that he would be _____ happy with a small breakfast.
2. He _____ enjoyed orange juice and a roll.

456   Appendix C: Writing Skills Review

3. Then, he _____ ate scrambled eggs, bacon, and another roll.
4. He couldn't be _____ through until he had drunk three cups of coffee with milk and _____ sugar.
5. After eating all this food, he was _____ full.

Using Connector Words (Pronouns, Articles, Prepositions, and Conjunctions)

Pronouns are words like *I, you, my,* and *her* that can take the place of a noun. For example: *He* is cooking *it* for *them.*

This sentence sounds a little mysterious because we may not know *who* he is, *what* he is cooking, or *for whom* he is cooking. Adding more specific description will solve the mystery for your reader. For example: Raoul is cooking paella for his two great aunts. Watch out for unclear pronouns—and try to replace "them" with more descriptive nouns.

Subject pronouns can take the place of the subject.

Paul went to the movies.	He went to the movies.
Roger and Roberta are alone.	They are alone.
Homework can be time consuming.	It can be time consuming.

Is *it* clear enough in that last sentence? Sometimes when we write a rough draft, we use *it* as a "short cut" to expressing our ideas, but our readers may not easily understand us. Most readers would probably prefer *homework* rather than *it* in the sample sentence.

Object pronouns are used when the person is receiving something. For example, give the book to *him;* don't give the book to *me;* take *us* to the movies; or let's take *them* with *us.*

Possessive pronouns show ownership. For example, what have you done with *my* car keys? Can I use *your* car? If I can't use *your* car, could I use *his* car, or do I have to take *their* van? Use Exhibit C.1 to study the three cases of pronouns: subject, object, and possessive.

Decide whether to use an *object* pronoun or a *subject* pronoun by asking these questions:

- Who or what is the sentence about? Is the subject doing something, as in *he* is reading? Use a *subject* pronoun!

- Is the person receiving something? Is there a preposition right before the pronoun, as in give the book *to* him? Use an *object* pronoun!

- Does the person, place, or thing belong to someone? Is the pronoun right before the noun, as in *my* book? Use a *possessive* pronoun!

Understanding Pronouns   Exhibit C.1

	Subject	Object	Possessive	
	Singular			
First person	I	(to) me	my	mine
Second person	you	(to) you	your	yours
Third person	he, she, it	(to) her, him, it	his, her	his, hers, its
	Plural			
First person	we	(to) us	our	ours
Second person	you	(to) you	your	yours
Third person	they	(to) them	their	theirs

Using Pronouns   Exercise C.6

Fill in the missing pronoun(s) in the following sentences. Circle whether the answer is (SP) a subject pronoun, (OP) an object pronoun, or (PP) a possessive pronoun.

1. Give _____ the customer order. (SP), (OP), or (PP)
2. George and _____ have worked here over seven years. (SP), (OP), or (PP)
3. Paula and _____ went to San Diego (SP), (OP), or (PP) with _____ (SP), (OP), or (PP) to work with _____ new customers. (SP), (OP), or (PP)
4. Just between _____ and _____, (SP), (OP), or (PP) the trip was very expensive.
5. Where is _____ expense report? (SP), (OP), or (PP)
6. Get the monthly report from _____. (SP), (OP), or (PP)
7. _____ would like (SP), (OP), or (PP) to hear more about _____ promotion. (SP), (OP), or (PP)

More Practice with Pronouns   Exercise C.7

The following paragraph has 11 pronouns. Circle each one, noticing how they are used. To improve clarity, should any pronouns be changed to nouns? List the number of the sentence and revise any sentence that needs clearer nouns instead of pronouns.

(1) My favorite movies are old romantic comedies. (2) Most of the best of these black and white films were made in the 1930s, but since I don't subscribe to cable TV, I don't often get to see them. (3) I like them for how they show life in the "old days," when everybody dressed in fancy clothes for dinner, when detectives solved all the crimes, and when the movie ends, the girl gets the guy. (4) Of course, they were popular so that people could stop worrying about the Great Depression for a little while. (5) People who were struggling with unemployment liked to fantasize about living a rich life. (6) They didn't really solve any problems, but for an hour or so, they were happy, rich, and loved. (7) Maybe that's why for many of us, such romantic adventure movies as *The Princess Bride* or *Jewel of the Nile* are still popular today.

458 Appendix C: Writing Skills Review

An **article** is a word like *a, an,* or *the* that shows whether the word that follows is a general word (like *a* park, meaning any park) or a very specific word (like *the* park, meaning one special park).

If you were to say, "I want to go to *a* park," you are saying you don't really care which park, you just want to go to any park. If you say, "I want to go to *the* park, you are saying you want to go to one particular park, and you assume the reader knows which park.

Do you need to say, "I want to go to the Yellowstone National Park?" Most times, using a proper name means you omit the article. For example, "I am going to New York City," or "I want to go to Yellowstone Park."

Plural nouns can be a little troublesome. If you say, "I have *a* very good supervisor," the article "a" stands for "one" as in "I have one very good supervisor." If you are referring to all your supervisors, omit the article! You would say, "I have very good supervisors." Is *the* used correctly in this next sentence?

I want to go to all *the* national parks in the United States.

Using "the" in this sentence points to all the national parks, so it is correct to use *the* in front of some plural nouns.

Sometimes we have difficulty knowing when to use "a" or when to use "an." Use "an" in front of a noun that starts with any of the letters *a, e, i, o,* or *u* (called vowels). For example:

Give me *an* apple, *an* ice cream bar, and *a* bag to put them in.

Using articles correctly can be very difficult for people who do not use these "pointer" words in their own language. One way to practice using articles is to notice how they are being used by circling them in a newspaper article.

Exercise C.8 Using Articles

Add articles to the following sentences. You will need to add more than one article!

1. Once upon time, old man lived in small house in woods by Carpathian mountains.

2. He was very respected old man because he could tell person's fortune by just looking at person's face.

3. Many people took train to woods to see old man, even though they said they didn't believe him.

4. After talking to old man, they were surprised by how much he knew about them.

A **preposition** in a word that, like a traffic signal, shows direction (*at, toward, on, under, above*) or relationships (*with, between*). Using prepositions precisely adds clarity to your writing. For example, Mark went *with* Susan *toward* the back *of* the bus.

Appendix C: Writing Skills Review   **459**

Some commonly used prepositions are listed here.

about	among	beside	for	of	to
above	around	between	from	on	toward
across	at	both	in	over	under
after	before	by	inside	past	with
against	behind	down	near	through	within
along	below				

Using Prepositions   Exercise C.9

Fill in the blanks with the preposition of your choice. Be prepared to say why this preposition works better than other prepositions you might have chosen.

1. Why can't we go _____ Paula to the airport?
2. I am waiting _____ Paula before we leave.
3. You'll find her briefcase _____ the table.
4. Put her suitcase _____ the back seat.
5. Please go _____ the ticket office to verify the flight.
6. Go _____ the street and _____ the waiting room.

A **conjunction** is a word that connects other words, phrases, clauses, or sentences together. You will very commonly use one of two kinds of conjunctions, a coordinating conjunction and a subordinating conjunction.

Coordinating conjunctions (words like *but, or, yet, so, for, and, nor*) let the reader know that the words being connected are roughly equal in value.

Rachel *and* James will commute to Los Angeles.
Rachel will study music, *but* James will study political science.

Some writers believe that you must use a comma every time you use a coordinating conjunction like "and." But conjunctions like "and" often connect words together. When "and" connects two complete sentences, a comma must be used. If either of the sentences is incomplete, a comma cannot be used.

Rachel *and* James will commute to Los Angeles. (No comma is used because "and" connects two subjects: Rachel and James.)

Rachel will study music, *but* James will study political science. (A comma is used because the two sentences are complete by themselves.)

Rachel plays the piano for many hours each day *and* somehow finds time to play the violin as well. (No comma is used because "and" connects two verbs that have the same subject.)

Subordinating conjunctions (words like *if, when, since, although, even though, as if, because, before, until, while*) let the reader know that the clause that begins with a subordinating conjunction is secondary to the main clause.

A clause that begins with a subordinating conjunction is considered a dependent clause or a fragment unless it is connected to an independent clause. This dependent clause can be placed at the beginning or at the end of the sentence.

When it is raining.	(fragment: What happens when it rains?)
Since last week.	(fragment: What happened last week?)
When it is raining, I stay home.	(complete sentence)
Four people have been hired since last week.	(complete sentence)

The dependent clause can be attached either at the beginning of the sentence or at the end of the sentence, depending on which ideas are most important. Which of the following do you prefer?

When it is raining, I stay home.	I stay home when it is raining.
Since last week, four people have been hired.	Four people have been hired since last week.

Notice how the conjunctions in these next sentences reinforce the meaning of the sentence. For example, using "and" adds to the discussion. Using "but" introduces an opposing idea. Using words like "when," "if," or "while" set a condition. Notice where the comma is used and where it is *not* used!

Murphy had just returned from China, *and* he wanted to talk about his trip.

Murphy had just returned from China, *but* he didn't want to talk about his trip.

If college were less expensive, I believe more people would attend.

When my cat wants to go outside, he sits patiently by the door.

We have many books to donate to the library *before* the fall sale begins.

Exercise C.10 Using Conjunctions and Subordinating Conjunctions

Add conjunctions to the following.

1. Charlotte _____ Roger were ready to go to dinner.
2. _____ Roger was very hungry, Charlotte had not eaten all day.
3. Roger was concerned about Charlotte _____ she had worked all day.
4. _____ Charlotte got to dinner, she decided she was hungry.
5. Charlotte ate a rib steak and a piece of chocolate cheesecake _____ the night was over.

Sentence Patterns

Understanding sentence patterns (how sentences are put together) can help you decide in what order you can best express your ideas. This section will help you understand the most common of sentence patterns. You will also see how the sentence pattern is linked to punctuation.

A complete sentence has two parts: a **subject** (what the sentence is about) and a **predicate** (the verb and all the words related to that verb that describe the subject). Notice how the subjects in the following sentences show what the sentence is about.

Subject	*Predicate*
The cabin	has air conditioning.
The small log cabin	can accommodate five people.
Mary and Martha	ran quickly from the cabin.

Sometimes the *subject* does not appear as part of the sentence. For example, you might say: "Could you please shut the door." Or you could say, "Shut the door!" The subject *you* is understood as part of the command: Shut the door!

Finding Subjects and Predicates   Exercise C.11

Circle the complete subject and underline the complete predicate in each of the following:

1. It is very humid today.
2. Kansas and Missouri have frequent thunderstorms.
3. Camping in the rain can be an unforgettable experience.
4. The mosquitoes flew around the tent, making us nervous.

Write two additional sentences that include a subject and a predicate.

Combining Sentences

All writing involves adding and subtracting words from a basic, simple sentence. You can make your writing more specific by adding words, phrases, and clauses to the subject or to the predicate.

Practicing adding and subtracting words, phrases, and clauses to your writing will help you develop ideas more fully and will help you control your punctuation.

Adding Words.   Adding adjectives and adverbs (also called modifiers because they "add to" or modify the original idea) is an important way to add more information to your writing. To summarize what our chapter has covered:

Nouns describe people, places, or things (for example: the *house*).

Adjectives add information about nouns and usually are placed **before** the noun (for example: the *blue* house).

Verbs describe action or a state of being (for example: She *ran* or she *is feeling* stronger today).

Adverbs modify verbs by describing **how,** usually come **after** the verb and often end with -ly (for example: She ran *slowly*).

Exercise C.12 Using Verbs, Adverbs, and Adjectives

Fill in the missing verbs, adverbs, and adjectives for the following exercise.

When I return to my _____ office, I see that _____ light _____ on my _____ telephone that tells me I have _____ messages waiting for me. Voice mail is an _____ tool, but people need to know how to leave messages. Three key ideas to _____ your use of voice mail are:

1. Keep your message _____. Try to _____ unessential information.
2. Talk as _____ as possible. Include important information. Avoid general wording.
3. Leave your phone number so the person can _____ call you back.

As the writer, you are always in charge of deciding where you want to add words. In this next example, *suddenly,* the adverb modifying *how* the tent collapsed, is put at the beginning and at the end of the sentence. Which sentence do you prefer and why?

Suddenly, the tent collapsed.
The tent collapsed *suddenly.*

Adding Phrases. Prepositional phrases can help you describe your subject more precisely or describe exactly where something happened. For example, which sentence tells you the whole story?

Base sentence: Rita lost her temper.
Expanded: Rita lost her temper in Cincinnati on Interstate 75.
   in Cincinnati   (preposition + proper noun)
   on Interstate 75   (preposition + proper noun)

When prepositional phrases introduce or begin the sentence, you need to use a comma. Prepositional phrases can be used anywhere else in a sentence *without commas,* unless you are listing or using the phrases in a series. Study these examples:

In the morning, we discovered the tent had collapsed.
We discovered the tent had collapsed *in the morning.*

I looked for the missing car keys *in the trunk, on the picnic table, inside the tent,* and *under the car.*

We finally found the car keys *on the ground under the car,* after looking for three hours.

One way to help you think of information that could be included in your writing is to answer these questions: *Who? What? Where? When? How? How much? Why?*

Using Prepositional Phrases   Exercise C.13

Write a paragraph about a vacation or camping trip that you have taken either recently or as a child. Write about seven sentences. After you have finished your draft, go over each sentence to make sure it has a prepositional phrase. Underline the prepositional phrases. Exchange papers with a classmate and proofread each other's paper.

Adding Clauses. A **clause** always has a subject and a verb. A clause is also known as a simple sentence or an **independent clause** if it can stand by itself. Study these examples of independent clauses:

It is raining.

John and Rita have finished eating breakfast.

Is it time to start driving?

If the clause begins with a modifier of some kind, it is called a **dependent clause.** A dependent clause cannot stand alone; it is an incomplete sentence, otherwise known as a **fragment.** Study these examples of dependent clauses:

If it is raining

Even though John and Rita have finished eating breakfast

When it is time to start driving

Adverbial and adjective clauses also add more information. Usually adverbial clauses tell more about *how* and adjective clauses add more information about the subject (*who* or *what*).

Base sentence:	Rita lost her temper.
Adverbial clause:	after her engine exploded
	(adverb + pronoun + noun + verb)

After her engine exploded, Rita lost her temper.
Rita lost her temper *after her engine exploded.*

Base sentence:	Rita lost her temper.
Adjective clause:	who had been driving for 13 hours
	(pronoun + verb phrase + prepositional phrase)

Rita, *who had been driving for 13 hours,* lost her temper.

464   Appendix C: Writing Skills Review

Exercise C.14   Using Adverbial and Adjective Clauses

Practice writing sentences with adverbial and adjective clauses by completing the following and identifying whether each sentence includes an *adjective clause* or an *adverbial clause*.

1. Write a sentence about your best friend, using *who*.
2. Write a sentence about a grandparent that begins with *after*.
3. Write a sentence about your parent beginning with *although*.
4. Write a sentence about a pet that includes a *that* clause.

Whether you are adding adverbial or adjective clauses, you as the writer decide where the clause should go. Which of the following sentences do you like best and why?

Even though the tent had collapsed, we still enjoyed our camping trip.
We still enjoyed our camping trip *even though the tent had collapsed.*

If you are telling a story, you'll decide at the very last second which sentence structure will work best to heighten the story for your audience. Sometimes how we present our ideas comes from our culture. For example, which of these sentences would you most likely say?

Yellowstone National Park has unique dangers.
Unique dangers exist in Yellowstone National Park.

The first sentence follows the pattern most North Americans use: Subject + Verb + Object.

(Yellowstone National Park)   (has)   (unique dangers).
    (subject)   *(verb)*   *(object)*

Exercise C.15   Revising Base Sentences

Add any of the words, phrases, and clauses that have been discussed in this section to the following base sentences. Your goal is to use very specific descriptions of nouns and verbs in your revised sentences. You may have to change the meaning so the sentence makes sense for you. Here's an example:

    Base sentence: I hate broccoli.
    Revised: I don't mind eating broccoli when it is covered with lemon and garlic salt.

1. Your car needs repair.
2. I like working with computers.
3. My first job was unforgettable.
4. My worktable is cluttered.
5. My dog likes cats.

Appendix C: Writing Skills Review **465**

SUMMARY EXERCISE

Complete the sentences that follow to write a first rough draft. Then revise your paragraph by adding words, phrases, and clauses to make it more descriptive. You may want to add more sentences or delete some ideas.

After you have finished your draft, exchange it with a classmate to make sure your story is complete and you have included as much specific description as possible. Turn your completed revision in to your instructor with all rough drafts.

My First Promotion

1. When I got my first promotion at:

2. I got my first promotion by:

3. My co-workers felt:

4. My friends said:

5. My supervisor was:

6. I felt that:

7. I immediately:

8. I'll never forget:

9. Today, promotions are:

BUILDING PUNCTUATION SKILLS

Some people like to think about punctuation as a way to let readers know when they can take a breath. However, the real reason we use punctuation is so that important parts of sentences don't run together, creating confusion for the reader.

This section will help you to review these important punctuation skills:

- Using periods, semicolons, and question marks   Go to page 466

- Using commas   Go to page 468
  Introductory commas
  Interrupter commas
  Ending commas

- Changing sentence patterns   Go to page 470
  Simple sentence
  Compound sentence
  Complex sentence
  Compound complex sentence

- Using apostrophes   Go to page 471

466   Appendix C: Writing Skills Review

Complete the exercises and the summary exercise as you work through this section, checking for correct answers at the end of this Appendix.

Using Periods, Semicolons, and Question Marks

No matter how long or short the sentence is, *a **period** gives the reader a clear signal that the sentence has ended.*

Not OK: Harry slept in
OK: Harry slept in.

Sometimes two sentences are so closely related that they can be connected by a **semicolon.** *A semicolon can replace a period because it shows the end of a sentence.*

Not OK: Harry slept in, Robert went hiking.
OK: Harry slept in; Robert went hiking.

Not OK: Harry had always dreaded camping, Robert had gone camping all of his life.
OK: Harry had always dreaded camping; Robert had gone camping all of his life.

Notice how choppy two short sentences can sound. That's why it may be better to use a semicolon or to combine the two sentences by adding a comma and conjunction (more about this in the next section).

What if two sentences were joined together without any punctuation at all? This problem is called a fused sentence. You can use a semicolon to connect two closely related, short sentences.

Not OK: The cafe was closed it was after midnight.
OK: The cafe was closed; it was after midnight.

Notice how the punctuation implies that the cafe was closed *because* it was after midnight.

Semicolons can also separate items in a complicated list. Here is a simple list that uses commas to separate the items:

Not OK: My favorite colors are blue; green; and purple.
OK: My favorite colors are blue, green, and purple.

Notice how the semicolon makes it easier to read this more complicated list:

Not OK: I have lived in Philadelphia, Pennsylvania, San Francisco, California, Phoenix, Arizona and Portland, Oregon. (In this sentence, cities and states run together.)
OK: I have lived in Philadelphia, Pennsylvania; San Francisco, California; Phoenix, Arizona; and Portland, Oregon. (In this sentence, cities and states are separated by semicolons.)

When we ask someone a question, our voices rise at the end of the question. The question mark shows this in writing.

OK: Are you ready to go to the subway?
OK: Have you finished the McKenzie report?
OK: Susan is ready, isn't she?

Using Periods, Semicolons, and Question Marks Exercise C.16

Change or add punctuation to the following:

1. Have you learned how to use the Internet yet.
2. I have just taken my first class on the Internet learning how to jump on the information highway was fascinating.
3. I have had to learn a new vocabulary words such as "gopher," "mosaic," and "web" have taken on new meanings.
4. My favorite people in the class were Mary a college librarian George a retired contractor and Martha a nurse.
5. My teacher was very understanding he helped us gain confidence by letting us practice on the computers.
6. When will you get your new computer.

More Practice with Punctuation Exercise C.17

Use each of these sentence patterns at least once in correcting the following:

Sentence, and Sentence	(Use comma plus coordinating conjunction.)
Sentence; Sentence	(Use semicolon.)
Dependent Clause, Sentence	(Turn one sentence into a dependent clause at the beginning of the sentence.)
Sentence, Dependent Clause	(Put dependent clause at end of sentence, using an optional comma.)
Sentence. Sentence	(Use period between two sentences.)

1. At first I was not comfortable using the computer to prepare my assignments but with practice my skills in revising are improving.
2. Our classroom only has three printers these DeskJet printers are so quiet that we can't hear when a document is being printed.
3. Using the new program was time-consuming but that didn't seem to matter we started to enjoy making changes to our writing.
4. I was not able to find the RETURN key in fact the ENTER key is the same as the RETURN key!
5. I could print my work out if only I could remember the print command!

Using Commas

One way to think about where we put commas is to consider how they are placed after introductory, interrupting, and ending sentence elements.

An *introductory word, phrase,* or *clause* that appears at the start of a sentence is separated from the rest of the sentence with a comma:

Word: *Suddenly,* the roof fell in.

Phrase: *In the beginning,* no one wanted to watch TV.

Clause: *After watching the evening news,* we ate dinner.

Exercise C.18 Using Commas

Write three sentences about your first job that use the following introductory word, clause, or phrase:
1. Use *quickly* as your introductory word.
2. Use *on* to start your introductory phrase.
3. Use *although* to start your introductory clause.

Exercise C.19 Adding Introductory Commas

Add commas to the following sentences:
1. Quickly Martha opened the door.
2. Even though it was very early it was still dark.
3. Because it was the first day of her vacation Martha was ready to go.
4. If it had been raining she would have been disappointed.
5. At first she had just hoped for a camping trip but there on the table were five airplane tickets to Hawaii.

An *interrupter* word, phrase, or clause can appear in the middle of the sentence, usually between the subject and the verb. *If the interrupter is a word or phrase, commas are not needed. If the interrupter is a clause, commas are needed.*

Word: The roof *suddenly* fell in.

Phrase: No one *in the living room* wanted to watch TV.

Clause: John, *after watching the evening news,* ate dinner.

Clause: John, *having lived in Arizona for five years,* was ready to move to Alaska.

Clause: Roberta, *who likes to eat pizza with anchovies,* was ready for dinner.

Clause: Driving, *which is considered an essential skill,* requires concentration and common sense.

Sometimes a clause gives essential information to the sentence. When this happens, commas are left out. You can tell if the information is essential by asking this question: If I take the clause out, does the sentence still make sense? *If the clause gives essential information, leave commas out!*

OK: The sweater that has blue roses on it is mine.
OK: That sweater is mine. ("That" points to my sweater.)
Not OK: The sweater is mine. (Which sweater?)

OK: The woman who lives in apartment 49 got a job.
Not OK: The woman got a job. (Which woman?)

Sometimes commas are used to show whether or not the information is essential.

OK: My brother, Stewart, loves dogs. (I have one brother.)
OK: My brother Stewart loves dogs. (I have several brothers.)

Sometimes key connector words are left out. When this happens, a comma is essential!

OK: I traded my station wagon in for a sports car, which was a drastic change.
OK: I traded my station wagon in for a sports car, a drastic change.

Punctuating "Interrupter" Clauses   Exercise C.20

Write four sentences about the weekend that use an *interrupting* word, clause, or phrase immediately after the subject.

1. Use *happily* as your interrupting word.
2. Use *at* to start your interrupting phrase.
3. Use *since* to start your interrupting clause.
4. Use *which* to start your interrupting clause.

A *word, phrase,* or *clause* can also be added at the end of the sentence. *Most of the time, commas are not used when adding words, phrases, or clauses at the end of a sentence.*

If the clause is very long, you may want to use a comma so the reader can easily understand you. Sometimes you can add emphasis to the ending word, phrase, or clause by using a dash.

Word: The roof fell in *suddenly.*
Word with dash for emphasis: The roof fell in—*suddenly.*

Phrase: No one wanted to watch TV *in the beginning.*

Clause: John ate dinner *after watching the evening news.*
Clause: Mary, Roberta, and John stayed for dinner, hoping they would be served sizzling rice soup.

Exercise C.21 Using Commas with Ending Elements

Write six sentences about working with customers that include an *ending* word, clause, or phrase. Use commas when needed.

1. Use *quietly* as your ending word.
2. Use *sadly* as your ending word.
3. Use *under* to start your ending phrase.
4. Use *in* to start your ending phrase.
5. Use *even though* to start your ending clause.
6. Use *since* to start your ending clause.

Changing Sentence Patterns

Another way to think about where to put the commas is to consider the types of sentences we write. We use four basic patterns of sentences:

Simple Sentence = *Independent clause* (complete sentence).
Example: I turned on the light.

Compound Sentence = *Independent clause + Independent clause.*
Example: I turned on the light, but I still could not see.
Example: I turned on the light, and I could see everything.

Complex Sentence = *Dependent clause + Independent clause.*
Example: When it grew dark, I turned on the light.
Example: I, who fear the dark, turned on the light.

Compound Complex Sentence = *Dependent clause + Independent clause + Independent clause.*
Example: When it grew dark, I turned on the light; the chair was gone.
Example: When it grew dark, I turned on the light; however, no lights came on.

When we add clauses to sentences, we make them more interesting for readers rather than writing one simple sentence after another. Notice how short, choppy sentences interfere with the content of this next example.

My daughter is a teenager. She doesn't like to clean her room. Her room is messy. It is very messy. It is so messy that she can't walk through it without stepping on something.

Revised: My teenage daughter doesn't like to clean her very messy room, even though she can't walk through it without stepping on something.

Combining sentences also helps us practice punctuation.

Changing Sentence Patterns Exercise C.22

Please combine the following sentences, emphasizing *sentence variety!* Watch for transitions between ideas. Delete words that are repeated and don't add meaning. Try to use each of the four sentence types at least once. Use the word in parentheses to combine the two sentences.

1. I have saved all of the money from my summer job. I'm ready to pay my tuition. (therefore)
2. I have some money left after paying tuition. I will stand in line again to buy my books. (if)
3. Spring break and daylight savings time are coming soon. We will have to get up an hour earlier. (but)
4. Paul is the smartest person I have ever met. Paul spends three hours every day reading newspapers or books. (who)
5. Combine these short sentences into *one* sentence:
   My husband is a sports fan.
   He doesn't like to watch television.
   The television is too loud.
   The television has too many commercials.
   The announcer's voice seems to drone on forever.
6. Combine these short sentences into *one* sentence:
   My husband refuses to miss even one game.
   It doesn't matter which sport.
   It doesn't matter what time of year it is.
   It doesn't matter what time of day it is.
   It doesn't matter if it's a holiday.
   The game must be played through!

More Practice with Commas Exercise C.23

Add commas to the following sentences where needed.

1. The woman who was arrested by the police was charged with arson.
2. The cat's favorite hiding place is under the bed a quiet spot that is also cool and dark.
3. Although our television was broken we still heard the game by listening to the radio.
4. This version of "Stardust" which is fast and has a steady beat would be good to dance to.
5. The waiter took a coffee break while the customers lingered over dessert.
6. Happily Mary took the increase in pay.
7. Aurancia who just won the tennis competition at Wimbledon grew up in Spain.
8. I found it hard to decide whom to favor in this tennis match even though one of the players was from the United States.

Using Apostrophes

Apostrophes can show possession or ownership, like *Joe's place* or the *students' study hall.* Apostrophes can also be used to show contractions, where the apostrophe fills in for missing letters. *Martha's ready to go* instead of *Martha is ready to go.*

Possessive apostrophes show ownership. Most of the time, the apostrophe is placed between the end of the *base word* and before the *s*.

The dog that belongs to <u>Paul</u>	becomes	<u>Paul's</u> dog
The books that belong to the <u>children</u>	becomes	the <u>children's</u> books
The parking lot of the <u>workers</u>	becomes	the <u>workers'</u> parking lot

People often have problems putting the apostrophe in exactly the right place when the base word is plural. You should start with the base word (children), and then add the 's (children's).

Contractions are also used when the apostrophe takes the place of missing letters. Contractions are usually considered informal wording.

I cannot	becomes	I can't
it is	becomes	it's
you are	becomes	you're
Paul is hungry.	becomes	Paul's hungry.

Note that "it's" with an apostrophe should be used *only* when the word is a contraction for "it is." In the following sentence, there's no contraction: *The dog lost its bone.* If "it's" is used, the sentence would mean, *"The dog lost it is bone,"* which is incorrect.

Exercise C.24 Using Contractions

Add apostrophes to the following sentences.

1. Her friends dont know that she's working at a competitors store.
2. The first dishwashers job was to scrape and sort the customers dishes.
3. Theyre visiting their parents farm to study crop rotation.
4. A persons mind doesnt work like a tape recorder.
5. Its the parents responsibility to make sure that they dont buy childrens toys with sharp edges that can cut a youngsters tender skin.
6. Isnt it a shame that football players salaries have gotten so high that the teams owners arent able to keep ticket prices at last years level?

Exercise C.25 More Practice Using Apostrophes

Add apostrophes where they are needed in the following sentences:

1. The shop supervisor lost the womens paychecks.
2. My brothers associates took a trip last summer.
3. Its too late to go for an hours drive.

4. Marias manager kept track of one employees work schedule.
5. The managers office was bigger than the typists' offices.
6. The secretary made a copy of this years tax return.
7. We could hear the childrens voices outside.

Summary Exercise—Sentence-Combining Review

Expand each sentence by following the directions shown in parentheses after each sentence. Your goal is to practice different sentence patterns and add more information to the sentence. Check your answers at the end of this Appendix.

Example: The girls were standing upstairs. (add other *verbs*)
Revised: The girls *were standing* upstairs and *shivering* in the cold.

1. Carl **had repainted** the boat. (add verbs)
2. We loaded our **fishing gear** onto the boat. (add direct objects)
3. The boat tipped. (add adverbs)
4. The **boys** were waiting by the pier. (add other subjects)
5. **It was very early in the morning.** (add another independent clause)
6. Everyone was ready for the trip **to the lake.** (add more prepositional phrases)
7. Peter began to eat lunch. (add adjectives)
8. **Peter** was not happy about eating lunch on the boat. (change the subject to a pronoun)
9. Rhonda started the motor. (add an adjective clause)
10. Combine the following into *one* sentence:
    Reynard found a map.
    The map was mysterious.
    The map was titled, "The Quest."
    The map showed a castle.
    The map had a note.
    The note was written hastily.
    The note said, "Come quickly or all is lost."
    The note was signed with the initial "A."

REVISING FOR CORRECT PRONOUNS

This section will help you use subject and object pronouns correctly, revise unclear pronoun references, and use "you" and "they" with more precision. Work through the exercises as you read these sections. Answers are at the end of this Appendix.

Using subject and object pronouns	Go to page 474
Proofreading for pronoun agreement	Go to page 476
Clarifying pronoun reference	Go to page 477

474 Appendix C: Writing Skills Review

Using Subject and Object Pronouns

Most of the time, we use the correct pronoun by unconsciously copying what others say and write. This process has been going on since we learned to talk. However, what we have heard may not always be correct. Many of us use incorrect grammar when we talk or write informally.

The following chart shows both subject pronouns and object pronouns. Notice that the form (or case) of a pronoun changes depending on whether it is used as a subject or as an object. Pronouns can be singular or plural, depending on the word they replace or refer to.

Subject Pronouns	Object Pronouns
I	me
you	you
he	him
she	her
it	it
we	us
they	them
who	whom

Examples of pronouns used correctly:

OK: *Paul* wants to give the book to *Maria.*
OK: *He* wants to give the book to *her.*

OK: *Maria* wants to go to the movies with *Paul.*
OK: *She* wants to go to the movies with *him.*

OK: Maria and Robert want to go with *Paul* and *Albert.*
OK: *They* want to go with *them.*

OK: Who is calling?
OK: You want me to give the book to whom?

Notice that pronouns can be singular or plural, depending on the word they refer to.

Using Subject and Object Pronouns.   Problems occur when we mix up subject pronouns with object pronouns. Sometimes this happens because the wording just sounds better.

NOT OK: Her wants to go to the movies.
OK: She wants to go to the movies. (subject pronoun used correctly)

| NOT OK: | Mary and her want to go to the movies. |
| OK: | Mary and she want to go to the movies. (subject pronoun used correctly) |

Notice that a preposition is used just before the object pronoun: *to* her, *with* him, *with* them. You can decide whether to use an object pronoun by looking for a preposition. Give the book to? Give the book to **her.**

| NOT OK: | Just between you and I, I'm confused. |
| OK: | Just between you and me, I'm confused. |

Here *between* signals an object pronoun used correctly.

The following chart includes some common prepositions. Notice how many of these prepositions point out the direction or location of either a thing or a person.

about	at	down	over
above	before	for	past
across	behind	from	through
after	below	in	to
against	beside	inside	toward
along	between	near	under
among	both	of	with
around	by	on	within

Use the Subject Pronoun for Comparisons. When we are comparing someone or some thing to another, we need to use a subject pronoun.

NOT OK:	She is taller than me.
OK:	She is taller than I am.
OK:	She is taller than I. ("Am" is implied.)
OK:	I am taller than she. ("Is" is implied.)

In this case, we are comparing two equals. The subject "she" can be exchanged for the subject complement "I". This kind of comparison always uses a subject pronoun.

Selecting the Correct Pronoun Exercise C.26

When we check to see if we use subject pronouns in place of subjects, or object pronouns in place of objects, we are checking for correct use of the case of the pronoun.

Circle the correct subject or object pronoun(s) in the following exercise, checking your answers at the end of Appendix C.

1. My roommate and (I) (me) became close friends.
2. The meal was prepared especially for (they) (them) and (we) (us).

3. For (we) (us) students, attendance is required.
4. A problem has come up between you and (I) (me).
5. The river rose after the Browns and (us) (we) had safely crossed.
6. (Who) (Whom) is on the phone?
7. The audience did not know (who) (whom) to applaud.
8. The audience wanted to applaud (she) (her).
9. Either (he) (him) or (I) (me) will wait for the shuttle.
10. The music drew Sylvia and (I) (me) closer.
11. The substitutes were (her) (she) and (I) (me).
12. (He) (Him) and (she) (her) were admitted to the hospital.
13. Sylvia is taller than (I) (me).
14. Ronald can type faster than (her) (she).

Proofreading for Pronoun Agreement

When the pronoun is singular and the word it refers to is plural (or vice versa), we call this a pronoun agreement problem.

NOT OK:	Each bird has their own cage.
OK:	*All* of the birds *have their* own *cages*.
NOT OK:	Each of the companies had hired all their employees.
OK:	*Each* of the companies had hired all *its* employees.
NOT OK:	Mildred and Paula didn't want her own office.
OK:	*Mildred and Paula* didn't want *their* own office.
OK:	*Neither Mildred nor Paula* wanted *her* own office.

Notice with this last example, "and" leads the reader to think of Mildred and Paula as "they" sharing one office, while using "neither . . . nor" leads the reader to think of the two as each having her own office. Once you decide which wording most accurately describes the situation, you can then choose your pronouns.

Exercise C.27   Editing for Pronoun Agreement

Read the following sentences to make sure the pronouns agree with the singular or plural words they refer to. Circle the correct answer, checking your answers at the end of this Appendix.

1. If the workers (needs) (need) a ride to the workshop, (he) (she) (they) can call the Motor Pool Department.
2. Yesterday, one of the workers had (his) (the) (their) assembly line shut down while (his) (the) (their) machine was replaced.
3. (He) (She) (They) lost four hours of production time while (his) (her) (their) machines were down.

4. One of the supervisors attended a workshop and then left for (her) (their) vacation.
5. The supervisor gave overtime to all of the workers, making (him) (her) (them) very happy.
6. No one wanted to turn in time sheets until (his or her) (the) (their) time sheets were correct.
7. Roberto and Paul wanted (his) (their) own office.
8. Neither Roberto nor Paul wanted to share (his) (their) workstation.
9. Give Roberta and Paula (her) (their) receipt.
10. Neither Roberta nor Paula wanted (her) (their) receipt.

Clarifying Pronoun Reference

Your constant challenge is to write so that readers will easily understand what you mean—the first time they read your work.

Pronouns Should Agree with Their Antecedents. When you use a pronoun, double-check that the pronoun clearly refers to the word that it is linked to.

UNCLEAR:	Mary told her she couldn't work late.
STILL UNCLEAR:	Mary told Susan she couldn't work late.
OK:	Mary told Susan, "I can't work late."
OK:	Mary told Susan, "You can't work late."
OK:	When Mary learned she couldn't work late, she told Susan.

Pronouns Should Not Offend the Reader. You might write, "Jack wanted his promotion before March," or "Jacqueline took her report to her supervisor." These uses of the pronoun are quite correct!

However, using "he," "his," or "him" when you mean someone in general may seem biased to your audience. Most of the time, you can write your way around this problem by using plural pronouns:

NOT OK:	Each supervisor should have his own office.
OK:	All supervisors should have their own offices.

Pronouns Should Be Used with Precision. Sometimes we use "code words" to take shortcuts in our writing. Instead of writing, "The Department of Health, Education, and Welfare of the U.S. federal government has decided not to fund the school lunch program," we will say, "They have decided not to fund the school lunch program."

This approach works if our audience can correctly define the nonspecific "they" or "you" that holds our discussion together. This kind of writing often shows up at the drafting stage and needs revision. Making sure your reader understands exactly what you are writing about is especially important in workplace writing.

Exercise C.28 Clarifying Pronoun Reference

Read the following sentences to make sure the pronouns are used as clearly as possible. Make any revisions needed and check your answers at the end of this Appendix.

1. After taking the key out of the envelope, it was taken to the lab.
2. Under the Bill of Rights, each citizen is guaranteed his rights to privacy.
3. Anyone should have job security when he goes to work.
4. A few minutes after Susan discovered Barbara had the missing key, she left the lab.
5. After taking the draft out of the lab notebook, I turned it in to my supervisor.
6. As I read each page of the final report, I worked hard to correct the errors, marking it with a yellow marker.
7. They really appreciate hard workers here.
8. Each applicant turned in his completed application and waited to be called for an interview.
9. If it just involves filing, Sam files faster than I.
10. It is agreed that we should switch to four shifts a day.

Pronouns Should Clarify Point of View. Sometimes when we are writing or talking, we speak from our personal point of view. This first-person point of view emphasizes "I" or "we" statements and is commonly used in the workplace for trip reports or lab reports that describe what "I" did. The "I" point of view emphasizes the *speaker*.

I have just completed 100 hours of overtime.

I would like to go to the movies tonight.

We are ready to buy a new car.

When we write instructions, business letters, or some sales materials, we switch to the "you" point of view. This second-person point of view emphasizes the *audience*.

You will enjoy owning the new Turbojet engine 800.

Before you can use your new printer, you must install the printer driver.

When we want to emphasize the subject we are writing or talking about, third person is most appropriate. The "it" point of view emphasizes the *subject* or *topic*.

The classroom is empty today.

The Jensen report was finally finished.

Using more than one point of view can confuse the reader. This next exercise will help you practice editing to use a single, consistent point of view.

Appendix C: Writing Skills Review

Exercise C.29 Clarifying Point of View

Revise the following paragraph where needed so that it is consistently written from the students' point of view. As you revise, make sure you change all words that are affected by your change in point of view. Compare your final version to the revision at the end of this Appendix.

Their registration system at this college creates frustration. Everybody has to preregister before they go to registration, but then you still have to spend hours waiting in a long line before they tell you that the courses you wanted to take are filled. You then have to sign up for another section. Then we have to wait in another line. This drives me crazy. A student goes to college to learn; they don't go to college to stand in line. The administration is thinking of setting up telephone registration. I favor this change!

Pronouns Should Avoid Confusing Shifts in Person. As the previous exercise shows, sometimes we shift to a different pronoun without realizing the effect this can have on the clarity of our writing. Confusing shifts can occur when we move from singular to plural, when we switch verb tenses, subjects, or sentence structure (also called parallel form).

NOT OK:	We entered the museum, not knowing you were supposed to pay an admission fee.
OK:	*We* entered the museum, not knowing *we* were supposed to pay an admission fee.
NOT OK:	I enjoy camping, hiking, and the Internet fascinates me.
OK:	I enjoy camping, hiking, and learning about the Internet.

Revising Confusing Shifts Exercise C.30

Read the following sentences to make sure there are no confusing shifts in subjects, verb tense, pronouns, or sentence structure. Revise any errors, checking against the answers at the end of this Appendix.

1. We went to the employment office, not realizing that you needed to register as you entered the building.
2. She wanted to know whether to take the job and did it pay well.
3. We planned to commute by car, but bus was found to be cheaper.
4. Oliver had no way of knowing we were there until he walks through the door.
5. When people get mad, you can sometimes act foolishly.
6. The waiter scowled at the diners because he left a small tip.
7. A person should always have professional playing experience before they coach.
8. To revise a paper, read it through for errors, and then you should look at its logical organization.
9. Each member of the squad looked to me as their leader.
10. I had to decide where to go on my vacation and could I afford it.

480   Appendix C: Writing Skills Review

REVISING FOR EFFECTIVE VERB USE

Verbs describe action and bring precision to writing. By understanding how to revise verbs to emphasize action, your writing can be clearer and less wordy. This section will help you proofread for agreement between subjects and verbs and check for consistency in verb tense throughout a piece of writing.

Understanding verbs	Go to page 480
Reviewing verb tenses	Go to page 481
Looking at irregular verb forms	Go to page 483
Checking for subject/verb agreement	Go to page 486
Using active and passive verbs	Go to page 487

Understanding Verbs

Verbs can show **action:**

He *ran.*
They *scrambled* out of the burning car.

Verbs also show conditions or a state of being; sometimes these are called "to be" verbs.

She *is* on time.
They *are* very tired.

Verb tenses change when you change the time referred to in the sentence.

Mary *works* in Chicago.	(present)
Last year, Mary *worked* in Chicago.	(past)
Next year, Mary *will work* in San Francisco.	(future)

Complete verbs can have one or more "helping" or auxiliary verbs in addition to a central verb:

We *have been driving* for 15 hours.
Even though it *has* not *been raining* for very long, the humidity *has dropped* by 10 points.

Note that words like **not** or **often** can be used in the middle of a verb phrase, even though these words are not verbs.

Verb phrases can be used as either subjects or objects.

Waiting for Fred *upset* me.   **-ing verb** used as a subject + primary verb
    (Waiting for Fred = **verb phrase** used as a subject)
    (upset = **primary verb** for this sentence)

I *wanted to talk* to my supervisor.   Subject + primary verb + **infinitive used as object**
(I = subject)
(wanted = verb)
(to talk = infinitive used as object)
(to my supervisor = prepositional phrase)

Often, more than one verb will be used in a sentence. The verb that appears in the independent clause is called the **primary verb.**

Having not eaten for three hours, I *found* a small delicatessen near the office and *wolfed* down a Rueben sandwich.
(*All italicized words* = verbs)
(*found* and *wolfed* = **primary verbs**)

Running for the bus and *catching* it *meant* that I *would be* on time for work.
(*All italicized words* = verbs)
(*meant* = **primary verb**)

Finding Verbs   Exercise C.31

Underline all *verbs* in the following sentences. Put parentheses around the (*primary verb*).

1. I will see you after attending the meeting.
2. Architects came from Chicago to redesign our building.
3. Our building was surrounded by modern, high-rise skyscrapers that seemed to reach to the sky.
4. Without remodeling, our building would have been old-fashioned.
5. We could have modernized the building less expensively if we had limited the budget to replacing carpeting, rewiring electrical circuits, and repainting offices.

Reviewing Verb Tenses

Verbs can describe action or a "state of being" in the past, present, or future. The **tense** of a verb describes the **time** these actions or states of being occurred. The **form** of the verb changes to show the past, present, or future. There are three **simple** tenses in English:

Simple

Past:	She talked.	(in the past)
Present:	She talks.	(now)
Future:	She will talk.	(in a future time)

When we want to show a *completed* action or state of being, we add the **perfect** form to one of these three **simple** tenses. A few verbs (like *be, did,* and *have*), can be "helping" or auxiliary verbs as well as main verbs (She *is looking* for her book, or She *has been looking* for Fred).

Perfect

Past Perfect:	She had talked.	(before yesterday)
Present Perfect:	She has talked.	(before now)
Future Perfect:	She will have talked.	(before a future time)

Exercise C.32 Changing Present to Past Tense

Change all of the present-tense verbs in the following paragraph to past-tense verbs. Start your revision with: When I *worked* at the college, I *saw* . . .

When I work at the college, I see lines of students waiting to be helped with registration at the college. Just before each term begins, I meet with students, check their schedules, talk about their career plans, and help them register for the new term. Sometimes students do not have very much time to think about what they want to do, and they feel stressed out. Sometimes students know what they want to do, but they don't know which classes would be best for them or which classes are required for graduation. A very few students see me during the term to talk about their future. Although students register by telephone or computer, I still like to talk directly with students—especially when they need help!

If the action or state of being is *continuing,* we can add a **progressive** form to either the **simple** tense of past, present, or future tenses or to the **perfect** form of past perfect, present perfect, or future perfect.

Progressive

Past Progressive:	She was talking.	(and will continue)
Present Progressive:	She is talking.	(and will continue)
Future Progressive:	She will be talking.	(and is expected to continue)

The *present progressive* tense shows an action that began in the past and that is expected to continue into the future. For example:

She *has* been talking now for several hours, and I do not know when she will stop.

He *has* been feeling ill for some time and plans to see the doctor next week.

The *future progressive* tense describes some future action that will be completed by a specific time in the future. For example:

By the time you park the car, I *will have purchased* our tickets for tonight's basketball game.

Before the bakery opens tomorrow morning, Mary *will have baked* 500 sticky cinnamon buns.

Perfect Progressive

Past Perfect Progressive: She had been talking. (and stopped)

Present Perfect Progressive: She has been talking. (and stopped)

Future Perfect Progressive: She will have been talking. (and stopped)

We use the *past progressive* tense to describe an action in the past that was expected to continue but something (either an action or a state of being) happened to prevent the action from continuing. For example:

She *had* been talking, but she stopped before lunch.

She *has* been talking; however, she was just interrupted by the moderator.

By the end of this shift, she *will have been* working 19 hours without a break.

Notice that in each use of the progressive tense, some kind of specific condition is in effect.

Using Perfect Tenses   Exercise C.33

Write a paragraph of about eight sentences to describe something memorable that happened to you during your first job. Be sure you use *past perfect, present perfect,* or *future perfect* verbs in each sentence. Proofread your paper and turn it in to your instructor. Starting with one of the following sentences may be helpful:

Before I began my first job, I had been . . .

I had never worked before when I was hired as a . . .

I had been tired before the shift began, but then . . .

Looking at Irregular Verb Forms

Because most of us learn our vocabulary by listening to others, we may not know the rules that shape our language. Often, these rules have exceptions and contradict each other. An area that seems particularly difficult to understand is when to use the standard verb form and when to switch to an irregular verb form.

You can form the past tense or the past participle of most verbs by adding *-d* or *-ed* to the root of the infinitive form.

Infinitive:	*to complete*
Present:	She *completes* the test on Friday.
Past tense:	She *completed* the test.
Past participle:	She *had completed* the test last Friday.
Infinitive:	*to install*
Present:	Roberta *installs* the new air conditioner.

484 Appendix C: Writing Skills Review

Past tense:	Roberta *installed* the new air conditioner.
Past participle:	She *had installed* the air conditioner just before the heat wave began.

Infinitive:	*to suppose*
Present:	She *supposes* that everyone will be on time.
Past tense:	She *supposed* Mary was late.
Past participle:	She *had supposed* that her grades would have been mailed by last Friday.

For the past tense of most regular verbs, we just add *-d* or *-ed*. However, irregular verbs routinely use one of the following six common patterns for the past and past participle forms.[1]

1. *Vowels change.* (*Example:* to swim, swam, had swum)
    Past tense: I swam all day yesterday.
    Past participle: I had swum in last year's invitational swimming competition.

2. *-en added.* (*Example:* to eat, ate, had eaten)
    Past tense: He ate everything he could before his plate was taken.
    Past participle: He had not eaten for three days.

3. *-ee changes to -e.* (*Example:* to feel, felt, had felt)
    Past tense: She felt quietly happy.
    Past participle: Last time she went to the dentist, she had felt no pain.

4. *-d changes to -t.* (*Example:* to send, sent, had sent)
    Past tense: Roger sent me a copy of his grades.
    Past participle: He had sent the wrong copy, so he was racing to make the deadline.

5. *Changes to -ught.* (*Example:* to bring, brought, had brought)
    Past tense: Lisa brought her lunch to work.
    Past participle: Lisa had brought her lunch to work until she met Sam and Mary.

6. *No change.* (*Example:* to read, read, had read)
    Past tense: He read the homework before class.
    Past participle: He has read five books this summer.

Exhibit C.2 shows some common irregular verb forms organized by these patterns.

---

[1]Hodges, John C., and Mary E. Whitten. *Harbrace College Handbook.* New York: Harcourt Brace Jovanovich (1982), p. 544. Used by permission.

Appendix C: Writing Skills Review   **485**

Common Irregular Verb Forms   Exhibit C.2 ∎

Infinitive	Past Tense	with Participle
1. *Vowels change* (*Example:* to swim, swam, had swum)		
to do	did	had done
to drink	drank	had drunk
to fly	flew	had flown
to go	went	had gone
to know	knew	had known
2. *-en added* (*Example:* to eat, ate, had eaten)		
to break	broke	had broken
to fall	fell	had fallen
to speak	spoke	had spoken
to take	took	had taken
to write	wrote	had written
3. *-ee changes to -e* (*Example:* to feel, felt, had felt)		
to sleep	slept	had slept
to weep	wept	had wept
4. *-d changes to -t* (*Example:* to send, sent, had sent)		
to build	built	had built
5. *changes to -ught* (Example: to bring, brought, had brought)		
to buy	bought	had bought
to catch	caught	had caught
to think	thought	had thought
6. *No change* (Example: to read, read, had read)		
to set	set	had set

Practicing Verb Tense   Exercise C.34

To practice your understanding of verb tenses overall, fill in the following chart with the verb forms for third person for the following two verbs: *to call* and *to fly*.

Tense	First Person	Third Person
Present	I call	She _____
Past	I called	She _____

Future	I will call	She _____
Present Progressive	I am calling	She _____
Past Progressive	I was calling	She _____
Present Perfect	I have called	She _____
Past Perfect	I had called	She _____
Future Perfect	I will have called	She _____

Tense	First Person	Third Person
Present	I fly	They _____
Past	I flew	They _____
Future	I will fly	They _____
Present Progressive	I am flying	They _____
Past Progressive	I was flying	They _____
Present Perfect	I have flown	They _____
Past Perfect	I had flown	They _____
Future Perfect	I will have flown	They _____

Checking for Subject/Verb Agreement

Verbs need to agree with their subjects. To check for subject/verb agreement, ask if the subject is singular. If it is, use a singular form of the verb.

My cat sleeps.	Singular subject (cat) + Singular verb (sleeps)
The cats were asleep.	Plural subject (cats) + Plural verb (were asleep)

When sentences are longer, it can be challenging to find the "true" subject of the sentence and to check to make sure the verb matches the subject, whether it is singular or plural. For example:

My *supervisor,* who works on the night shift with five other supervisors, *is* well liked.

The *interviews* with Mary *are* scheduled for 2 PM today.

The *report* that Ron has and that Paula is planning to take on her trip *needs* proofreading.

To check for subject/verb agreement:

- Find the subject.

- Cross out any dependent clauses or prepositional phrases.

- Check to see the verb matches the subject, whether it is singular or plural.

When you start a sentence with **this** or **there,** check the verb to make sure it agrees with the subject it is pointing to, or rewrite to clarify the true subject.

This   is   the first time that I have used computers.
This + Verb + Singular Subject
Revised: Today is the first time I have used computers.

There   *are   five new people*   working on our shift.
There + Verb + Plural Subject
Revised: Five new people are working on our shift.

Editing for Subject/Verb Agreement   Exercise C.35

Circle the correct verb for each of the following sentences.

1. The roses in the pot on the balcony (need) (needs) watering.
2. The final set of drawings (was) (were) ready to be picked up.
3. Five boxes of computer paper (is) (are) all we need.
4. The computers and one printer (was) (were) stolen.
5. A good set of golf clubs (cost) (costs) about $400.
6. This (is) (are) the last time I will see you.
7. There (is) (are) only five seconds left.
8. Many small companies located in San Diego (serve) (serves) tourists.
9. People who live in San Diego (is) (are) more likely to have a sun tan.
10. Putting flowers in all of our offices (is) (are) expensive.

Using Active and Passive Verbs

Most of the time, we write using **active** verbs, for example:

He jumps. She ran. I saw. You looked.

Passive verbs emphasize the *objects* instead of the *subjects* (I, you, we, they). The result can be wordy, unclear, and confusing to your readers. You can check your use of passive verbs by asking "who" questions and then deciding if the information needed is important.

The report has been read.	(Who read the report?)
They have been seen.	(Who saw them?)
It has been stated that the car was speeding.	(Who said this? Whose car was speeding?)

Notice how passive verbs can be used to hide the identity of the person who is responsible for the action. Sometimes in controversial or negative situations, we need to downplay the subject. Which of the next two sentences does Maria prefer?

Active: Maria completed the late report on Tuesday.

Passive: The late report was completed on Tuesday.

Exercise C.36 Using Passive Voice

Rewrite the following sentences, revising active verbs to passive verbs. Decide for each sentence whether active or passive voice is the "best" choice.

1. Larry chaired the meeting.
2. John, Robert, and Sheila knew the meeting was important.
3. Robert asked Susana and Sheila to go to Mexico City to attend a conference on computers.
4. Susana said she would go to the conference, but Sheila said she had made other commitments.
5. Sheila installed the new computers that arrived on Monday.

Summary Exercise—Verb Review

Use this final exercise to review your progress in using verbs correctly. Circle the word that should be used in the following sentences.

1. They've (went) (gone) to work already.
2. There (is) (are) many new students this fall.
3. Even this supervisor with two assistants (has) (have) too much work to do.
4. The dog has (ran) (run) himself to death.
5. She (has) (have) been working here since last January, and she (fits) (fit) very well on our team.
6. She (has been) (had been) working here since last January, but she (was transferred) (had been transferred) to the new building.
7. The report on my desk behind the computers (need) (needs) to be retyped.
8. Roger (has been) (had been) looking forward to his vacation, but he (has been) (had been) (is) planning to leave in early June.

PROOFREADING

Begin building your proofreading skills by methodically checking work before it leaves your desk. Try to notice *the pattern of errors* that occurs most in your writing. Try the following suggestions each time you complete a document:

Appendix C: Writing Skills Review **489**

- Check for problems with *spelling*, after you run a spelling checker. Know which words you most often misspell and check for these.

- Look for key problems in *grammar*—especially subject/pronoun agreement.

- Double-check *punctuation*. Know which kinds of comma errors you're most likely to make and double-check these as well.

- Reread for any *typing errors*. If the document is very technical, work with a colleague to slow down your reading speed, or read it out loud.

- Look again at the overall *format* of the paper once all changes have been made. Make any final adjustments to ensure your document will be easy to read.

You can use this section to review your spelling skills and commonly confused words.

Building spelling skills	Go to page 489
Using commonly confused words	Go to page 490

Building Spelling Skills

Some people believe "Good spellers were born, not made." Improving your spelling can be difficult. You can improve your spelling overall by keeping a list of the words you misspell most often. Some of these may include the commonly confused words listed in the next section. By keeping a list, you can then work on memorizing those key words you use most often—so that in time you can spell them correctly. Here are some additional suggestions:

- Use a dictionary. Look up any words you are uncertain about.

- Make a list of the words you have difficulty with. Circle the problem area.

- Listen to how you pronounce the words you have difficulty with. Sometimes we don't say what our eyes see (for example, *could of* instead of *could have*).

- Read as much as you can. This builds your vocabulary and helps with spelling.

- Use a spelling checker—but also proofread to catch those words the spelling checker doesn't catch (for example, *thee* instead of *the*).

Most people have learned a few spelling rules over the years. Since English has "borrowed" many words from other languages, spelling rules are equally complicated. Here is a list of some key ideas to help with spelling.

When you're proofreading for spelling, double-check for:

Contractions (I've, he's, won't, couldn't).

Silent letters (pneumatic, psychology).

490 Appendix C: Writing Skills Review

Words that use ie or ei (science, grieve, conceive).

Words that end with -ary or -ery (boundary, gallery).

Words that end with -ent or -ant: (dependent, reliant).

Words that end with -ize, -ise, or -yze (realize, revise, analyze).

Words that end with -ible or -able: (defensible, dependable).

*Double consonants (*bookkeeping, recommend).

When you're proofreading in general, look for:

Missing or extra words.

Missing or extra letters.

Incorrect words (The computer was an the desk.).

Commonly confused words (homynyms).

Exercise C.37   Starting a Spelling List

As part of your writing for this class, set a goal of learning 10 correctly spelled words that are difficult for you. Start a spelling list of the word correctly spelled and the word misspelled with the error circled. Review this list each week. Add new words that you need to memorize. Turn your list in with your final project and a final comment that summarizes whether this approach works for you.

Using Commonly Confused Words

Some words are confusing because they sound alike but are spelled differently. Exhibit C.3 lists pairs of words (called *homonyms*) that sound alike but have different definitions. Concentrate on those that are most confusing to you.

Exercise C.38   Reviewing Commonly Confused Words

Choose three sets of words from Exhibit C.3 and write a sentence for each word to practice your understanding of the differences between the two words. Add these words to your spelling list.

Appendix C: Writing Skills Review   **491**

Common Homonyms   Exhibit C.3 ■

I accept your offer.	*Accept/except*	Everyone except him
I advise you to do this.	*Advise/advice*	Take my advice.
You have affected me.	*Affect/effect*	This stage effect is just right.
We are all ready.	*All ready/already*	I've been there already.
The brakes don't work.	*Brake/break*	Don't break the glass.
Cite your sources.	*Cite/site*	Visit the construction site.
My conscience tells me to.	*Conscience/conscious*	He may be conscious.
I counsel you to do this.	*Counsel/council*	The city council met.
Eat the sweet dessert.	*Dessert/desert*	Cross the sandy desert.
It flew from its cage.	*Its/it's*	It's too late.
Hear the music.	*Hear/here*	Put the book here by me.
Don't lose your money.	*Lose/loose*	I have loose change.
Think of the past events.	*Past/passed*	He passed this way.
It's too personal.	*Personal/personnel*	Go to the personnel office.
I'm taller than she is.	*Than/then*	She then left the store.
Their family is waiting.	*Their/they're/there*	They're often over there.
She threw the football.	*Threw/through*	It flew through the window.
I too am ready to go.	*Too/two/to*	The two of us are too hot to wait.
Where are you going?	*Where/were/wear*	What were you going to wear?
Whether you want to or not	*Whether/weather*	What's the weather?
Who's on first?	*Who's/whose*	Whose book is this?
You're on time!	*You're/your*	Your ride is here.

ANSWERS TO APPENDIX C EXERCISES

Answers for Understanding Sentence Structure

C.1 Writing with Nouns (Answers will vary.)

Roger, Marilyn, and *I* went to *Sadie's Restaurant* last *Thursday night* to find the *chef* was overworked. We had to wait for 45 *minutes* before our *dinners* arrived; however, when we finally started eating, we realized the *wait* had been worthwhile. The *rib steak* was tender, the *fetuccini* creamy, and the *shrimp* freshly tossed with a *garlic butter sauce.* We'll return next *Saturday night,* despite the *wait.*

C.2 Using Adjectives

1. A *horrible, evil-smelling* gas was leaking from the hot water heater.
2. The *foul-smelling* odor was *very* strong.

492 Appendix C: Writing Skills Review

3. I called my *absentee* landlord.
4. She said this was not a *big* problem.
5. The evil smell was seeping into my *small, but beautiful* bathroom.
6. Every time I took even a *short* shower, the water smelled like *rotten* broccoli.

C.3 Finding Verbs

1. *Open* the window, please.
2. In Mexico, windows *are always opened* early in the morning.
3. Every window *has* an iron grill in front of it.
4. Nearly every house *is built* around a patio.
5. The patios *are filled* with many different plants and flowers, each one in its own earthen pot.

C.4 More Practice Finding Verbs

1. "To be or not to be," *cried* Hamlet.
2. Trying out for a play *can be* stressful.
3. After working hard to memorize his lines, Fernando *was* ready for the audition.
4. Fernando *wanted* to be selected for a part.
5. The director selecting the main characters *chose* Fernando.

C.5 Using Adverbs

1. Jorge said that he would be *very* happy with a small breakfast.
2. He *thoroughly* enjoyed orange juice and a roll.
3. Then, he *slowly* ate scrambled eggs, bacon, and another roll.
4. He couldn't be *completely* through until he had drunk three cups of coffee with milk and sugar.
5. After eating all this food, he was *extremely* full.

C.6 Using Pronouns

(SP) a subject pronoun, (OP) an object pronoun, (PP) a possessive pronoun.

1. Give *him* the customer order. (OP)
2. George and *she* have worked here over seven years. (SP)
3. Paula and *she* went to San Diego (SP) with *him* (OP) to work with *their* new customers. (PP)
4. Just between *you* and *me*, (OP) the trip was very expensive.
5. Where is *their* expense report? (PP)
6. Get the monthly report from *her*. (OP)
7. *He* (SP) would like to hear more about *her* promotion. (PP)

Appendix C: Writing Skills Review   **493**

C.7 More Practice with Pronouns

(1) (My) favorite movies are old romantic comedies. (2) Most of the best of these black and white films were made in the 1930s, but since (I) don't subscribe to cable TV, (I) don't often get to see (them.) (3) (I) like (them) for how (they) show life in the "old days," when everybody dressed in fancy clothes for dinner, when detectives solved all the crimes, and when the movie ends, the girl gets the guy. (4) Of course, (they) were popular so that people could stop worrying about the Great Depression for a little while. (5) People who were struggling with unemployment liked to fantasize about living a rich life. (6) (They) didn't really solve any problems, but for an hour or so, (they) were happy, rich, and loved. (7) Maybe that's why for many of (us,) such romantic adventure movies as *The Princess Bride* or *Jewel of the Nile* are still popular today.

Revised:

(3) I like old movies for how they . . .

(4) Of course, these movies were . . .

(6) These movies didn't . . . the theater-goers were happy . . .

C.8 Using Articles

1. Once upon *a* time, *an* old man lived in a small house in the woods by *the* Carpathian mountains.
2. He was *a* very respected old man because he could tell *a* person's fortune by just looking at *the* person's face.
3. Many people took *the* train to *the* woods to see *the* old man, even though they said they didn't believe him.
4. After talking to *the* old man, they were surprised by how much he knew about them.

C.9 Using Prepositions

1. Why can't we go *with* Paula to the airport?
2. I am waiting *for* Paula before we leave.
3. You'll find her briefcase *beside* the table.
4. Put her suitcase *under* the back seat.
5. Please go *to* the ticket office to verify the flight.
6. Go *across* the street and into the waiting room.

C.10 Using Conjunctions and Subordinating Conjunctions

1. Charlotte *and* Roger were ready to go to dinner.
2. *Although* Roger was very hungry, Charlotte had not eaten all day.
3. Roger was concerned about Charlotte *because* she had not eaten all day.
4. *Once* Charlotte got to dinner, she decided she was hungry.
5. Charlotte ate a rib steak and a piece of chocolate cheesecake *before* the night was over.

494 Appendix C: Writing Skills Review

C.11 Finding Subjects and Predicates

1. (It) *is very humid today.*
2. (Kansas and Missouri) *have frequent thunderstorms.*
3. (Camping in the rain) *can be an unforgettable experience.*
4. (The mosquitoes) *flew around the tent, making us nervous.*
5. Answers will vary.
6. Answers will vary.

C.12 Using Verbs, Adverbs, and Adjectives

When I return to my *small* office, I see that *red* light *flashing* on my *desk* telephone that tells me I have *too many* messages waiting for me. Voice mail is an *excellent* tool, but people need to know how to leave messages. Three key ideas to *improving* your use of voice mail are:

1. Keep your message *short*. Try to *avoid* unessential information.
2. Talk as *specifically* as possible. Include important information. Avoid general wording.
3. Leave your phone number so the person can *easily* call you back.

C.13 Using Prepositional Phrases (Answers will vary.)

When I was a child, my family spent every weekend camping *in the woods*. I can remember the time we plotted to take down the tent while my sister was still *inside the tent!* We were giggling so hard that we were sure that she would hear us as we worked *outside the tent. At first*, she was surprised. She laughed as hard as the rest *of us*, but later, she told our parents. She got an ice cream cone and sat *in the front seat of the car*. Even though the trip home seemed long as we sat *in the back seat* and dreamed *of ice cream cones*, we still have our memories.

C.14 Using Adverbial and Adjective Clauses (Answers will vary.)

1. My best friend, *who grew up in New York City*, visits there every year. (adjective clause)
2. *After I spent the day with my grandfather,* listening to his stories, I couldn't believe he really was 100 years old. (adverbial clause)
3. *Although my mother sang well*, she never sang any opera. (adverbial clause)
4. The cat *that has two black spots over its eyes* is my favorite. (adjective clause)

C.15 Revising Base Sentences (Answers will vary.)

1. Your 1986 Toyota Camry needs its transmission repaired.
2. I like learning new software (like the HTML Assistant) on my new Pentium computer.
3. My first job manufacturing and selling rubber stamps was unforgettable.
4. My worktable is impossibly cluttered with papers, stacks of books, and a forgotten cup of tea.
5. Despite generations of a different history, my Dalmatian loves cats of any kind.

Appendix C: Writing Skills Review **495**

Answers for Building Punctuation Skills

C.16 Using Periods, Semicolons, and Question Marks

1. Have you learned how to use the Internet yet?
2. I have just taken my first class on the Internet; learning how to jump on the information highway was fascinating.
3. I have had to learn a new vocabulary; words such as "gopher," "mosaic," and "web" have taken on new meanings.
4. My favorite people in the class were Mary, a college librarian; George, a retired contractor; and Martha, a nurse.
5. My teacher was very understanding; he helped us gain confidence by letting us practice on the computers.
6. When will you get your new computer?

C.17 More Practice with Punctuation

1. At first, I was not comfortable using the computer to prepare my assignments, but with practice, my skills in revising are improving.
2. Our classroom only has three printers; these DeskJet printers are so quiet that we can't hear when a document is being printed.
3. Using the new program was time-consuming, but that didn't seem to matter. We started to enjoy making changes to our writing.
4. I was not able to find the RETURN key; in fact, the ENTER key is the same as the RETURN key!
5. I could print my work out, if only I could remember the print command! (optional comma)

C.18 Using Commas (Answers will vary.)

1. Quickly, Mark started the car.
2. On the table, Mark could see his car keys.
3. Although this was his first car, he was a very good driver.

C.19 Adding Introductory Commas

1. Quickly, Martha opened the door.
2. Even though it was very early, it was still dark.
3. Because it was the first day of her vacation, Martha was ready to go.
4. If it had been raining, she would have been disappointed.
5. At first, she had just hoped for a camping trip, but there on the table were five tickets to Hawaii.

C.20 Punctuating "Interrupter" Clauses (Answers will vary.)

1. Martha, happily, took her checkbook with her.
2. Robert, at the start of his vacation, was a little nervous about flying on an airplane.

496   Appendix C: Writing Skills Review

3. Robert, since he had not had breakfast, stopped for a quick bite at the airport.
4. The breakfast, which Robert ate, was excellent.

C.21 Using Commas with Ending Elements (Answers will vary.)

1. Martha sat in her assigned seat quietly.
2. "Do I have to give up the window seat?" she asked sadly.
3. She put her reading bag and new magazine under the seat.
4. She had been eager to start the trip in the morning
5. The tray of hot lunch looked very appetizing even though she wasn't hungry.
6. She liked the strawberry shortcake best since she didn't eat dessert often.

C.22 Changing Sentence Patterns

1. I have saved all of the money from my summer job; therefore, I'm ready to pay my tuition.
2. If I have some money left after paying tuition, I will stand in line again to buy my books.
3. Spring break and daylight savings time are coming soon, but we will have to get up an hour earlier.
4. Paul, who spends three hours every day reading newspapers or books, is the smartest person I have ever met.
5. Even though my husband is a sports fan, he doesn't like to watch television because it's too loud, has too many commercials, and the announcer's voice seems to drone on forever.
6. My husband refuses to miss even one game; it doesn't matter which sport, what time of year, what time of day, or if it's a holiday because the game must be played through!

C.23 More Practice with Commas

1. The woman who was arrested by the police was charged with arson.
2. The cat's favorite hiding place is under the bed, a quiet spot that is also cool and dark.
3. Although our television was broken, we still heard the game by listening to the radio.
4. This version of "Stardust," which is fast and has a steady beat, would be good to dance to.
5. The waiter took a coffee break while the customers lingered over dessert.
6. Happily, Mary took the increase in pay.
7. Aurancia, who just won the national tennis competition at Wimbledon, booked a flight to New York.
8. I found it hard to decide whom to favor in the tennis match, even though one of the players was from the United States. (optional comma)

Appendix C: Writing Skills Review   497

C.24 Using Contractions

1. Her friends (don't) know that (she's) working at a (competitor's) store.
2. The first (dishwasher's) job was to scrape and sort the (customers') dishes.
3. (They're) visiting their (parents') farm to study crop rotation. (assumes two parents)
4. A (person's) mind (doesn't) work like a tape recorder.
5. (It's) the (parents') responsibility to make sure that they (don't) buy (children's) toys with sharp edges that can cut a (youngster's) tender skin.
6. (Isn't) it a shame that football (players') salaries have gotten so high that the (teams') owners (aren't) able to keep ticket prices at last (year's) level?

C.25 More Practice Using Apostrophes

1. The shop supervisor lost the (women's) paychecks.
2. My (brother's) associates took a trip last summer. (assumes one brother)
3. (It's) too late to go for an (hour's) drive.
4. (Maria's) manager kept track of one (employee's) work schedule.
5. The (manager's) office was bigger than the (typists') offices.
6. The secretary made a copy of this (year's) tax return.
7. We could hear the (children's) voices outside.

Summary Exercise—Sentence-Combining Review (Answers will vary.)

1. Carl **had sanded and repainted** the boat.
2. We loaded our **lunch, our fishing gear, and maps** onto the boat.
3. The boat tipped **slowly over on its side.**
4. The **boys and girls** were waiting by the pier.
5. It was very early in the morning, **but the time was just right to go fishing.**
6. Everyone was ready for the trip **to the lake in the boat by the dock.**
7. Peter began to eat his **soggy and cold** lunch.
8. **He** was not happy about eating lunch on the boat.
9. Rhonda, **who had been working at the lodge for three summers,** started the motor.
10. Reynard found a mysterious map titled, "The Quest" with a note written hastily on it; signed with the initial "A," the note said, "Come quickly or all is lost."

C.26 Selecting the Correct Pronoun

1. My roommate and (I) became close friends.
2. The meal was prepared especially for (them) and (us).
3. For (us) students, attendance is required.
4. A problem has come up between you and (me).
5. The river rose after the Browns and (we) had safely crossed.
6. (Who) is on the phone?

498   Appendix C: Writing Skills Review

7. The audience did not know (whom) to applaud.
8. The audience wanted to applaud (her).
9. Either (he) or (I) will wait for the shuttle.
10. The music drew Sylvia and (me) closer.
11. The substitutes were (she) and (I).
12. (He) and (she) were admitted to the hospital.
13. Sylvia is taller than (I).
14. Ronald can type faster than (she).

C.27 Editing for Pronoun Agreement

1. If the workers (need) a ride to the workshop, (they) can call the Motor Pool Department.
2. Yesterday, one of the workers had (the) assembly line shut down while (his) machine was replaced.
3. (They) lost four hours of production time while (their) machines were down.
4. One of the supervisors had attended a workshop and then left for (her) vacation.
5. The supervisor gave overtime to all of the workers, making (them) very happy.
6. No one wanted to turn in time sheets until (his or her) time sheets were correct.
      NOTE: (the) would also be correct!
7. Roberto and Paul wanted (their) own office.
8. Neither Roberto nor Paul wanted to share (his) workstation.
9. Give Roberta and Paula (their) receipt.
10. Neither Roberta nor Paula wanted (her) receipt.

C.28 Clarifying Pronoun Reference

1. After taking the key out of the envelope, (the key) was taken to the lab.
2. Under the Bill of Rights, (all citizens are) guaranteed (their) rights to privacy.
3. Anyone should have job security when (he or she) goes to work.
4. A few minutes after Susan discovered Barbara had the missing key, (Susan) left the lab.
5. After taking the draft out of the lab notebook, I turned (the notebook) in to my supervisor.
6. As I read each page of the final report, I worked hard to correct (the errors, marking each one) with a yellow marker.
7. (The management at Ryerson Computers) really appreciate hard workers here.
8. Each applicant turned in (a) completed application and waited to be called for an interview.
9. If (the job) just involves filing, Sam files faster than I.
10. (The Board) agreed that we should approve four shifts a day.

Appendix C: Writing Skills Review **499**

C.29 Clarifying Point of View

(This exercise can be revised for any point of view; the revision needs to be consistent all the way through.)

The registration system at this college creates frustration. All students have to preregister before they go to registration, but then they still have to spend hours waiting in a long line before the admitting clerks tell the students that the courses they wanted to take are filled. Students then have to sign up for another section and wait in another line. This drives all the students crazy. Students go to college to learn; they don't go to college to stand in line. The administration is thinking of setting up telephone registration. Students favor this change!

C.30 Revising Confusing Shifts

1. We went to the employment office, not realizing that (we) needed to register as (we) entered the building.
2. She wanted to know whether to take the job and (whether it paid) well.
3. We planned to commute by car, but (we found the bus) to be cheaper.
4. Oliver had no way of knowing we were there until he (walked) through the door.
5. When people get mad, (they) can sometimes act foolishly.
6. The waiter scowled at the diners because (they) left a small tip.
7. (People) should have professional playing experience before they coach.
8. To revise a paper, read it through for errors, and then look at its logical organization.
9. (The members) of the squad looked to me as their leader.
10. I had to decide where I should go on my vacation and whether I could afford it.

C.31 Finding Verbs

1. I (*will see*) you after *attending* the meeting.
2. Architects (*came*) from Chicago *to redesign* our building.
3. Our building (*was surrounded*) by modern, high-rise skyscrapers that *seemed to reach* to the sky.
4. Without remodeling, our building (*would have been*) old-fashioned.
5. We (*could have modernized*) the building less expensively if we *had limited* the budget to *replacing* carpeting, *rewiring* electrical circuits, and *repainting* offices.

C.32 Changing Present to Past Tense

When I *worked* at the college, I *saw* lines of students waiting to be helped with registration at the college. Just before each term *began,* I *met* with students, *checked* their schedules, *talked* about their career plans and *helped* them register for the new term. Sometimes students *did not have* very much time to think about what they *wanted to do,* and they *felt* stressed out. Sometimes students *knew* what they *wanted to do,* but they *didn't know* which classes *were* best for

500   Appendix C: Writing Skills Review

them or which classes *were required* for graduation. A very few students *saw* me during the term to talk about their future. Although students *registered* by telephone or computer, I still *liked* to talk directly with students—especially when they *needed* help!

C.33 Using Perfect Tenses

Paragraphs will vary.

C.34 Practicing Verb Tense

1. She calls, called, will call, is calling, was calling, has called, had called, and will have called.
2. They fly, flew, will fly, are flying, were flying, have flown, had flown, and will have flown.

C.35 Editing for Subject/Verb Agreement

1. The roses in the pot on the balcony (need) watering.
2. The final set of drawings (was) ready to be picked up.
3. Five boxes of computer paper (are) all we need.
4. The computers and one printer (were) stolen.
5. A good set of golf clubs (costs) about $400.
6. This (is) the last time I will see you.
7. There (are) only five seconds left.
8. Many small companies located in San Diego (serve) tourists.
9. People who live in San Diego (are) more likely to have a sun tan.
10. Putting flowers in all of our offices (is) expensive.

C.36 Using Passive Voice

1. The meeting was chaired by Larry. (Not acceptable)
2. It was decided the meeting was important. (Not acceptable)
3. Susana and Sheila were asked by Robert to go to Mexico City to attend a conference on computers. (Acceptable; active voice preferred.)
4. It was said by Susana that she wanted to go to the conference, but it was said by Sheila that she had made other commitments. (Not acceptable)
5. The new computers that arrived on Monday were installed by Sheila. (Acceptable; active voice preferred.)

Summary Exercise—Verb Review

1. They've (gone) to work already.
2. There (are) many new students this fall.
3. Even this supervisor with two assistants (has) too much work to do.
4. The dog has (run) himself to death.

Appendix C: Writing Skills Review **501**

5. She (has) been working here since last January, and she (fits) very well on our team.
6. She (had been) working here since last January, but she (was transferred) to the new building.
7. The report on my desk behind the computers (needs) to be retyped.
8. Roger (has been) looking forward to his vacation, but he (is) planning to leave in early June.

C.37 Starting a Spelling List

Lists will vary.

C.38 Reviewing Commonly Confused Words

Lists will vary.

INDEX

A

Accounting systems, computerized, 11
Adjective, 453, 454; *see also* Sentence structure
Adverb, 453; *see also* Sentence structure
Agenda; *see* Meetings
Apostrophe, 465, 471–472
Application letter, 338, 356; *see also* Jobs
  checklist for, 343–345
  writing, 339–343, 345–346
Article, 453, 456, 458; *see also* Sentence structure
Association of Records Managers and Administrators (ARMA International), 129, 130, 131
Audience, 32, 33, 35, 36, 38, 42, 47, 63, 76, 132, 153, 204, 229, 234, 258, 265 271; *see also* Forms; Presentations
  analyzing, 11
  performance evaluations, 151–152
  primary, 33, 34
  range of, 34
  secondary, 33, 34
  solving problems for, 22, 38
  workplace, 18
  writer and, 33–34
  writing to, 22, 38

B

Bar-coding systems, 186
Before Going Toll-Free, Consider Who Might Call (Norman), 242
Bibliography, 55, 77, 447, 449; *see also* Documentation; Reports
  annotated, 68
  formats for, 69–71
  in-text citations, 68, 77
  preparing, 67–71
  publication information and, 68
Blue Collared (Crockett), 362
Blyth, John, 316
Bulletin boards, 186

C

Career planning, 306–319, 360, 386, 388; *see also* Jobs
Cautions; *see* Design, document; Manuals, technical
CD ROM databases, 63, 64, 65; *see also* Library
  bibliographies and, 68
Certificate of Origin for the North American Free Trade Agreement, 137
Checklists, 198
  checklist for, 234–235
  designing, 232–234
Classes, formal, 9

Code of Professional Responsibility (Cunningham), 130
Coles, Robert, 16
Combining Computers and People to Build Flexibility (Upton), 278
Comma, 465, 468–469, 471
Communication skills, 8, 9, 19, 23; *see also* Listening skills; Social skills; Speaking skills; Writing skills;
  crises and, 305
  refining, 304
Communications technology, 11; *see also* Computers; E-mail
Comparisons, 46
Computers, 11, 31, 54, 186, 282, 283, 284, 306; *see also* CD ROM databases; E-mail; Graphics; Internet; Word processing
Conclusion, 28, 40, 101
Conferences, 9
Conflict, 17
Conjunction, 453, 456, 459–450; *see also* Sentence structure
Content, 37, 47–48, 51, 76, 114, 120, 186, 198, 204, 226, 230, 234, 245, 258, 265, 271
  document design and, 37
  organization and, 37

504 Index

Content—*Cont.*
style and, 37, 38
Content map, 39, 40
Contractions, 472
Co-workers, 2; *see also* Informing;
Instructions
writing for, 192–236
Crockett, Roger O., 362, 366
Culture, corporate, 286–287, 288
Cunningham, Patrick, 130
Customer complaints, 29
responding to, 261–262
Customers, 236, 240; *see also*
Letters
loyalty, 241
phone and, 266
selling to, 267–270
service and, 245
solving problems for, 241

D
Dahir, Mubarak, 55, 56, 73
Davidson, Keav, 51
Definitions, technical
expanded, 95, 97–98
expanding techniques, 99
formal, 95, 96–97, 98
parenthetical, 95, 96
stipulative, 95, 101
Descriptive writing, 6
Design, document, 37, 38, 48, 77,
86, 115, 120, 198, 200, 205,
226, 230–231, 235, 258, 265,
271
cautions, 216, 217
headings, 213–214
instructions and, 210–219
lists, 214–216
notes, 215
readability and, 112–114
warnings, 216, 217, 218, 219
Devil's advocate, 16, 17
Directions; *see* Informing
Discovery draft; *see* Writer's block
Discussion, 28, 40, 101
Documentation
APA, 444–447
MLA, 442–444

Documentation—*Cont.*
numbered system, 448–449
Drafting; *see* Writing process

E
Editing, 19, 49, 78, 122, 168, 205,
231, 235, 236, 258, 265, 271
*Educating Engineers for the Real
World* (Mubarak), 56–57
E-mail, 11, 167, 173–176
checklist for, 185–186
day to day use of, 180–184
definition of, 176–177
ethical issues and, 184
privacy and, 184
problems with
for receivers, 180–182
for senders, 182–183
technology and, 183–184
reading, 182
writing skills and, 184–184
writing with, 176–179
Electricity and Automation
(Macauley), 88–89
*Entry-Level Workers Can Buy Only
a Wafer-Thin Slice of the
American Pie* (Crockett), 366
Errors, correcting, 29
Evaluation, performance, 375–377
checklist for, 385–386
of employees, 384–385
evaluating, 386–387
forms for, 377–381
self-evaluation, 381, 383
Examples, 5, 28, 42, 44

F
Facts, 5, 28, 41, 44
Fax, 177, 186
Feedback, 9
Field reports, 29, 157–159, 170
Flowcharts, 31, 128, 168, 169
organizational, 160
process, 163–165
Footnotes, 55, 443, 446, 448–449;
see also Documentation;
Reports

Format, 37, 122, 186, 232, 246
Forms, 29, 117, 128–129,
132–134, 141, 254; *see also*
Field reports; Flowcharts; Raw
data; Tables
for analyzing tasks, 160–165
for appointments, 128
audience and, 141–143
billing, 136
computerized, 136
ethics and, 129–132
fill-in-the-blank, 134, 135
from the field, 133
from the floor, 133
headings, 145
invoices, 134
lists, 144
logs, 137, 168–169
order, 134
patient charts, 137, 140, 143
for performance evaluations,
377–380
for projects, 128, 165–167
raw data and, 144–152
routine, 137, 142
for schedules, 128, 160
Formula, 40

G
Goals
personal, 370
professional, 371–372
Grammar, 19, 49, 78; *see also*
Sentence structure
Graphics, 18, 77, 99; *see also*
Design, document;
Instructions; Reports
computer-generated, 110
cut-away drawings, 109
designing, 110–111
exploded drawings, 109
line drawings, 108
photographs, 107
for resumes, 329, 332
in technical descriptions,
106–112
Groups, 20; *see also* Teams
individual attitudes in, 14, 15

Index **505**

Groups—*Cont.*
  life cycle of, 14, 15, 16
  process of, 16
  roles in, 13, 14
  stages of, 14–15

H

Harvey, Jeanne Walker, 173, 174
Headings, 128; *see also* Design,
  document; Forms; Raw data

I

Informative writing, 153; *see also*
  Informing; Letters; Memos
  audience and, 153
  direct pattern, 153, 254–255
  indirect pattern, 153, 255, 257
Informing, 35, 36, 192; *see also*
  Letters; Memos
Instructing, 86
Instructions, 192, 193, 198, 206
  checklist for, 229–231
  document design for, 210–219,
    226
  effective, 206
  format for, 207
  graphics for, 221, 224, 225
  layout for, 221–222
  narrative style and, 207
  parts of, 209–210
  planning, 219–221
  process for writing, 226–227
  product manual, 208
  readers' needs and, 227–229
Internet, 54, 63, 64, 66–67, 177;
  see also E-mail; Library
Interruptions, 11
Interview, employment, 347–354
  questions at, 350–354
  types of, 348–350
Introduction, 28, 40, 101
Inventory
  control systems, electronic, 11,
    283
  just-in-time (JIT) systems,
    283–284
  management, 283–284
Invoices; *see* Forms

J

Job application; *see* Application
  letter; Jobs
Jobs; *see also* Application letter;
  Career planning; Evaluation,
  performance; Goals;
  Interview, employment;
  Resume
  applying for, 319–326, 338
  career plan, 372–375
  offers of, 354
  promotion, writing for, 369
Journal, 2, 3, 5, 8, 22, 53, 55, 57,
  86, 128, 132, 170, 173, 189,
  193, 239, 241, 277, 311,
  314–315, 359, 361
  double entries, 73, 74
  reaction entries, 73

L

Lacitis, Eric, 3, 4, 5, 6
Leadership skills, 15
Learning Skills Inventory, 285
Learning styles, 285–286
Letters, 11, 240; *see also*
  Application letter; Customer
  complaints; Customers; Jobs
  block, 397, 398
  body of, 249
  checklist for, 265
  closing, 249–250
  collection, 262–264
  customers and, 246–272
  format of, 248–249, 250–252
  greeting style, 249
  indented, 399
  informative, 236, 254–258
    checklist for, 258
  persuasive, 236, 259
  quick format for, 394–396
  sales, 267–271
    checklist for, 271
  thank-you, 355–356
  "you" focus, 252–253
Library, 63; *see also* Internet;
  Research

Library—*Cont.*
  call numbers, 64, 65
  card catalog, computerized, 64,
    68, 69
  computer searches, 64, 65
  encyclopedias in, 65
  interlibrary loan, 66
  journals in, 65–66
  reference section, 64, 65
  resources, 64–67, 82–83
Library of Congress Subject Index,
  64
Line drawings; *see* Graphics
Listening skills, 6, 14
Lists, 128; *see also* Design,
  document; Forms; Raw data
Logs; *see* Forms

M

Macauley, David, 87
Main idea, 5, 28, 60, 75; *see also*
  Paragraphs; Topic sentences
Manuals, technical
  cautions in, 91
  notes in, 91
  organization of, 92
  reading, 92, 93
  warnings in, 91
Maps, reading, 52
Math skills, 3
Mechanisms; *see* Tools
Meetings; *see also* Presentations
  agenda for, 298–300
  checklist for, 305–306
  formal, 291
  hostile questions in, 300
  informal, 291
  leading, 301
  managing, 291–292
  minutes, writing, 302
  personal relationships and,
    289–290
  physical factors and, 289
  planning factors and, 289
  preparing for, 292–294
  purpose of, 288
  "rules" for, 290

506 Index

Memos, 6, 11, 12, 128, 153, 198, 394; *see also* Informing; Instructions; Reports
checklist for, 119–120, 204–205
confidential, 35
to inform, 200–203
parts of, 203–204
quick format for, 400–401
Metaphors, 46, 99, 105
Meyer-Briggs Personality Test, 285
Minutes, writing; *see* Meetings
Mobility Trap, The (Blyth), 316

N

Nelton, Sharon, 23, 24–27
Networks, personal, 9
Norman, Jan, 241, 242
Notes; *see* Design, document; Manuals, technical
Note taking, 9
Nouns, 453; *see also* Sentence structure
Numbers, 128, 154
breakdown comparison, 150
growth percentage, 149
Nurturing Diversity (Nelton), 24–27

O

Operating skills, 6, 7
Opinion; *see* Topic sentences
Orders, 11, 29
Organization, 37, 38, 47–48, 51, 76, 114, 120, 198, 204, 226, 230, 235, 246, 258, 265, 271
operating order, 102–103
order of assemble, 104
order of importance, 104
patterns of, 153–154
spatial order, 102
Ownership of written work, 11–12

P

Paragraphs, 18, 41, 42, 49
supporting, 44

Paraphrasing, 72, 132
Patient charts, 29; *see also* Forms
Percentages; *see* Numbers
Period, 465, 466
Personal integrity, 12
Persuading, 35, 36, 118; *see also* Letters, persuasive
Photographs; *see* Graphics
Plagiarism, 55
Pouliot, Janine S., 193, 194
Prepositions, 453, 456, 458–459; *see also* Sentence structure
Presentations, 292–294
audiences and, 294–296
technical information, 296–297
visual aids and, 297–298
Primary reader; *see* Audience, primary
Privacy for Consumers and Workers Act, 184
Problem solving, *see* Audience
Processes, 86, 87, 99
describing, 104–106
steps in, 128
Production; *see* Writing process
Product manual; *see* Instructions
Progress reports, 29
Projects, 2
Promotion; *see* Jobs
Proofreading, 19, 30, 38, 49, 78, 122, 168, 186, 205, 231, 235, 236, 258, 265, 271, 452, 488–489
Pronoun, 452, 456–457, 473; *see also* Sentence structure
agreement, 476
object, 474–475
reference of, 477–479
subject, 474–475
Proposals, 6
Psychology classes, 7
Punctuation, 19, 49, 78, 451, 465–472

Q

Quality circles, 3
Question mark, 465, 466, 467
Quotations, 28, 46, 72

R

Raw data
headings and, 144
interpreting, 148
lists and, 144
reporting, 144
summarizing, 148–149
tables and, 146
Reaction; *see* Journal; Reading skills; Summary
Reading; *see also* Reading skills
for workplace writing, 54, 58
Reading skills, 3, 5, 28, 57, 90, 197, 244, 281, 318, 368
reading for content, 5–6, 28, 58, 90, 132, 175, 197, 244, 181, 318, 368
reading for reaction, 6, 28, 58, 72–74, 90, 132, 176, 197–198, 244, 281, 318, 369
technical materials and, 54, 60, 61
workplace and, 58
Record keeping, 8, 9
Reports, 6, 8, 12, 402–403; *see also* Documentation
bibliography, 436, 442
body, 416, 420
conclusion, 430
cover memos, 40
footnotes, 442
graphic elements, 418
introduction and background, 412–414
recommendations section, 432
summary, 410
survey section, 418, 438
table of contents, 408
title page, 406
Request for proposal (RFP), 115
Research, 48; *see also* Writing process
library, 35, 48, 54, 55
writing from, 54, 55
Resume
checklist for, 337–338
editing, 337
format for, 326–329, 336

S

Schools, 6
Secondary reader; *see* Audience, secondary
Selling; *see* Customers
Semicolon, 465, 466
Sentences; *see also* Sentence structure
  adding clauses, 463–464
  adding phrases, 462–463
  combining, 461–462
  patterns in, 461, 470–471
  revising, 464
Sentence structure, 19, 49, 78, 451; *see also* Sentences
  clauses, 452
  connector words, 456
  content words, 453
  parts of speech, 452, 453
  phrases, 452
  predicate,
  subject, 461
Set Your Rules on E-Mail (Harvey), 174
Shift logs, 29
Signature, 11, 12
Similes, 46
Slang, 302
Social skills, 6, 7–8, 11, 14
Speaking skills, 6
Specifications, 29; *see also* Tools and equipment
  preparing, 115–117
Speech classes, 7
Spelling, 19, 49, 78, 489–491
Statistics, 5, 28, 42
Stories, 5, 28
Style, 37, 38, 48, 77, 114, 120, 122, 132, 186, 198, 200, 204,

Resume—*Cont.*

graphic features, 329, 332
layout for, 329, 336
objective, 332
Returns of merchandise, 29
Reviewing; *see* Writing process
Revising; *see* Writing process
  checklist for, 47–48

Style—*Cont.*

226, 230, 235, 246, 258, 265, 271
concise, 155–157
Summary, 71–72, 77, 78; *see also* Reports
  checklist for, 76
  writing, 54, 55, 60, 61, 75–76
Summary statement, 42
Supervisors, 2, 11, 272
  approval of written work, 11

T

Tables, 128; *see also* Forms; Raw data
  introducing, 146
  interpreting, 147
Teams, 2, 3, 9, 272, 282; *see also* Groups; Meetings; Teamwork
  conflict in, 17
  corporate culture and, 286–287
  crises and, 305
  learning styles and, 285–286
  performance standards and, 151–152
  working with, 284–285
  writing with, 12–13, 18, 32
Teamwork; *see also* Groups; Meetings; Teams
  encouraging, 16–17
  inventory and, 283
  manufacturing and, 282–283
Technical articles
  finding, 63
  reading, 60–61
Technical description, 117–118; *see also* Graphics; Tools and equipment
  checklist for, 114–115, 121–122
Technical information, 55; *see also* Technical articles; Vocabulary
Technical journals, 63
Technical objects; *see* Tools and equipment
Technical skills, 6, 7, 14
Telecommuters, 186
Theories of Motion (Davidson and Williams), 51–52

Thesis statement, 42
Thinking skills, 3
Tone, 28
Tools and equipment; *see also* Manuals, technical
  defined, 93, 95
  definitions of, 95–101
  describing, 77, 87, 90, 94–104
Topic sentences, 18, 41–44, 45
  analyzing, 43
Trade associations, professional, 59
Trade journals, 58–59
Transitional words and phrases, 45–46
Transition statement, 42

U

Upton, David M., 278
Usage, 19
User profile, 227

V

Verb, 452, 453, 455, 480; *see also* Sentence structure
  active, 487–488
  agreement with subject, 486–487
  irregular, 483–486
  passive, 487–488
  tenses of, 481–483
Verbal communication, 276; *see also* Meetings; Slang
Video conferences, 186
Video phones, 186
Visual aids; *see* Presentations
Visual Approach to Employee Safety, A (Poulist), 194
Vocabulary technical, 61–63
Voice mail, 11, 177, 186
Volunteering, 9

W

Warnings; *see* Manuals, technical
Way Things Work, The (Macauley), 87
Why Job Seekers Don't Make the Cut (Lacitis), 4–5

Index

Williams, A. R., 51
Word processing, 18, 30
Workshops, 9
Writer's block, 38–41
Writing process, 3, 153–157; *see also* Audience; Content
analyzing, 29–32
describing, 10–18
developing, 22–48
drafting and, 10, 18, 30, 38, 41, 155, 226

Writing process—*Cont.*
planning and, 10, 18, 30, 38
production, 30, 31, 38
purpose and, 35–36, 37, 38, 42, 47, 76, 204, 229, 234, 258, 265, 271
researching and, 18, 30, 38
reviewing, 18, 30, 38
revising and, 10, 18, 30, 38, 40, 157
teams and, 12–13, 18

Writing process—*Cont.*
workplace conditions and, 11–12
workplace versus academic, 35
Writing skills, 3, 6, 8; *see also* Writing process
review of, 49